WC S36 JIR 2008

Hepatitis C Virus Disease

Emilio Jirillo

Editor

Hepatitis C Virus Disease

Immunobiology and Clinical Applications

 Springer

Editor
Emilio Jirillo
University of Bari and
National Institute for Digestive Disease
Castellana Grotte, (Bari), Italy

Library of Congress Control Number: 2007932062

ISBN: 978-0-387-71375-5   e-ISBN: 978-0-387-71376-2

9 8 7 6 5 4 3 2 1

springer.com

Preface

Hepatits C virus (HCV) infection represents a worldwide disseminated disease, but despite numerous studies, its pathogenesis and medical treatment have not been completely elucidated.

As far as HCV pathogenesis is concerned, HCV genotypes and viral antigens have been investigated with the aim to find a correlation with disease severity and response to treatment. On the other hand, according to current literature, immunological response is implicated in disease progression rather than in host protection. Finally, use of interferon (IFN)-alpha alone or in combination with other antiviral drugs, e.g., ribavirin (RIB), at the moment represents the most effective treatment in chronic HCV disease, even if the percentage of cured patients is still low.

On these grounds, the present book, entitled *Hepatis C Virus Disease: Immunobiology and Clinical Applications*, will emphasize the most recent advances in HCV infection, moving from basic research to clinical application. In particular, in the first chapters of this volume, the full spectrum of immune responses to HCV is analyzed, taking into account either innate or adoptive immunity involvement. In this respect, the role of antigen-presenting cells (macrophages and dendritic celsl) and Toll-like receptors and that of T helper, T cytotoxic, natural killer, and T regulatory cells will be discussed in the course of HCV disease.

At the same time, deficits of innate immunity at the peripheral level with an easier access of microbes into the host will be described also in view of a putative interference of microbial products with IFN trreatment.

In the last part of this volume, a series of contributions elucidates the state of the art of IFN-alpha treatment in HCV patients and the effectiveness of therapy also in relation to HCV genotypes. Besides the combined treatment with IFN-alpha and RIB, the use and applications of pegylated IFNs aree the object of intensive speculation in specific chapters. Finally, the complicated HCV disease and its treatment are discussed.

In summary, this volume, written by various scientists with specific expertise in the field of HCV infection, should represent an efficacious up-to-date on the state of the art of HCV disease in diifferent geographical areas. Moreover, a clear description of disease pathogenesis, a detailed clarification of immune mechanisms, and a deep elucidation of the pharmacology of antiviral drugs should be very useful for a large readership, even including medical students who may wish to learn basic principles of HCV infection.

Emilio Jirillo, M.D.
University of Bari and National Institute for Digestive Disease
Castellana Grotte (Bari), Italy

Contents

List of Contributors

Luigi Amati
National Institute for Digestive Diseases
Castellana Grotte (Bari), Italy

Rafael Bárcena
Service of Liver-Gastroenterology
Universidad de Alcala de Henares
Madrid, Spain

Eleanor Barnes
Nuffield Department of Medicine
University of Oxford
Oxford, UK

Patrick Bertolino
AW Morrow Gastroenterology and
Liver Centre
Institute of Cancer Medicine and Cell
Biology
Sidney, Australia

David G. Bowen
AW Morrow Gastroenterology and
Liver Centre
Institute of Cancer Medicine and Cell
Biology
Sidney, Australia

Antonio Colecchia
Department of Internal Medicine
and Gastroenterology
University of Bologna
Bologna, Italy

Wan-Long Chuang
Department of Internal Medicine
Kaohsiung Medical University Hospital
Kaohsiung, Taiwan

Srikanta Dash
Department of Pathology and
Laboratory Medicine
Tulane University Health Science
Center
New Orleans, LA USA

Eva Dazert
Department of Molecular Virology
University of Heidelberg
Heidelberg, Germany

Davide Festi
Department of Internal Medicine
and Gastroenterology
University of Bologna
Bologna, Italy

Michael Frese
School of Health Sciences
University of Canberra
Canberra, Australia

James Fung
Department of Medicine
University of Hong Kong
Hong Kong SAR

Norihiro Furusyo
Department of Environmental Medicine
and Infectious Diseases and
Department of General Medicine
Kyushu University Hospital
Fukuoka, Japan

Robert F Garry
Department of Microbiology and
Immunology
Tulane University Health Sciences
Center
New Orleans, LA USA

Sidharta Hazari
Department of Pathology and
Laboratory
Medicine Tulane University Healthy
Sciences Center New Orleans, LA
USA

Jun Hayashi
Department of Environmental Medicine
and Infectious Diseases
and Department of General Medicine
Kyushu University Hospital
Fukuoka, Japan

Norio Hayashi
Department of Gastroenterology and
Hepatology
Osaka University Graduate School of
Medicine
Osaka, Japan

Emilio Jirillo
Immunology
University of Bari, Bari Italy
and National Institute for Digestive
Diseases
Castellana Grotte (Bari), Italy

Paul Klenerman
Nuffield Department of Medicine
University of Oxford
Oxford, UK

Ching-Lung Lai
Department of Medicine
University of Hong Kong
Hong Kong SAR

Francesca Lodato
Department of Internal Medicine and
Gastroenterology
University of Bologna
Bologna, Italy

S. James Matthews
School of Pharmacy
Bouve College of Health Sciences
Boston, MA USA

Giuseppe Mazzella
Department of Internal Medicine and
Gastroenterology
University of Bologna
Bologna, Italy

Geoffrey W. McCaughan
AW Morrow Gastroenterology and
Liver Centre
Institute of Cancer Medicine and Cell
Biology
Sidney, Australia

Christopher McCoy
Department of Infectious Diseases
Beth Israel Deaconess Medical Center
Boston, MA USA

Ana Moreno
Service of Infectious Diseases
Universidad de Alcala de Henares
Madrid, Spain

Masayuki Murata
Department of Environmental Medicine
and
Infectious Diseases and Department
of General Medicine
Kyushu University Hospital
Fukuoka, Japan

Vittorio Pugliese
National Institute for Digestive Diseases
Castellana Grotte (Bari), Italy

Carmen Quereda
Service of Infectious Diseases
Universidad de Alcala de Henares
Madrid, Spain

Tertsuo Takehara
Department of Gastroenterology and
Hepatology

Osaka University Graduate School of
Medicine
Osaka, Japan

Ming-Lung Yu
Department of Internal Medicine
Kaohsiung Medical University Hospital
Kaohsiung, Taiwan

Man-Fung Yuen
Department of Medicine
University of Hong Kong
Hong Kong SAR

Innate Immunity in Type C Hepatitis

Tetsuo Takehara and Norio Hayashi

Hepatitis C Virus and Hepatocellular Carcinoma

As early as the days of Hippocrates, hepatitis has been described as a disease that occurs in the young and shows the cardinal symptom of jaundice, which sometimes develops into a critical condition. Ironically, research on hepatitis progressed rapidly during World War II because injuries and the terrible sanitary conditions of the battlefields caused serious hepatitis epidemics. People recognized that hepatitis could be classified into two types: infectious and serumal. The former became known as hepatitis A and the latter as hepatitis B. After the war, the hunt for hepatitis viruses had begun. First, the hepatitis B virus (HBV) was identified in 1967 by Blumberg, who was awarded a Nobel Prize in recognition of his discovery. Next, the hepatitis A virus (HAV) was discovered in 1973. These discoveries were thought to have clarified the causes of hepatitis, but by the following year, it was acknowledged that many cases of hepatitis were not caused by either HAV or HBV (Prince et al., 1974). Later, the hepatitis D virus (HDV) and the hepatitis E virus (HEV) were discovered in 1977 and 1983, respectively, but they were not the cause of hepatitis non-A, non-B, which is associated with blood transfusion. In 1989, HCV was identified by a molecular biological method where researchers induced the expression of cDNAs obtained from the blood plasma of a chimpanzee with hepatitis non-A, non-B and screened them with convalescent serum (Choo et al., 1989). HCV was the first virus to be discovered not by the previously used virological methods, but by a molecular biological method.

The discovery of HCV had a great impact on the treatment and prevention of liver diseases (Hayashi & Takehara, 2006). It turned out that not only did most patients who had been diagnosed as hepatitis non-A, non-B actually have hepatitis C, but also that there were quite a few hepatitis C patients among those who had been thought to have alcoholic liver disease or autoimmune hepatitis. As the natural history of hepatitis C was clarified, it became clear that the disease is a major risk factor for hepatocellular carcinoma (HCC) (Figure 1). The infection route of HCV is via the blood. Some patients who are exposed to the virus develop overt liver disease, but most of them remain in a latent state. Within six months of being infected, 30% of the patients expel the virus naturally while the remaining 70% enter a phase of persistent infection. Once patients enter in this latter phase, it is very rare for the

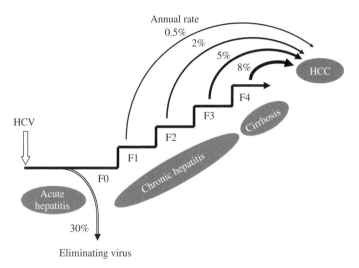

Fig. 1 Natural history of HCV infection and its incidence of HCC according to liver fibrosis stage

virus to be expelled naturally, with the estimated annual rate being less than 0.2% at most. Many of the patients with persistent infection show the medical conditions of chronic hepatitis and develop cirrhosis in 20 to 30 years. Patients with cirrhosis develop HCC at a very high annual rate of 8%, while patients with early chronic hepatitis do so at an annual rate of only 0.5%. The estimated number of patients with HCV is about 1.7 million in Japan and about 1.7 billion in the world. It is a serious public health problem as many of these patients belong to a high-risk group for HCC.

HCV does not fit into the classical definition of a tumor virus. The mechanism of carcinogenesis in patients with persistent infection of HCV is not fully understood, but it is usually explained from the virus and the inflammation viewpoints. From the virus viewpoint, the HCV core protein has an effect on the mitochondrial electron transport system and prompts the production of reactive oxygen species (ROS) (Moriya et al., 1998; Okuda et al., 2002). This process is thought to cause damage to the host gene. It has also been reported that the expression of HCV core protein activates $bcl\text{-}x_L$ transcription via the MAP kinase pathway as well as the activation of STAT3 (Otsuka et al., 2002). We have shown that high expression of $bcl\text{-}x_L$, observed in about one-third of HCC cases, is involved in the apoptosis resistance of cancer cells (Takehara et al., 2001; Takehara & Takahashi, 2003). From the inflammation viewpoint, it is thought that the inflammation itself induces oxidative stress and that hepatic regeneration in patients with hepatic disorder has an influence on the fixation of accumulated mutations (Kato et al., 2003). In any case, the larger clone size of the transformed hepatic cells caused by these factors leads to overt HCC. Generally, the innate immune system recognizes abnormality in autologous cells *in vivo*, and the immunological expulsion mechanism starts to function. Increasing evidence supports the possible involvement of innate immunity in the carcinogenesis of HCC and its development in patients with HCV infection.

Toll-like Receptor: Impact on the Study of Innate Immunity

The biological defense mechanism of higher organisms, including humans, is generally divided into innate immunity and adaptive immunity (Dranoff, 2004). In the adaptive immune response, gene rearrangement by T cells and B cells enables the establishment of a defense mechanism of high specification against "the molecular microstructure of a foreign substance," and this mechanism is immunologically memorized. However, as it takes a few days to induce adaptive immune responses, innate immunity works as an early defense mechanism. Innate immunity has existed as a biological defense mechanism from the earliest stages of evolution: for example, insects have only innate immunity as a defense mechanism. Cells involved in innate immunity include macrophages, neutrophils, NK cells, NKT cells, and γδT cells. Important humoral factors include complements, lectins, and interferons (IFNs) (Biron, 2001). Key notions in the paradigm of modern immunology are the presentation of the antigen by antigen-presenting cells (APCs) and the development of adaptive immune responses. Most research in immunology in a narrow sense has been focused in this area. In contrast, innate immunity has not been the focus of much attention because it is mainly involved in nonspecific phagocytosis and toxicity, which have been regarded as primitive immune responses. Recently, however, innate immunity has been receiving considerable attention for two major reasons. One is the discovery of Toll-like receptors (TLRs), beginning in 1996 (Lemaitre et al., 1996), and the other is a growing recognition that innate immune responses play a critical role not only in early immunity but also in determining the magnitude and direction of the subsequent adaptive immune responses.

TLRs are specific to structures peculiar to microbes, including bacterial and fungal compounds (such as LPS and flagellin) and microorganism-origin nucleic acids (such as double-stranded RNA and CpG DNA) (Akira et al., 2006). TLRs are molecules whose expression has been observed in non-hematopoietic cells as well as in hematopoietic cells such as dendritic cells (DCs). The discovery of TLRs was important because it showed that, at the molecular level, a living body recognizes the entrance of pathogens as "pathogen-associated molecular patterns" (PAMPs). This has shown that the innate immune system discriminates between self and not-self and recognizes abnormality via a mechanism that is different from the gene rearrangement of the adaptive immune system. As for the development of adaptive immunity, it had been thought to be a simple scheme where APCs (the most potent APC *in vivo* is a DC) trap antigens in peripheral tissues and present the antigens to T cells in secondary lymph nodes. However, it is now known that T cells cannot be activated by DCs without the process of DC maturation and that typical signals to induce DC maturation are sent by TLRs (Kaisho & Akira, 2003). Most of the adjuvants loosely recognized as activating factors of immunity have turned out to be TLR ligands. Thus, it can be said that the TLR has revealed the importance of innate immunity in adaptive immunization.

What is now clear is the existence of a scheme in which the recognition of pathogens by TLRs and DC maturation/activation are followed by adaptive immunization in immune responses to invading microbes. The questions then arise of how abnormality *in vivo* is recognized in the process of carcinogenesis and how this recognition can lead to adaptive immunization. The development of cancer is a process in which a normal cell becomes abnormal. The mechanism of recognizing "abnormality in autologous cells" cannot be explained by means of TLRs, which recognize "pathogen-associated molecular patterns." We need to consider another important system.

NK Receptors: System of Recognizing Abnormality in Autologous Cells

Abnormality that occurs in autologous cells *in vivo* is generally reflected in the decreased expression of, or a deficiency of, MHC class I molecules (Smyth et al., 2002). NK cells are a group of cells originally defined based on their nonspecific cytotoxic activity to transformed cells (Trinchieri, 1989). The cytotoxic activity of NK cells to transformed cells has been suggested to depend on the decrease of MHC class I expression in those cells (missing-self hypothesis) (Karre et al., 1986; Ljunggren & Karre, 1990). As a series of inhibitory receptors expressed in NK cells has been identified recently, the molecular mechanism is now understood as follows (Figure 2) (Ravetch & Lanier, 2000). Inhibitory receptors of NK cells inhibit NK activity in normal cells, recognizing MHC class I molecules, which are constantly expressed in normal cells. For cells in which MHC class I expression has decreased, such as in tumor cells, the inhibition is released and the NK cells can display their cytotoxic activity. Inhibitory receptors are generally divided into two types according to their structures: the immunoglobulin superfamily of type I transmembrane proteins and C-type lectins, or type II transmembrane proteins. Inhibitory

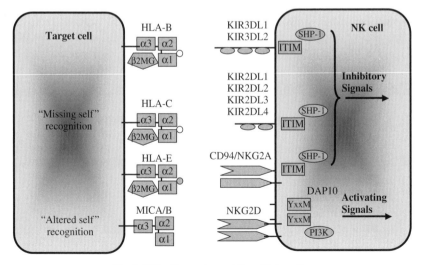

Fig. 2 NK-cell receptor and ligand interaction

receptors that belong to the immunoglobulin superfamily are called "killer cell immunoglobulin-like receptors" (KIRs), and more than 30 human cDNAs have been isolated. A number of these KIRs recognize HLA class I molecules present at the loci of HLA-B and -C, specifically genetic polymorphisms (for example, KIR2DL1 recognizes HLA-Cw4 and KIR2DL2 recognizes HLA-Cw3). Heterodimer receptors consisting of CD94 and NK group 2 (NKG2) are known to exist, being typical members of the C-type lectin family. Among them is CD94/NKG2A (NK group 2, member A), whose ligand is known to be HAL-E. CD94/NKG2A is thought to monitor the entire translation volume of HLA class I in a target cell by recognizing the leader sequence of HLA class I antigens presented by HLA-E. Inhibitory receptors belonging to the immunoglobulin superfamily and the C-type lectin family have a structure called "immunoreceptor tyrosine-based inhibitory motifs" (ITIM) in their intracellular domains. Tyrosyl residues of ITIM are phosphorylated by cross linking of ligands, and thus inhibitory signals are transmitted to NK cells.

Recently, NK cells have been shown to have activating receptors, which activate NK cell functions, as opposed to inhibitory receptors (Raulet, 2003). Among the molecules belonging to the NKG2 family of C-type lectin-like receptors, NKG2D (NK group 2, member D) is an unusual receptor. It has low homology with other NKG2 proteins. Structurally, it forms a homodimer with NKG2D and does not form a heterodimer with CD94. NKG2D forms a complex noncovalently with an adaptor protein called DAP10 and, through cross linking, recruits the p85 subunit of PI3 kinase, thus transmitting the activating signals to NK cells. The NKG2D-activating receptor has attracted attention because its expression in almost all NK cells presumably signifies its importance and because a ligand for it has been identified but not for many other activating receptors. Either MHC class I-related chain A or B (MICA or MICB), both of which are MHC-related molecules, is a ligand for NKG2D in humans (Bauer et al., 1999). MICA or MICB is not expressed in normal cells, being induced by the transformation of cells. This means that NK cells not only recognize abnormality in autologous cells as the missing self but also positively recognize the abnormal self, which is not expressed in normal cells as the altered self to regulate NK cell functions. In this manner, NK cells function as a system by which MHC class I or MHC-related molecules are recognized by the mediation of various NK receptors, leading to the recognition of abnormality in autologous cells.

MICA Expression in HCC and NK-Cell Sensitivity

MIC (MHC class I-related chain) genes make up a gene family identified in the HLA class I region, and seven MIC loci, from A to G, have been confirmed (Figure 3). C to G are pseudogenes, while both MICA and MICB encode 43 kDa proteins. MICA/B are glycoproteins expressed on the cellular membrane, and the structures of the extracellular domains composed of α1, α2, and α3 are similar to those of classic HLA class I molecules. However, their functions differ from those of classic HLA class I molecules because they lack the antigen-presenting function because their structures have no domain for peptides due to the narrow grooves formed by

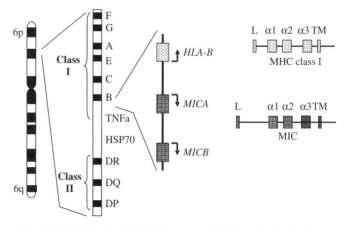

Fig. 3 MHC class I-related chain genes and their molecular structure

the α1 and α2 domains and because their expression is not induced by IFN as they do not need association with β2 microglobulin to be expressed on the membrane (Bahram et al., 1994). In general, MHC class I molecules are constantly expressed in cells while MICA/B are not expressed in normal cells, except in intestinal epithelial cells and some thymocytes. MICA/B expression is known to be induced in stressed cells and transformed epithelial cells. The functions of MICA/B have not been clarified since their discovery in 1994, although in 1999 they were found to be human ligands for NKG2D. Since then, their functional importance in the immune response has attracted attention (Jinushi et al., 2003a, 2003b).

The characteristics of cancer cells from various organs can be examined from the perspective of MHC class I molecule expression. For example, with colon cancer cells, the decreased expression or the deficiency of HLA class I molecules is observed in many mechanisms (Miyagi et al., 2003), and this is thought to be an immune evasion mechanism of colon cancer. By contrast, many hepatoma cells distinctively retain the expression of HLA class I molecules (Takehara et al., 1992). This works against hepatoma cells because of the development of an antigen-specific immune response, although this does help them evade NK cells. In order to clarify the molecular mechanism of NK-cell immune surveillance of HCC, we examined the MICA/B expression in HCC (Jinushi et al., 2003c). Immunohistological examination and PCR analysis of human HCC revealed that non-neoplastic liver tissue had no MICA/B expression, while about 50% of the HCC tissues had it. FACS analysis of hepatoma cell lines showed that many of them had MICA/B expression. When we examined the cytotoxic activity of CD56-positive cells (NK cells) separated from human peripheral blood, the target hepatoma cell lines were found to be susceptible to their cytotoxic activity in various degrees. What is important is that when anti-MICA/B antibody or anti-NKG2D antibody was added to mask these molecules, the cytotoxic activity of CD56-positive cells decreased. Therefore, it appears that the activation of NKG2D by MICA/B plays an important role in inducing the NK-cell sensitivity of HCC (Figure 4).

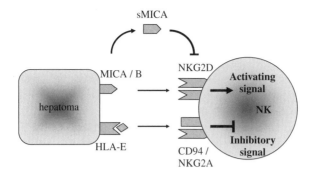

Fig. 4 HCC modulation of NK cells via NK inhibitory and activating receptors

NK Receptor Expression in Chronic Hepatitis C Patients

There is no established view of NK activity in patients with HCV: some researchers have reported decreased NK activity in HCV patients, while others have reported levels equivalent to those of healthy individuals (Ahmad & Alvarez, 2004; Golden-Mason & Rosen, 2006). In order to evaluate the function of NK cells in patients with HCV, we separated CD56-positive cells from the peripheral blood of patients with HCV and healthy donors and examined their cytotoxic activity. To K562 cells, a classic NK-sensitive target, CD56-positive cells in patients with HCV showed the same level of cytotoxic activity as those in healthy donors, but to hepatoma cell lines, the cytotoxic activity of CD56-positive cells decreased in patients with HCV. This suggests that the NK-cell receptor expression profile might differ between patients with HCV and healthy individuals. We next comprehensively analyzed NK receptor expression in CD56-positive cells using FACS. The results showed that for KIR, among the inhibitory receptors, there was no expression difference between the two groups, whereas for NKG2A and CD94, there was a significant increase of expression frequency in patients with HCV (Jinushi et al., 2004). On the other hand, as for NKG2D, one of the activating receptors, there was a trend toward decrease of expression frequency in patients with HCV, but it did not reach a significant level.

We further examined the expression of HLA-E, a ligand for CD94/NKG2A receptor. HLA-E was positive for primary hepatocytes and all hepatoma cell lines tested (HepG2, Hep3B, Huh7), while it was negative for K562. During the chromium release assay of NK-cell cytotoxic activity targeting hepatoma cell lines, we added anti-NKG2A neutralizing antibody, and the cytotoxic activity of NK cells was significantly increased. As for K562, by contrast, the addition of anti-NKG2A antibody made no difference in NK-cell sensitivity. Therefore, signals sent to NKG2A from HLA-E were thought to be inhibitory to NK-cell responsiveness in hepatoma cell lines. The addition of anti-NKG2A antibody led to a clearer increase in NK-cell sensitivity in patients with HCV. This demonstrated that the increase of NKG2A expression caused the decrease of NK-cell cytotoxic activity against hepatoma in patients with HCV.

These findings suggest that the responsiveness of human NK cells to hepatoma cell lines is regulated by a balance of activating signals from NKG2D and inhibitory signals from NKG2A (Figure 4) (Takehara & Hayashi, 2005). In patients with HCV, NKG2A expression frequency is increased and NK responsiveness to hepatoma cells is decreased. This suggests that NK cells, part of the innate immune system, act to recognize and expel hepatoma cells. Hepatocarcinogenesis in patients with HCV involves virus and inflammation factors, and it appears that the additional involvement of these immunological factors might result in a higher hepatoma incidence rate.

Control of DC Function by NK Cells

NK cells make up a cell family defined by an index of direct effector functions of being cytotoxic to transformed cells. Recently, NK cells have attracted attention for the possibility of having effects on the development of adaptive immunity through the modification of DC maturation, activation, and cell death (Gerosa et al., 2002). We have been conducting *in vitro* experiments on how NK cells are involved in DC maturation and activation (Jinushi et al., 2006). DCs that are inductively differentiated from peripheral blood monocytes of healthy donors using GM-CSF and IL-4 show an immature phenotype (IM-DC). In order to clarify how NK cells might be involved in the maturation of these IM-DCs, we conducted a mixed-culture test of IM-DCs and NK cells. The co-culture of IM-DCs with NK cells did not induce maturation, but a 48-hour co-culture of IM-DCs with hepatoma cell lines and NK cells led to increased expression of CD40, CD86, and HLA-DR in DCs and induced DC maturation. During the co-culture, we inserted a trans-well membrane between DCs and NK cells, but DC maturation was induced. This showed that DC maturation was induced via humoral factors, and not by direct cell-to-cell contacts. In fact, the stimulation of IM-DCs using a 24-hour mixed-culture supernatant of NK cells and hepatoma cells resulted in inducing maturation. This maturation was accompanied by functional activation, and allostimulatory capacity toward CD4-positive T cells from healthy donors was significantly enhanced compared with that of IM-DC.

Next, we examined DC maturation and functional activation resulting from co-culture with hepatoma cells and NK cells using NK cells from patients with HCV instead of those from healthy donors. The stimulation of IM-DCs using the supernatant of a 24-hour co-culture of NK cells from HCV patients with hepatoma cells resulted in suppressed DC maturation and allostimulatory capacity compared with the case in which we used NK cells from healthy donors. In order to examine whether NKG2A signals from HLA-E during a mixed culture of hepatoma cells and NK cells are involved in this inhibition of DC maturation and activation, we conducted an inhibition experiment by adding anti-NKG2A antibody during the mixed culture. DC maturation and activation resulting from culture supernatant stimulation were enhanced by adding anti-NKG2A antibody either in the case of

NK cells from healthy donors or in the case of those from patients with HCV. However, DC maturation and activation were more notably enhanced when NK cells from patients with HCV were used. Levels of various cytokines present in the culture supernatant were quantitatively analyzed in each case, and the levels of IFNγ and TNFα were high when NK cells from healthy donors were used, whereas the levels of IL-10 and TGFβ were high when those from patients with HCV were used. Results from experiments where a neutralizing antibody was added for each cytokine into the culture supernatant suggested that the change of the cytokine balance affected DC maturation and activation.

These observations suggest that in patients with HCV, excess NKG2A signals in NK cells not only have an inhibitory effect on the direct effector activity of NK cells but also have a negative effect on the subsequent DC maturation and activation. With improved methods to detect specific T cell responses, such as the ELISPOT assay, there have been reports of some T cell responses specific to cancer antigens in the case of HCC as well. Down-regulated adaptive immune responses from NK cells to DCs might inhibit development of the adaptive immunity.

Secretion of Soluble MICA into Serum in HCC and NKG2D Expression in NK Cells

Recently, it has been reported that some MICA expressed in tumor cells are truncated and their extracellular domains are secreted into culture solutions as soluble forms (Groh et al., 2002; Salih et al., 2002). It is known that soluble MICA (sMICA) is detected in the serum of patients with prostate cancer, colon cancer, brain neoplasm, and leukemia. The importance of this phenomenon is that MICA expression in tumor cells is decreased due to cleavage and NK responsiveness is decreased due to induced NKG2D internalization. These events might be involved in the ability to evade the immunomechanism.

In order to examine the importance of sMICA in liver disease, we conducted ELISA quantitation of serum sMICA from healthy donors and patients with chronic HBV/HCV and HCC (Jinushi et al., 2005). Only a small amount of sMICA was detected in a small number of cases of healthy donors and patients with chronic hepatitis, while notably larger amounts of sMICA were detected in some patients with HCC. We also conducted the test according to the HCC stage and observed that the number of sMICA positive cases was notably more frequent for advanced HCC than for early HCC. We conducted FACS analysis of NKG2D expression in CD56-positive cells from healthy donors and patients with HCC (sMICA positive/negative) and chronic HCV. The results showed that the level of NKG2D expression in patients with hepatitis or HCC (sMICA negative) was the same as that in healthy donors, while the level of NKG2D expression in patients with HCC (sMICA positive) was decreased. Next, in order to examine whether sMICA is involved in the decreased NKG2D expression, we treated CD56-positive cells from healthy donors

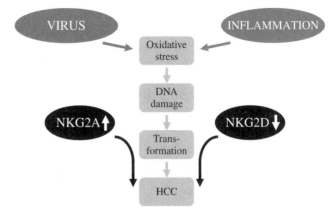

Fig. 5 Possible mechanisms of HCV-related liver carcinogenesis

with 10% patient serum for 48 hours and then determined the NKG2D expression. A significant decrease in NKG2D expression was noted after treatment with serum from patients with HCC (sMICA positive), while the expression remained the same after treatment with serum from other patients. Furthermore, when we added patient serum treated with an antibody that recognized $\alpha 1$ and $\alpha 2$ domains of MICA to NK cells, the decrease of NKG2D expression caused by the serum treatment was canceled. This showed that sMICA of the patient serum was involved in the decrease of NKG2D expression. Using the chromium release assay, we compared hepatoma cell line-specific cytotoxic activity of CD56-positive cells in healthy donors and patients with HCC (sMICA positive/negative) and found that the cytotoxic activity decreased most in patients with HCC (sMICA positive). These observations suggest that an increase of HCC tumor size leads to the release of sMICA into the blood, which triggers a decrease in the NKG2D expression in NK cells and the responsiveness of NK cells to hepatoma cells, thus further suppressing the immune response to tumors (Figure 5).

Immunotherapy for HCC

If HCC is in a limited area, hepatectomy or medical ablation treatment can be prescribed. If the reserve capacity of the liver is sufficient, the technical problems are resolved and the treatment can be conduced with an adequate margin of safety, then radical therapeutic measures can be taken. However, HCC possesses biological characteristics of multiple occurrence in both time and space dimensions, making its treatment extremely difficult. Topical treatment has no effect on multiple HCC spreading over both hepatic lobes. In such a case, transcatheter arterial embolization or arterial injection chemotherapy can be an option, but the effectiveness is limited. Even if topical treatment is successful, distant recurrence is likely to occur in many cases. Thus, there is an urgent need to establish ways to prevent recurrence. In order to improve the prognosis of progressive HCC and the recurrence-free survival rate after topical treatment, we need to develop whole-liver treatment based on a new point of view. It is in this context that there are high expectations for immunotherapy to treat HCC (Butterfield, 2004; Palmer et al., 2005; Avila et al., 2006).

Immunotherapy for cancer has been developing from nonspecific to specific, from those using unknown mechanisms to those where the mechanism has been clarified. To holistically activate immune responses *in vivo*, immunomodulating therapy using bacterial compounds has been replaced by cytokine therapy. Adoptive immunotherapy using lymphokine-activated killer cells (LAK) has been replaced by therapy using tumor-infiltrating lymphocytes (TIL) that are tumor-specific, and nonspecific immunostimulation has been replaced by specific treatment using DC or tumor-specific antigens. In addition, conventional approaches have been reevaluated from an up-to-date point of view: bacterial compounds used for immunostimulation have been found to be ligands for TLRs, which activate innate or adaptive immunity by inducing DC maturation. To treat HCC, possibilities being explored include the application of cytokine therapy, adoptive immunotherapy, DC therapy, and tumor-derived peptide therapy.

The knowledge of the hepatoma recognition mechanism mediated by NK receptors should be useful for developing the immunotherapy for HCC based on new strategies. One possibility lies in the exploration of the method to induce MICA/B expression in hepatoma cells. We demonstrated that all-*trans*-retinoic acid inducement increases MICA/B expression in hepatoma cells and makes them more susceptible to NK cells (Jinushi et al., 2003c). The fact that this phenomenon is lost in the presence of synthetic retinoid, which functions as a competitive inhibitor of all-*trans*-retinoic acid receptors, shows that it is a specific action of all-*trans*-retinoic acid mediated by receptors, not a nonspecific response. It has been reported recently that DNA toxic antitumor agents (Gasser et al., 2005) or inhibitors of histone deacetylating enzyme (Armeanu et al., 2005; Skov et al., 2005) similarly induce MICA/B expression in neoplastic cells. It is also known that with respect to NKG2D expression in NK cells, cytokines such as IL-15 have the inducing capacity. As for the combined therapy of chemotherapeutic agents and cytokines, various combinations have been explored for the treatment of progressive HCC. New combinations should be examined from the perspective of NK receptors and the expression of their ligands.

Concluding Remarks

HCV infection and subsequent hepatocarcinogenesis and HCC progression can be summarized from the perspective of receptor expression in NK cells as follows. HCV patients show increased NKG2A expression, and patients with advanced HCC display decreased NKG2D expression under the influence of sMICA. There is a link between the staged change of NK-cell phenotypes and the decreased cytotoxic activity of NK cells to HLA-E positive/MICA positive hepatoma cell lines. This might have a disadvantageous effect on a living body in the ablation of transformed liver cells or the growth of tumor. It appears that apart from well-known factors from the virus viewpoint and the inflammation viewpoint, the modulation of innate immunity like this is involved in the high rate of hepatocarcinogenesis and the following progression in patients with HCV (Figure 5).

In vivo, DCs are the most potent APC to activate naïve T cells, but to initiate adaptive immunity in this manner, DCs should be mature and activated. In the case of microbe infection, the recognition of molecular structures peculiar to microbes by TLRs induces DC maturation. In the process of carcinogenesis, on the other hand, NK receptors expressed in NK cells only recognize abnormality in autologous cells (altered self or missing self) and activate NK cells. Thus, NK receptors are involved in direct resistance to tumor cells. In addition, NK-cell activation has an influence on DC maturation via humoral factors. Therefore, NK receptors as well as TLRs play their part as an interface in transmitting information about abnormality arising *in vivo* to the immune systems (Figure 6). It appears that the aberrant expression of NK-cell receptors in patients with HCV or progressive HCC might have a negative influence on the development of not only innate immune responses but also adaptive immune responses.

The expression of NK-cell receptors and their ligands changes dynamically during the process from HCV infection to hepatocarcinogenesis to the progression of HCC. The expression kinetics of these molecules might have a close connection with the process. The modification of the expression of these molecules by drugs or cytokines might lead to the development of cancer immunotherapy based on a new perspective. There are great expectations for further progress of research in this field.

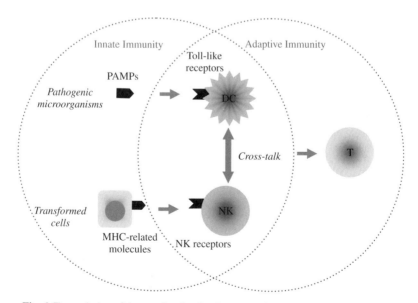

Fig. 6 Transmission of danger signals of pathogens and tumors to the immune system

References

Ahmad, A., Alvarez, F. (2004). Role of NK and NKT cells in the immunopathogenesis of HCV-induced hepatitis. *Journal of Leukocyte Biology*, 76: 743–759.

Akira, S., Uematsu, S., Takeuchi, O. (2006). Pathogen recognition and innate immunity. *Cell*, 124: 783–801.

Armeanu, S., Bitzer, M., Lauer, U.M., Venturelli, S., Pathil, A., Krusch, M., Kaiser, S., Jobst, J., Smirnow, I., Wagner, A., Steinle, A., Salih, H.R. (2005). Natural killer cell-mediated lysis of hepatoma cells via specific induction of NKG2D ligands by the histone deacetylase inhibitor sodium valproate. *Cancer Research*, 65: 6321–6329.

Avila, M.A., Berasain, C., Sangro, B., Prieto, J. (2006). New therapies for hepatocellular carcinoma. *Oncogene*, 25: 3866–3884.

Bahram, S., Bresnahan, M., Geraghty, D.E., Spies, T. (1994). A second lineage of mammalian major histocompatibility complex class I genes. *Proceedings of the National Academy of Sciences of the United States of America*, 91: 6259–6263.

Bauer, S., Groh, V., Wu, J., Steinle, A., Phillips, J.H., Lanier, L.L., Spies, T. (1999). Activation of NK cells and T cells by NKG2D, a receptor for stress-inducible MICA. *Science*, 285: 727–729.

Biron, C.A. (2001). Interferons alpha and beta as immune regulators—a new look. *Immunity*, 6: 661–664.

Butterfield, L.H. (2004). Immunotherapeutic strategies for hepatocellular carcinoma. *Gastroenterology*, 127: S232–S241.

Choo, Q.L., Kuo, G., Weiner, A.J., Overby, L.R., Bradley, D.W., Houghton, M. (1989). Isolation of cDNA clone derived from a blood-borne non-A, non-B viral hepatitis genome. *Science*, 244: 359–362.

Dranoff, G. (2004). Cytokines in cancer pathogenesis and cancer therapy. *Nature Reviews Cancer*, 4: 11–22.

Gasser, S., Orrulic, S., Brown, E.J., Raulet, D.H. (2005). The DNA damage pathway regulates innate immune system ligands of the NKG2D receptor. *Nature*, 436: 1186–1190.

Gerosa, F., Baldani-Guerra, B., Nisii, C., Marchesini, V., Carra, G., Trinchieri, G. (2002). Reciprocal activating interaction between natural killer cells and dendritic cells. *Journal of Experimental Medicine*, 195: 327–333.

Golden-Mason, L., Rosen, H.R. (2006). Natural killer cells: primary target for hepatitis C virus immune evasion strategies? *Liver Transplantation*, 12: 363–372.

Groh, V., Wu, J., Yee, C., Spies, T. (2002). Tumour-derived soluble MIC ligands impair expression of NKG2D and T-cell activation. *Nature*, 419: 734–738.

Hayashi, N., Takehara, T. (2006). Anti-viral therapy for chronic hepatitis C: past, present, and future. *Journal of Gastroenterology*, 41: 17–27.

Jinushi, M., Takehara, T., Kanto, T., Tatsumi, T., Groh, V., Spies, T., Miyagi, T., Suzuki, T., Sasaki, Y., Hayashi, N. (2003a). Critical role of MHC class I-related chain A and B expression on interferon α-stimulated dendritic cells in NK cell activation: Impairment in chronic hepatitis C virus infection. *Journal of Immunology*, 170: 1249–1256.

Jinushi, M., Takehara, T., Tatsumi, T., Kanto, T., Groh, V., Spies, T., Suzuki, T., Miyagi, T., Hayashi, N. (2003b). Autocrine/paracrine IL-15 that is required for type I IFN-mediated dendritic cell expression of MHC class I-related chain A and B is impaired in hepatitis C virus infection. *Journal of Immunology*, 171: 5423–5429.

Jinushi, M., Takehara, T., Tatsumi, T., Kanto, T., Groh, V., Spies, T., Kimura, R., Miyagi, T., Mochizuki, K., Sasaki, Y., Hayashi, N. (2003c). Expression and role of MICA and MICB in human hepatocellular carcinomas and their regulation by retinoic acid. *International Journal of Cancer*, 104: 354–361.

Jinushi, M., Takehara, T., Tatsumi, T., Kanto, T., Miyagi, T., Suzuki, T., Kanazawa, Y., Hiramatsu, N., Hayashi, N. (2004). Negative regulation of NK cell activities by inhibitory receptor CD94/NKG2A leads to altered NK cell-induced modulation of dendritic cell functions in chronic hepatitis C virus infection. *Journal of Immunology*, 173: 6072–6081.

Jinushi, M., Takehara, T., Tatsumi, T., Hiramatsu, N., Sakamori, R., Yamaguchi, S., Hayashi, N. (2005). Impairment of natural killer cell and dendritic cell functions by the soluble form of MHC class I-related chain A in advanced human hepatocellular carcinomas. *Journal of Hepatology*, 43: 1013–1020.

Jinushi, M., Takehara, T., Tatsumi, T., Yamaguchi, S., Sakamori, R., Hiramatsu, N., Kanto, T., Ohkawa, K., Hayashi, N. (2006). Natural killer cell and hepatic cell interaction via NKG2A leads to dendritic cell-mediated induction of CD4+ CD25+ T cells with PD-1-dependent regulatory activities. *Immunology*, in press.

Kaisho, T., Akira, S. (2003). Regulation of dendritic cell function through Toll-like receptors. *Current Molecular Medicine*, 4: 373–385.

Karre, K., Ljunggren, H.G., Piontek, G., Keissling, R. (1986). Selective rejection of H-2 deficient lymphoma variants suggest alternative immune defense strategy. *Nature*, 319: 675–678.

Kato, T., Miyamoto, M., Date, T., Yasui, K., Taya, C., Yonekawa, H., Ohue, C., Yagi, S., Seki, E., Hirano, T., Fujimoto, J., Shirai, T., Wakita, T. (2003). Repeated hepatocyte injury promotes hepatic tumorigenesis in hepatitis C virus transgenic mice. *Cancer Science*, 94: 679–685.

Lemaitre, B., Nicolas, E., Michaut, L., Reichhart, J.M., Hoffmann, J.A. (1996). The dorsoventral regulatory gene cassette spatzle/Toll/cactus controls the potent antifungal response in Drosophila adults. *Cell*, 86: 973–983.

Ljunggren, H.G., Karre, K. (1990). In search on the "missing self": MHC molecules and NK cell recognition. *Immunology Today*, 11: 7–10.

Miyagi, T., Tatsumi, T., Takehara, T., Kanto, T., Kuzushita, N., Sugimoto, Y., Jinushi, M., Kasahara, A., Sasaki, Y., Hori, M., Hayashi, N. (2003). Impaired expression of proteasome subunits and human leukocyte antigens class I in human colon cancer cells. *Journal of Gastroenterology and Hepatology*, 18: 32–40.

Moriya, K., Fujie, H., Shintani, Y., Yotsuyanagi, H., Tsutsumi, T., Ishibashi, K., Matsuura, Y., Kimura, S., Miyamura, T., Koike, K. (1998). The core protein of hepatitis C virus induces hepatocellular carcinoma in transgenic mice. *Nature Medicine*, 4: 1065–1067.

Okuda, M., Li, K., Beard, M.R., Showalter, L.A., Scholle, F., Lemon, S.M., Weinman, S.A. (2002). Mitochondrial injury, oxidative stress, and antioxidant gene expression are induced by hepatitis C virus core protein. *Gastroenterology*, 122: 366–375.

Otsuka, M., Kato, N., Taniguchi, H., Yoshida, H., Goto, T., Shiratori, Y., Omata, M. (2002). Hepatitis C virus core protein inhibits apoptosis via enhanced Bcl-xL expression. *Virology*, 296: 84–93.

Palmer, D.H., Hussain, S.A., Johnson, P.J. (2005). Gene- and immunotherapy for hepatocellular carcinoma. *Expert Opinion on Biological Therapy*, 5: 507–523.

Prince, A.M., Brotman, B., Grady, G.F., Kuhns, W.J., Hazzi, C., Levine, R.W., Millian, S.J. (1974). Long-incubation post-transfusion hepatitis without serological evidence of exposure to hepatitis-B virus. *Lancet*, 7875: 241–246.

Raulet, D.H. (2003). Roles of the NKG2D immunoreceptor and its ligands. *Nature Reviews Immunology*, 10: 781–790.

Ravetch, J.V., Lanier, L.L. (2000). Immune inhibitory receptors. *Science*, 290: 84–89.

Salih, H.R., Rammensee, H.G., Steinle, A. (2002). Downregulation of MICA on human tumors by proteolytic shedding. *Journal of Immunology*, 169: 4098–4102.

Skov, S., Pedersen, M.T., Andresen, L., Straten, P.T., Woetmann, A., Odum, N. (2005). Cancer cell become susceptible to natural killer cell killing after exposure to histone deacetylase inhibitors due to glycogen synthase kinase-3-dependent expression of MHC class I-related chain A and B. *Cancer Research*, 65: 11136–11145.

Smyth, M.J., Hayakawa, Y., Takeda, K., Yagita, H. (2002). New aspects of natural-killer-cell surveillance and therapy of cancer. *Nature Reviews Cancer*, 2: 850–861.

Takehara, T., Hayashi, N., Katayama, K., Ueda, K., Towata, T., Kasahara, A., Fusamoto, H., Kamada, T. (1992). Enhanced expression of HLA class I by inhibited replication of hepatitis B virus. *Journal of Hepatology*, 14: 232–236.

Takehara, T., Liu, X., Fujimoto, J., Friedman, S.L., Takahashi, H. (2001). Expression and role of Bcl-xL in human hepatocellular carcinomas. *Hepatology*, 34: 55–61.

Takehara, T., Takahashi, H. (2003). Suppression of Bcl-xL deamidation in human hepatocellular carcinomas. *Cancer Research*, 63: 3054–3057.

Takehara, T., Hayashi, N. (2005). Natural killer cells in hepatitis C virus infection: from innate immunity to adaptive immunity. *Clinical Gastroenterology and Hepatology*, 3: S78–S81.

Trinchieri, G. (1989). Biology of natural killer cells. *Advances in Immunology*, 47: 187–376.

Mechanisms of Interferon Action and Resistance in Chronic Hepatitis C Virus Infection: Lessons Learned from Cell Culture Studies

Srikanta Dash*, Sidhartha Hazari, Robert F Garry, and Fredric Regenstein

Abstract Alpha interferon, usually in combination with ribavirin, is currently the standard care for patients infected with hepatitis C virus. Unfortunately, a significant number of patients fail to eradicate their infection with this regimen. The molecular details concerning the failure of many patients to achieve sustained clearance of the virus infection after interferon therapy are currently unknown. The primary focus of this chapter is to provide an overview of interferon action and resistance against hepatitis C virus (HCV) based on our understanding developed from *in vitro* experiments. Interferon first binds to receptors on the cell surface; this initiates a cascade of signal transduction pathways leading to the activation of antiviral genes. Using a cell culture model, we determined that the activation of an interferon promoter (interferon inducible genes) is important for a successful antiviral response against HCV. The level of activation of the IFN promoter by exogenous interferon appears to vary among different replicon cell lines. It was observed that a replicon cell line showing low activation of the IFN promoter frequently develops resistant phenotypes compared to cell lines with higher activation. Furthermore, interferon-alpha, -beta, and -gamma are each found to inhibit replication of HCV in the cell culture. The antiviral action of interferon is targeted to the highly conserved 5' untranslated region (5'UTR) utilized by the virus to translate protein by an internal ribosome entry site (IRES) mechanism. This effect is the same among HCVs of other genotypes. Interferon inhibits translation of HCV by blocking at the level of formation of polyribosomes on the IRES containing mRNA. These *in vitro* studies suggest that differences in the regulation of IRES-mediated translation by interferon among hepatic cell clones may be directly related to the development of interferon resistance in chronic HCV infection.

* Corresponding Author: Srikanta Dash, Ph.D., Associate Professor, Department of Pathology and Laboratory Medicine, Tulane University Health Sciences Center, 1430 Tulane Avenue, SL 79, New Orleans, LA 70112, Tel.: 504-988-2519, Fax: 504-988-7389, E–mail: sdash@tulane.edu

Introduction

Hepatitis C virus (HCV) infection is a major public health problem. At present approximately 170 million people worldwide have been infected with HCV. It is the principal virus accounting for chronic liver disease in patients previously designated as having non-A, non-B hepatitis. The hepatitis C virus was cloned and sequenced by a team of investigators from the Chiron Corporation, California, USA, in 1989 (Choo et al., 1989). During subsequent years, significant progress has been made in areas of clinical and molecular virology, development of cell culture models, antiviral therapy, and viral pathogenesis. The hepatitis C virus is an enveloped virus of the *Flaviviridae family* containing a single-stranded, positive-sense RNA genome approximately 9,600 nucleotides in length (Francki et al., 1991; Rice, 1996). The viral genome is organized into a $5'$ untranslated region, followed by a large open reading frame and a $3'$ untranslated region. The HCV RNA genome directly binds to host cell ribosomes and is translated into a large polyprotein of 3,010 amino acids. This polyprotein is subsequently processed in the endoplasmic reticulum of the infected cell into structural proteins (core, El and E2, P7) and non-structural proteins (NS2, NS3, NS4A, NS4B, NS5A, and NS5B) (Reed & Rice, 2000). The structural proteins play important roles in the formation of complete virions, their export, and the infection of host cells. The nonstructural proteins provide the necessary enzymatic activities to replicate the HCV RNA genome. The viral genome persists in infected hepatocytes due to continuous replication of both positive- and negative-strand HCV RNAs in infected cells. The highly conserved structured RNA sequences located at the 5'UTR and 3'UTR are important in the viral translation and replication of the HCV genome (Friebe & Bartenschlager, 2002; Friebe et al., 2001; Yi & Lemon, 2003). Figure 1 demonstrates the structure of the positive-strand HCV genome and different mature proteins produced in the virus-infected cell.

Most people acquire HCV infection through direct contact with infected blood (e.g., blood transfusion, injection drug use). Following exposure to HCV, a robust host immune response is generated; however, in a majority of patients the response fails to eradicate the virus, leading to chronic infection. In chronically infected individuals, the virus preferentially replicates in the liver for prolonged intervals of time leading, both directly and indirectly, to potentially serious liver disease. It is now believed that long-standing chronic inflammation due to HCV infection triggers the development of hepatocellular carcinomas. The strong association between the chronic HCV infection and the development of hepatocellular carcinomas has been made in many parts of the world, including the United States, Japan, Australia, and Europe (El-Serag, 2004; Hoofnagle, 2004, Kiyosawa et al., 2004; Bosch et al., 2004). The mechanisms controlling the development of HCV into chronic infection and then evolving to cirrhosis and cancer appear to be complex.

Interferon-alpha alone, or in combination with ribavirin, is the standard therapy for acute and chronic HCV infection. Sustained virological response can be achieved in up to 90% of acute HCV infections and in approximately 50% of those chronic infections (Feld and Hoofnagle, 2005; Strader et al., 2004). HCV RNA

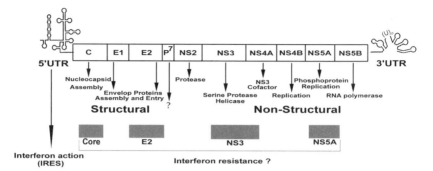

Fig. 1 Organization of the HCV RNA genome and protein translated from the single large open reading frame (ORF). The viral genome begins with a stretch of (1–341) untranslated sequences (5'UTR), followed by a large open reading frame (ORF), then another stretch of an untranslated region called 3'UTR. The 5'UTR that forms the complex secondary structure also initiates translation by directly binding to the host ribosome via internal ribosome entry site (IRES) mechanisms. Interferon specifically inhibits the IRES function in hepatic cells. Ten different mature proteins are generated due to the cleavage of large polyproteins. Among these (core, E1, E2) are structural proteins; NS2-NS5 are known as nonstructural proteins. The exact function of the p-7 protein is unknown. The core protein, envelope protein (E2), the nonstructural protein (NS5A), and the NS3/4A serine protease were reported to play an important role in the mechanisms of interferon resistance

levels in the blood and serum are used to monitor the response to interferon therapy in patients undergoing treatment. Approximately 60% of patients have undetectable levels of virus at the conclusion of therapy. Many chronic hepatitis C patients, particularly those infected with genotype 1, do not respond to interferon therapy. In most of these individuals, HCV RNA levels remain detectable throughout treatment. The reason why some patients become persistently infected, and why some respond to interferon therapy while others do not, is not clear. Understanding the mechanisms of interferon action and interferon resistance should open new directions to developing alternative strategies to improve the clinical efficacy of interferon therapy.

Interferon System

Interferons are the cytokines, which are produced initially to defend the host against infection, through mechanisms that inhibit the replication of a number of viruses. There are two main types of interferon. Type I interferons include interferon-alpha, interferon-beta, interferon-omega, and interferon-delta. Interferon-gamma is a Type II interferon. Interferon-alpha is mainly produced by leukocytes (dendritic cells and macrophages). Interferon-beta is produced by most of the epithelial cells and fibroblasts. Cells of immune system, including T cells and natural killer cells, produce interferon-gamma. Once stimulated by a viral infection, these cells go through a series of signaling events that leads to rapid production of interferons and other cytokines. Interferon is, therefore, an important cytokine during innate host defense against viral infection. It is believed that the interferon system is transcriptionally activated intracellularly within a few hours of virus infection through a cascade of signaling pathways that involve NF-kB, ATF2-c-Jun, and interferon-regulatory factors (IRF 3) and IRF-7 (Kawai & Akira, 2006). In humans, Type I interferons are encoded by 14 functional genes that form the interferon-alpha family. Single genes encode for interferon-beta and -omega, and three genes encodes for interferon-lambda (Sen, 2001; Bekisz et al., 2004). The biological significance of having multiple genes for interferon-alpha and only one for interferon-beta is not clear. The genes for different Type I interferons are all located together on human chromosome 9 (Diaz et al. , 1994), and the Type II interferon gene is located on chromosome 12 (Schroder et al., 2004). The commercially available recombinant interferon used against HCV is interferon-α2a, interferon-α2b, or a consensus interferon (Blatt et al., 1996). The consensus interferon is a recombinant protein that has the most common amino acid sequences derived from several natural interferon-alpha subtypes (Heathcote et al., 1998). All Type I interferons bind to the human interferon-alpha receptor (IFNAR), which consists of an IFNAR-1 and IFNAR-2 subunit (Uze et al., 1990, 1995; Novick et al., 1994; Colamonici et al., 1994; Domanski and Colamonici, 1996; Platanias et al., 1996). IFNAR-1 has a relative molecular weight of 110 kDa, while IFNAR-2 occurs as two forms due to differential splicing of the same gene. These include the IFNAR-2c protein of molecular weight 90–100 kDa and the IFNAR2b protein of molecular weight 51 kDa. There are two distinct interferon-gamma receptors (IFNGR-1 and IFNGR-2). IFNGR-1 has a major binding subunit protein with a molecular weight of 90 kDa; IRNGR-2, a 62-kDa protein, plays a minimal role in ligand binding and is important in downstream signaling pathways (Stark et al., 1998; Bach et al. 1997; Hemmi et al., 1994). All interferons activate a cascade of signal transduction pathways through its receptors that stimulate synthesis of numerous antiviral genes. The differences and similarities between the signaling pathways of Type I interferon and Type II interferon are summarized in Figure 2. Interferon binding to the cell surface receptors activates the intracellular signaling pathways, which involve Janus kinase (JAK1) and tyrosine kinase 2 (TYK2) and signal transducer and activator of transcription (STAT1 and STAT2) proteins. The JAKs phosphorylate the STAT proteins, which either homo- or heterodimerize and then translocate to the nucleus to induce the expression of the IFN-stimulated genes (ISG). The phosphorylated STAT1 and STAT2 combine with IRF-9 (interferon regulatory factor 9) to form a trimeric ISGF-3 complex. This complex enters the nucleus and binds to a consensus DNA sequence [GAAAN (N) GAAA] called the "interferon stimulated response element" (ISRE) (Goodbourn et al., 2000). This regulatory sequence is present upstream of most interferon-alpha and interferon-beta responsive genes. These cascades of molecular signaling are essential for stimulation of interferon-mediated gene transcription. In contrast, binding of interferon-gamma to its receptors leads to tyrosine phosphorylation of STAT1, but not STAT2. The phosphorylated STAT1 protein forms a homodimer called "gamma-activated factor" (GAF) that translocates to the nucleus and binds to a consensus sequence [TTNCNNNAA] called "gamma activation sequence" (GAS) elements. This DNA sequence is present in the upstream regulatory region of the interferon-gamma inducible genes. These cascades of biochemical reactions

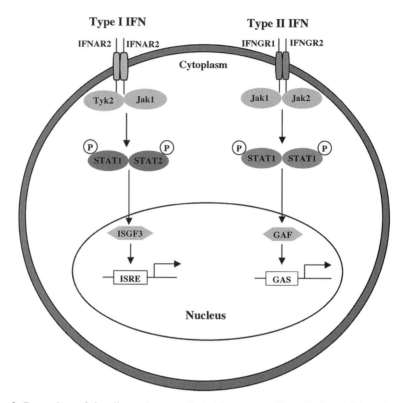

Fig. 2 Comparison of signaling pathways activated in a mammalian cell after addition of exogenous interferon. IFN-alpha/beta (type I IFN) and IFN-gamma (type II) binds to separate cell surface receptors. IFN-alpha or IFN-beta binding to their receptors activates two receptor associated tyrosine kinases, Jak1 and Tyk2, which then phosphorylate the STAT1 and STAT2 proteins. These two phosphorylated proteins combine with IRF-9 to form the trimeric ISGF3 complex. This complex enters the nucleus and binds to a regulatory consensus DNA sequence called ISRE (interferon sensitive response element) present in most of the type I interferon responsive genes, whereas IFN-gamma binding to its receptor leads to activation of Jak-1 and Jak-2 tyrosine kinases, resulting only in phosphorylation of STAT1 protein. The phosphorylated STAT1 protein forms a homodimer called "gamma-activated factor" (GAF). This complex enters the nucleus and binds to the consensus DNA sequence called the GAS (gamma activated sequence), which regulates the induction of type II responsive genes

occurring in normal cells due to interferon treatment have been termed the Jak-Stat pathways (Darnell, 1998). The Jak-Stat pathways activate a large number of genes in the IFN-treated hepatocyte, which are normally quiescent or present at low levels (William, 1991).

The roles of the interferon-stimulated genes have been well established while studying interferon action against different viruses (Katze et al., 2002). These include the double-stranded RNA-activated protein kinase PKR, which inhibits protein synthesis via eIF2alpha phosphorylation, the $2'$-$5'$ oligoadenylate synthatase ($2'$-$5'$ OAS) (which activates RNAse L to degrade viral RNA), the MX GTPase (which blocks viral transport inside the cell), p56 (which inhibits translation via

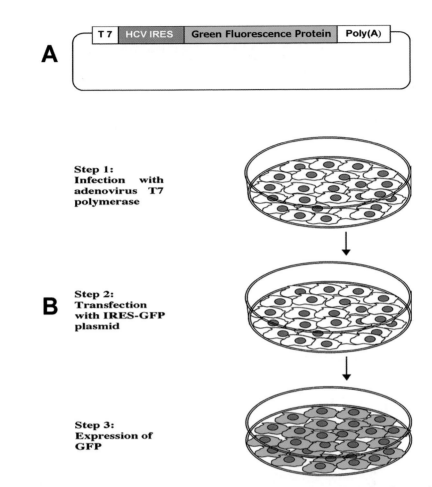

Fig. 3 Summarizes the expression strategy used in our laboratory to determine the interferon action on the HCV IRES translation. (a) Schematic representation of chimeric clone between HCV IRES sequence and green fluorescence protein (GFP). The chimeric clones were prepared by fusing the GFP coding sequences, including a poly (A) tail after the CCU sequence of the 5'UTR by overlapping PCR. (b) The experimental steps involved to express the IRES-GFP clone in hepatic cells. Cells were first infected with adenovirus T7 RNA polymerase. After 2 hours, transfected with plasmid DNA clone between HCV IRES-GFP using the FuGENE 6 transfection reagent. A fairly high-level expression of green fluorescence protein from the HCV IRES can be achieved within 24 hours in most of the cells that can be examined directly under a fluorescence microscope

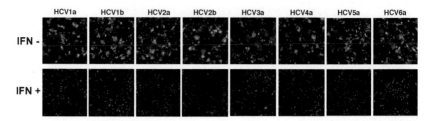

Fig. 4 Interferon alpha inhibits HCV IRES-GFP translation from six different HCV genotypes. Huh-7 cells were transfected with 1 og of HCV IRES-GFP chimera plasmid using the FuGENE 6 transfection reagent. Immediately after transfection, cells were treated with IFN alpha (1000 IU/ml). After 24 hours, cells were counterstained with DAPI (nuclear dye) and examined under a fluorescence microscope. Interferon treatment inhibits GFP expression from IRES of different HCV genotypes. Interferon has no effect on translation of red fluorescence protein in Huh-7 cells

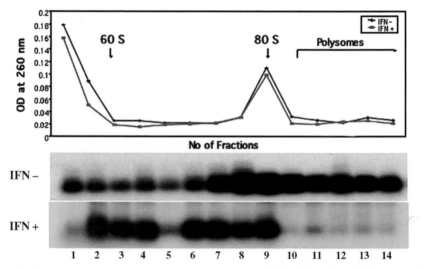

Fig. 5 Effect of interferon treatment on the translation of HCV IRES-GFP mRNA in Huh-7 cells. Huh-7 cells were transfected with HCV IRES-GFP chimera plasmid and then treated with IFN alpha (1000 IU/ml) for 24 hours. Cytoplasmic extracts were prepared and fractionated on 15–50% sucrose density gradients. Top panel shows that the separation of monosome and polysome profiles was not altered after interferon treatment. Total RNA from each fraction was isolated and separated on agarose gel electrophoresis. Distribution of HCV IRES-GFP mRNA in each fraction was determined by northern blot analysis. Results shown in the bottom panel indicate that IFN treatment prevented formation of polysome on the IRES containing GFP mRNA (lanes 10–14). Most of the IRES GFP mRNA is stuck in the fractions 1–9

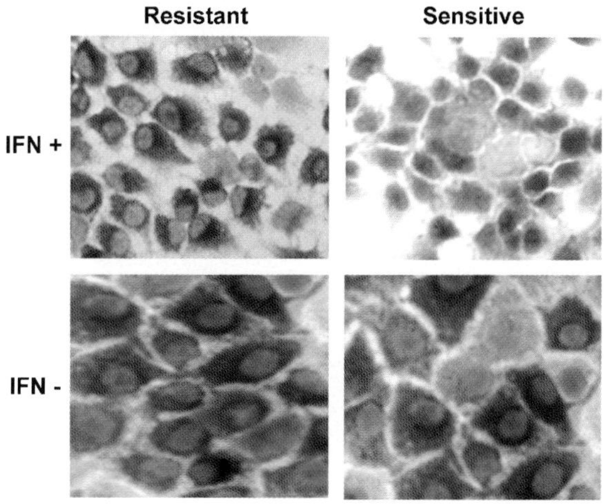

Fig. 6 Effect of interferon on the expression of HCV NS3 protein between replicon cell lines. One IFN-sensitive replicon cell line and one IFN-resistant replicon cell line were treated with interferon-alpha 2b at a concentration of 1000 IU/ml for 72 hours. Expression of NS3 protein was examined by an immunocyto chemical method using a monoclonal antibody. Only IFN-sensitive cells were negative for viral NS3 protein, not the resistant cells

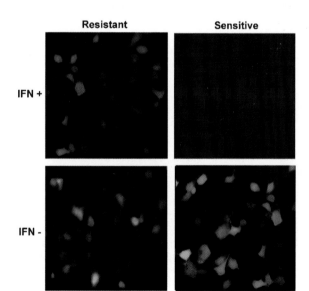

Fig. 8 Interferon effect on IRES-GFP translation between sensitive and resistant Huh-7 cells. One IFN-sensitive and one IFN-resistant Huh-7 cell line was transfected with IRES-GFP plasmid using the FuGENE 6 reagents. Then cells were immediately treated with IFN-a2b (1000 IU/ml). After 24 hours cells were examined for the expression of GFP under a fluorescence microscope. Interferon treatment inhibits HCV IRES in sensitive cells but not in resistant cell line

eIF3), and P-200 family proteins that impair cell proliferation through cellular factors such as NFkB, E2F, P53, and c-Myc. However, the exact mechanism by which interferon activates intracellular pathways to inhibit HCV replication is not fully understood. A detailed understanding of this intracellular signaling pathway is important to improve the success of interferon therapy against chronic HCV.

Mechanisms of Interferon Action

Our understanding of interferon action against the hepatitis C virus is possible due to the availability of HCV cell culture models. Work on this area began almost 10 years ago by Shimizu (1992, 1993, 1996), where HCV replication models were developed in lymphoid cell lines. Subsequently, full-length chimpanzee infectious clones for HCV were developed by the laboratories of Dr. Charles Rice, Rockefeller University (Kolykhalov et al., 1997), and Dr. Jens Bukh of NIH (Yanagi et al., 1997). An initial attempt to establish HCV replication models in hepatic cells was made using full-length RNA transfection (Yoo et al., 1995; Dash et al., 1997). The levels of HCV replication in these models remained low and required the RT-PCR method to detect HCV replication. These technical difficulties have demanded the development of a more reliable cell culture system for HCV. A significant advance in this area took place after development of a subgenomic replicon-based model by Lohmann and Bartenschalager (1999). This technology has allowed development of a stable Huh-7 cell line replicating HCV subgenomic RNA of different HCV genotypes (Miyamoto et al., 2006). Initially, this replicon model has been widely used to test interferon effects on virus replication. Recently, a cell culture model that supports full-length virus replication after natural infection has also been established (Lindenbach et al., 2005; Wakita et al., 2005; Zhong et al., 2005). These advances demonstrate that successful replication of HCV in cell culture can be achieved using cloned viral genome. Many investigators, including our own group, have utilized the replicon-based cell culture model and have shown that interferon-alpha inhibits HCV replication in cell culture (Guo et al., 2001, 2003; Zhu et al., 2003; Cheney et al., 2002; Vrolijik et al., 2003; Dash et al., 2005). The antiviral effect of interferon-alpha has been confirmed using full-length chimpanzee infectious clone for HCV 1a virus using a DNA-based replication model (Chung et al., 2001; Prabhu et al., 2004). Interferon-alpha inhibition of virus replication after natural infection has been also confirmed (Shimizu & Yoshikura, 1994; Lindenbach et al., 2005). Results of these investigations indicate that interferon-alpha successfully inhibits hepatitis C virus replication, and the amount of interferon used in all these experiments is between 10–1000 IU/ml. The maximum inhibitory effect was seen between 24 to 72 hours after interferon treatment. There are reports suggesting that other interferons (interferon-beta, interferon-gamma, interferon-lambda) can also inhibit replication of HCV (Frese et al., 2002; Cheney et al., 2002; Dash et al., 2005; Robek et al., 2005). It has also been reported that interferon inhibits the replication of HCV strain 2a in culture models (Kato et al., 2005). We have determined that replication of HCV is equally sensitive to a pegylated form of interferon-alpha, which has been very effective in the treatment of chronic hepatitis C virus infection (unpublished data). Taken together, it appears that interferon–alpha, –beta, and -gamma each can inhibit HCV replication in cell culture models.

The mechanism(s) by which each interferon inhibits hepatitis C virus replication is unknown. The antiviral genes induced by interferons directly or indirectly block at the level of viral translation and transcription (Landolfo et al., 1995). To address the mechanisms of interferon action, studies have been performed earlier to examine the effect of interferon at the level of HCV IRES translation using different subgenomic expression system. The HCV genome contains a 341-nucleotide untranslated sequence that binds to the host cell ribosomes and initiates translation by an internal ribosome entry site (IRES) mechanism (Ji et al., 2004; Otto and Pulglisi, 2004). Several laboratories have examined the action of IFN on HCV IRES–mediated translation (Kato et al., 2002; Guo et al., 2004; He et al., 2003; Shimazaki et al., 2002; Koev et al., 2002; Wang et al., 2003; Rivas-Estilla et al., 2002) with conflicting results. For example, studies by Shimazaki et al. (2002) suggested that interferon-alpha selectively inhibits IRES-mediated translation in a hepatic cell line stably transfected with a clone carrying a dicistronic cassette. This inhibition was due to reduced expression of La protein, but the protein kinase R (PKR) and the RNase L pathway were not involved. A report from Kato et al. (2002) used a dicistronic-based expression system and demonstrated that interferon-alpha inhibits cap- as well as HCV IRES-mediated translation without the involvement of La or PKR pathways. In contrast, Wang et al. (2003) used a similar dicistronic expression system and found that HCV as well as EMCV IRES-mediated translation was more sensitive to interferon-alpha than cap-dependent translation. A study by Koev et al. (2002) also suggested that cap-dependent translations were dramatically inhibited by interferon-alpha, whereas HCV IRES-mediated translations were only marginally inhibited in Huh-7 and HepG2 cells. In their study, IRES-mediated translation inhibition was observed only when the monocistronic IRES-luciferase expression system was used. Another study by Rivas-Estilla et al. (2002) suggested that interferon-induced PKR plays a direct role in the inhibition of protein synthesis from the subgenomic clone. These investigators found that PKR induces HCV IRES activity and inhibits EMCV IRES activity using a dicistronic-based expression system. Guo et al. (2004) also examined the effect of interferon on HCV IRES and EMCV IRES translation using a dicistronic-based expression. In their analysis, the antiviral activity of interferon was only a twofold inhibition on HCV IRES-directed translation when transfected with HCV replicon RNA, but not with dicistronic reporter construct. In summary, all these reports showed that HCV IRES is inhibited by interferon, while some studies demonstrated that cap-dependent translation was also inhibited. The reasons for these differences in the results from various laboratories regarding the IRES-mediated translation regulation are unknown. One possibility may be that the inhibition of both cap- and IRES-mediated translation inhibitions may reflect the use of the dicistronic-based expression systems. Another possible explanation may be the differences in the cell line or the relative sensitivity of the assay systems used in these studies.

To resolve this conflicting evidence, we have taken a different approach. We used a green fluorescence protein-based expression system that allows the effect of interferon on HCV translation to be directly determined in a reliable way under a fluorescent microscope (Kalkeri et al., 2001). An inducible expression system was established that allowed high-level expression of GFP from HCV IRES clone in the majority of Huh-7 cells using T7 RNA polymerase (Figure 3). Standard polymerase chain reaction (PCR) and cloning methods were used to construct chimeric clones between green fluorescence protein (GFP) and IRES of different genotypes of HCV. This model allows us to examine cells expressing GFP directly under a fluorescent microscope, without the requirement for immunological detection pro-

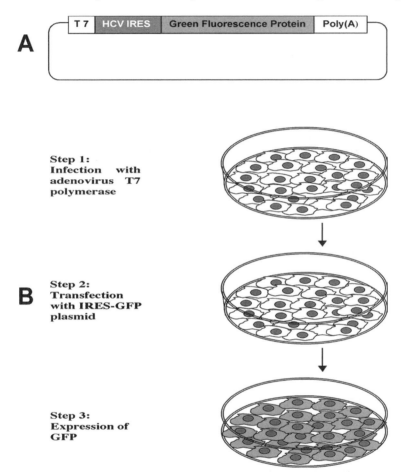

Fig. 3 Summarizes the expression strategy used in our laboratory to determine the interferon action on the HCV IRES translation. (a) Schematic representation of chimeric clone between HCV IRES sequence and green fluorescence protein (GFP). The chimeric clones were prepared by fusing the GFP coding sequences, including a poly (A) tail after the CCU sequence of the 5'UTR by overlapping PCR. (b) The experimental steps involved to express the IRES-GFP clone in hepatic cells. Cells were first infected with adenovirus T7 RNA polymerase. After 2 hours, transfected with plasmid DNA clone between HCV IRES-GFP using the FuGENE 6 transfection reagent. A fairly high-level expression of green fluorescence protein from the HCV IRES can be achieved within 24 hours in most of the cells that can be examined directly under a fluorescence microscope (See color insert.)

cedures or luciferase-based assay. To examine the effect of interferon treatment on non-IRES translation, Huh-7 cells were co-transfected with 1 μg of IRES-GFP plasmid DNA along with 1 μg of pDsRed2 plasmid (BD Biosciences Clontech, Palo Alto, CA) using the FuGENE 6 (Roche Molecular Biology, Indianapolis, IN) transfection reagent. The effect of interferon on IRES and non-IRES translation in a hepatic cell line was determined after 24 hours by examining the expression of green fluorescence or red fluorescence protein under a fluorescence microscope. Transfected cells were examined using a fluorescence microscope (Olympus) at 484 nm for the expression of green fluorescence, 563 nm for the expression of red fluorescence, and 340 nm for DAPI (nuclear dye). The percentage of GFP-positive Huh-7 cells was quantitatively measured using Cell Quest computer software and by flow analysis. Using this GFP-based subgenomic expression system, we demonstrated that interferon-alpha inhibits translation from the HCV IRES-GFP clone in a dose-dependent manner by fluorescence microscopy, Western blot analysis, and flow cytometry (Dash et al., 2005). In this expression system, cap-dependent, non-IRES translation of red fluorescence was not affected by interferon. Interestingly, we found that this effect also occurs for other positive-stranded RNA viruses that require IRES for protein translation such as encephalomyocarditis virus (EMCV) and classical swine fever virus (CSFV). Interferon-beta and -gamma also have a similar effect on IRES-mediated translation, indicating that there may be a common pathway by which different interferons inhibit IRES-mediated translation.

In patients being treated for chronic hepatitis C, it has been well established that individuals infected with genotype 1 respond less often than those infected with genotypes 2 and 3 (Davis & Lau, 1997; Davis et al., 1998). Our group has determined that the differential response to IFN therapy is not at the level of IRES-mediated translation. We showed that IFN treatment directly inhibited translation of GFP mediated by IRES sequences of HCVs derived from other genotypes (Figure 4) (Hazari et al., 2005b). We also found that inhibition of IRES-mediated translation of HCV was not sensitive to other cytokines, including IL-1, IL-6, or TNK-alpha. These studies establish that interferon-alpha, -beta, and -gamma each efficiently inhibit replication of HCV subgenomic RNA in cell culture. This inhibitory effect is targeted to the 5′ untranslated region (5'UTR), which the virus used to translate its genome by an internal ribosome entry site (IRES) mechanism.

The molecular mechanism by which interferon treatment selectively inhibits HCV IRES-GFP mRNA-mediated translation in Huh-7 cells is unknown. To understand the cause of IRES-GFP mRNA translation inhibition by interferon, we examined the involvement of RNA and protein degradation pathways and loading of ribosome on HCV IRES-GFP containing mRNA. Involvement of the RNA degradation pathway in interferon-treated cells was confirmed by comparing the stability of IRES-GFP mRNA in the transfected Huh-7 cells by northern blot analysis. Induction of the RNase L pathway in IFN-treated cells was detected by measuring levels of 2′-5′ oligoadenylate synthetase 3 (OAS3) protein by Western blot analysis. These studies indicate that RNAse L pathways are induced in interferon-treated cells. We did not notice a significant difference in the stability of HCV-IRES-GFP mRNA in Huh-7 cells treated with or without IFN-alpha 2b (Hazari et al., 2005a). These results suggest that the 2-5A-endonuclease system does not discriminate between

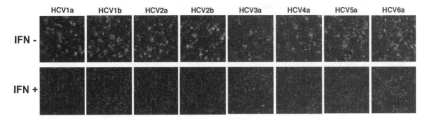

Fig. 4 Interferon alpha inhibits HCV IRES-GFP translation from six different HCV genotypes. Huh-7 cells were transfected with 1 μg of HCV IRES-GFP chimera plasmid using the FuGENE 6 transfection reagent. Immediately after transfection, cells were treated with IFN alpha (1000 IU/ml). After 24 hours, cells were counterstained with DAPI (nuclear dye) and examined under a fluorescence microscope. Interferon treatment inhibits GFP expression from IRES of different HCV genotypes. Interferon has no effect on translation of red fluorescence protein in Huh-7 cells (See color insert.)

viral and cellular mRNAs, indicating that RNaseL plays a minimal role in translational inhibition. These findings also suggest that the cause of translation inhibition may not be due to extensive degradation of IRES-GFP mRNA.

Interferon is known to activate the proteasome, which is a ubiquitously expressed multi-subunit complex that degrades proteins. To examine the possibility that IFN treatment induced the proteasome that specifically degrades GFP translated from the IRES construct, inhibition of GFP expression from IRES clones by interferon was assessed in the presence of two proteasome inhibitors: lactacystin and epoxomicin (Fenteany & Schreiber, 1998; Meng et al., 1999). Inhibition of GFP expression from the IRES clones by IFN treatment was not affected when Huh-7 cells were pretreated with the proteasome inhibitors, indicating that interferon action on the IRES inhibition is not due to activation of proteasome pathways. We also examined the possibility that interferon treatment impaired the loading of polyribosome with IRES containing mRNA in the transfected Huh-7 cells. This was accomplished by examining the distribution of IRES containing GFP mRNA in polysome fractions by northern blot analysis. The progression of mRNA association from monosomes to polysomes is an indication of increased ribosome loading and increased translation. An observed increase of mRNA into monosome fractions and reduction in polysome fractions is suggestive of a decreased loading of mRNA to ribosomes. To test this possibility, Huh-7 cells were transfected with IRES-GFP plasmid and then treated with IFN-alpha 2b (1,000 IU/ml) for 24 hours. Cytoplasmic extracts were prepared and applied to 15–45% (w/v) sucrose density gradients. The gradients were fractionated, and a polysome profile was generated. Total RNA from each fraction was isolated and analyzed by agarose gel electrophoresis. The amount of HCV IRES-GFP mRNA and capped GFP mRNA in the monosome or polysome fractions was determined by northern blot analysis (Figure 5) (Hazari et al., 2004). The association of IRES and capped messages in the ribosome fraction in Huh-7 cells was determined in experiments with or without interferon-alpha. Interferon treatment did not affect the basal distribution of monosome and polysomes but selectively prevented the loading of polyribosome with HCV IRES containing GFP mRNA. In two different control experiments, interferon treatment did not alter the distribution of either GFP mRNA or GAPDH mRNA, which express proteins by a

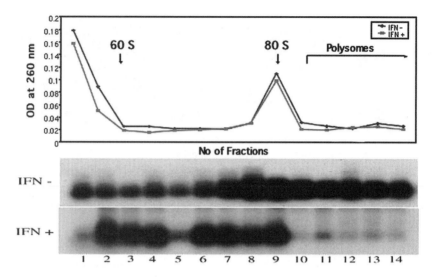

Fig. 5 Effect of interferon treatment on the translation of HCV IRES-GFP mRNA in Huh-7 cells. Huh-7 cells were transfected with HCV IRES-GFP chimera plasmid and then treated with IFN alpha (1000 IU/ml) for 24 hours. Cytoplasmic extracts were prepared and fractionated on 15–50% sucrose density gradients. Top panel shows that the separation of monosome and polysome profiles was not altered after interferon treatment. Total RNA from each fraction was isolated and separated on agarose gel electrophoresis. Distribution of HCV IRES-GFP mRNA in each fraction was determined by northern blot analysis. Results shown in the bottom panel indicate that IFN treatment prevented formation of polysome on the IRES containing GFP mRNA (lanes 10–14). Most of the IRES GFP mRNA is stuck in the fractions 1–9 (See color insert.)

non-IRES-dependent mechanism. These results suggest that RNA or protein degradation pathways do not contribute to the interferon action on IRES-mediated translation inhibition. Rather, interferon treatment specifically blocks the loading of the ribosome to the HCV IRES containing mRNA. In summary, we have found that interferon inhibits HCV replication by specifically blocking the translation at the level of loading of ribosome to the IRES containing mRNA.

Mechanisms of Interferon Resistance

The nucleotide sequences of HCV genomes isolated from patients from different parts of the world are quite heterogeneous. Six major genotypes of HCV virus show 30–50% variation in their nucleotide sequences (Simmonds, 2004). More than 50 subtypes of HCV have also been described, showing 15–30% difference in their nucleotide sequences. Isolates of HCV from a single patient can show a 1–5% difference in their nucleotide sequences (Hoofnagle, 2002). The sequence variability suggests that the HCV genome mutates frequently during replication and circulates in the serum as a population of quasispecies. Interferon therapy, usually in combination with ribavirin, is the standard treatment for chronic HCV infection throughout the world. HCV genotype is the most important predictor of treatment outcome. Sustained virologic responses can be achieved in up to 82% of patients infected with genotypes 2 and 3, whereas a substantially lower response rate, around 40–50%, is achieved in patients with genotype 1 (Manns et al., 2001; Fried et al., 2002; Hadziyannis et al., 2004). This clinical observation leads to speculation that viral factors may be a determinant to the response to interferon therapy. Additional viral factors, including viral load and viral heterogeneity (the more heterogenous, the less the response), have also been associated with response to interferon therapy (Pawlotsky, 2000, 2003). To address the role of virus heterogeneity in interferon resistance, a study by Enomoto and coworkers (1995, 1996) cloned and sequenced the entire HCV genomes from multiple interferon-sensitive and interferon-resistant chronic hepatitis C patients. They found that patients who respond to interferon had multiple amino acid substitutions in the NS5A gene between 2,209–2,248 compared to nonresponders, leading to the conclusion that the 40 amino acid sequence in the NS5A protein is the interferon sensitivity-determining region (ISDR). Patients exhibiting multiple substitutions in this region were sensitive to interferon, and patients who do not show any change in the amino acid sequence were nonresponders. In their study, the sequences of HCV derived from patients were compared to prototype HCV genotype 1b. A number of clinical studies have been performed in different countries following this initial observation, and results are inconclusive (Paterson et al., 1999; Hoffmann et al., 2005; Schinkel et al., 2004; Macquillan et al., 2004; Nousbaum et al., 2000).

To understand the role of the virus in the mechanisms of interferon resistance, several molecular studies have been performed using a cell line suggesting that individual proteins of HCV including core, envelope, NS3 protease, and NS5A proteins can block the antiviral action of interferon. Initially, many investigations have focused on one antiviral protein called "protein kinase R" (PKR) (Langland et al., 2006). This protein is induced by interferon and inhibits protein synthesis by phosphorylating eukaryotic translation initiation factor, eIF-2α. It is believed that phosphorylation of eIF-2α on S51 results in potent inhibitor of eIF2B, the nucleotide exchange factor necessary for recycling of eIF2α and ribosome loading to mRNA. Gale et al. (1997, 1998) reported that NS5A of HCV directly interacts with protein kinase R (PKR) and inhibits its function. It was found that NS5A inhibits PKR function in an ISDR-dependent mechanism. They showed that introduction of mutations seen in IFN-sensitive patients abolished the PKR-dependent suppression of HCV translation. Another report by Taylor et al. (1999) correlates the association of IFN resistance with the HCV E2 protein. They found that a 12-amino acid sequence (276-287) of E2 protein of IFN-resistant HCV strain is homologous to a PKR-eIF-2α phosphorylation site. Because of this sequence homology in the E2 gene, virus strains from the IFN-resistant patients may inhibit PKR activity and block the antiviral action of interferon. This observation provides a potential explanation for why the majority of genotype 1 HCV develops resistance to interferon. Following these reports, there were studies documenting that the NS5A protein can block interferon action by inducing expression of the pro-inflammatory cytokine IL-8. Polyak et al. (2001) found that IL-8 levels were significantly higher in chronic HCV patients who did not respond to IFN therapy. IL-8 has been found to block interferon action by reducing the expression of $2'-5'$-A oligoadenylate synthetase activity (Khaber et al., 1997). Additional reports suggest that the core protein of HCV can also interfere with interferon-induced signaling pathways in cultured cells. Studies performed by Lin et al. (2005) suggest that the core protein can bind and degrade the STAT1 protein, preventing IFN signaling to the nucleus via the Jak-STAT pathway. A report of Basu et al. (2001) indicates that the core protein can modulate formation of GAF and ISGF3 complex but does not interfere with the IFN-stimulated activation of IRF-1 genes. This effect may be due to the low abundance of STAT1 protein and its translocation to the cell nucleus. Another report by Lucas et al. (2005) suggests that core protein inhibits interferon-induced transcription of antiviral genes by decreasing binding of ISGF3 to the ISRE promoter. There is evidence suggesting that the expression of various structural and nonstructural proteins of HCV could inhibit the JAK-STAT pathway and prevent antiviral action of interferon alpha (Francois et al., 2000; Blinderbacher et al., 2003; Keskinen et al., 2002).

Recently, the laboratory of Michael Gale and Stanley Lemon published results of a series of studies indicating that the viral NS3 protease may be another player responsible for the persistent nature of HCV infection (Foy et al., 2003, 2005; Li et al., 2005; Sumpter et al., 2005; Ferreon et al., 2005). They found that the NS3 serine protease of HCV can stymie the antiviral state in Huh-7 cells by blocking intracellular production of interferon. These findings explain the reason why HCV frequently develops a chronic persistent infection in human. Studies emerging in these areas now suggest that natural virus infection cells maintain an antiviral state by inducing the expression of a large number of cellular antiviral defense genes, including the type 1 alpha/beta interferon, interferon-stimulated genes, cytokines, and pro-inflammatory cytokines (Karin et al., 2006). There are now reports suggesting that products of viral replication, such as double-stranded RNA, lead to IRF-3 and NF-kB activation through two distinct and independent pathways. One pathway involves the engagement of Toll-like receptor 3 (TLR-3) by viral double-stranded RNA, and the other involves recognition of structured viral RNA by cellular DexH/D RNA helicase and retinoic acid-inducible gene 1 (RIG-1) (Akira et al., 2006; Garcia-Sastre & Biron, 2006). These investigators have observed that the NS3 protease of HCV can disrupt each pathway and inhibit intracellular production of interferon-beta. These findings now provide a clear rationale for the chronic, persistent nature of HCV infection in humans. The inhibitory action of each protein of HCV on the interferon signaling pathways is summarized in Table 1. Taken together, these studies suggest that several different proteins encoded by HCV can interact with IFN signaling pathways and inhibit antiviral action against the virus.

Studies performed in our laboratory have provided evidence that cellular factors can also play a role in developing resistance to exogenous interferon. We used replicon-based stable cell clones since they represent a close homology of hepatocytes chronically infected with the hepatitis C virus. We initially tested whether the IFN signaling pathways were functional in different Huh-7 cell clones replicating the HCV subgenomic RNA. This was done in transient transfection experiments using a luciferase reporter assay attached to interferon-sensitive response elements (pISRE-Luc). Regulation of ISRE-mediated expression of firefly luciferase

Table 1 Inhibition of IFN Action by Different HCV Proteins

HCV Proteins	Cellular Targets	Mechanisms of Inhibition
Core protein	STAT1	Block JAK-STAT signaling
E2 protein	PKR, eIF-2α	Block translation
NS3/NS4A	IRF-3	Block IFN-β production
NS5A	PKR	Block translation
NS5A	IL-8,	RNA degradation

by interferon-alpha was studied by transfecting this clone directly into different replicon cell lines. We found that interferon treatment activates ISRE-mediated expression of luciferase in all three replicon cell lines, indicating that the pathway is functional in Huh-7 cells. We also noticed that the level of activation of the ISRE promoter (interferon promoter) varied among different replicon stable cell lines. The activation of this promoter is not exclusively due to replication of HCV, since differences are seen even after eliminating HCV replication from these cells. We found that the differential activation of the interferon promoter is due to the clonal nature of the replicon cell lines. The significance of ISRE-mediated transcriptional activation was studied in a replicon cell line by pretreatment of cells with actinomycin D, which inhibits cellular DNA-dependent RNA transcription. We determined that inhibition of the ISRE-mediated transcription of luciferase by actinomycin D makes HCV replication totally resistant to interferon-alpha (Pai et al., 2005). These *in vitro* studies suggest that activation of interferon-inducible genes is important in mounting a successful antiviral response against HCV. The level of IFN-promoter activation can be influenced by the clonal nature of cells, not by the replication of HCV subgenomic RNA clone used in our experiments. The above findings also suggest that the presence of nonstructural proteins (NS3, NS4A, NS4B, NS5A, and NS5B) did not inhibit activation of the ISRE promoter by IFN-alpha 2b. We have also showed that interferon treatment inhibits replication of full-length infectious HCV 1a clone in which both structural and nonstructural proteins were expressed (Prabhu et al., 2004). These observations suggest that activation of Jak-Stat pathways and induction of interferon-inducible genes are important in mounting a successful antiviral response against HCV.

In subsequent studies, we determined that differential activation of this IFN promoter among cell clones is also related to the development of an interferon-resistant phenotype in the cell culture (Hazari et al., 2005a). Two Huh-7 clones, 5–15 (higher activation) and Con-15 (low activation), replicating HCV were studied after treatment with interferon-alpha for an extended period of time in the culture. The development of cell colonies that are resistant to interferon-alpha action was examined. We found that interferon-resistant cell colonies were developed only in low inducer cell clones, and no resistant clones present in the cell clones exhibited higher activation of IFN promoter. Using this approach, we have now prepared several replicon cell clones in which HCV replication and translation are totally resistant to interferon-alpha action. Intracellular HCV expression was not altered in these replicon cells by interferon-alpha treatment (Figure 6). To examine whether interferon signaling is functional in these cell clones, HCV replication from

Fig. 6 Effect of interferon on the expression of HCV NS3 protein between replicon cell lines. One IFN-sensitive replicon cell line and one IFN-resistant replicon cell line were treated with interferon-alpha 2b at a concentration of 1000 IU/ml for 72 hours. Expression of NS3 protein was examined by an immunocyto chemical method using a monoclonal antibody. Only IFN-sensitive cells were negative for viral NS3 protein, not the resistant cells (See color insert.)

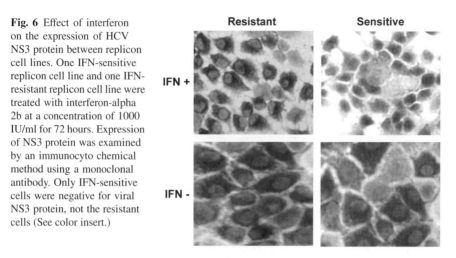

each of these interferon-resistant Huh-7 cell lines was eliminated by treatment with cyclosporin-A. Activation of ISRE-mediated expression of luciferase was measured in these cured and uncured cells after interferon treatment. We found that all the cured cell clones had lower interferon signaling and lower activation of ISRE promoter (Figure 7). To ensure that translational inhibition was a specific effect to the interferon treatment, we examined the effect of IFN on HCV IRES-GFP translation in both interferon-sensitive and interferon-resistant cells. Results shown in Figure 8

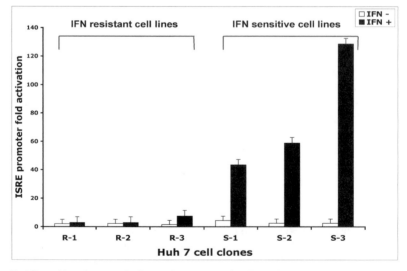

Fig. 7 Effect of interferon on firefly luciferase expression from ISRE promoter between sensitive and resistant cell clones. Huh-7 cells were transfected with 1 μg of reporter plasmid (pISRE-Luc) using the FuGENE 6 transfection reagent and then treated with IFN-alpha 2b (1000 IU/ml). After 24 hours cells were lysed, and equal amounts of protein lysates were counted for luciferase activity. R-1, R-2, R-3: Three different IFN-resistant Huh-7 cell lines developed in our laboratory. S-1, S-2, S-3: Three different IFN-sensitive Huh-7 cell lines. All three IFN-resistant cell clones show reduced expression of luciferase as compared to the sensitive cell lines

Fig. 8 Interferon effect on IRES-GFP translation between sensitive and resistant Huh-7 cells. One IFN-sensitive and one IFN-resistant Huh-7 cell line was transfected with IRES-GFP plasmid using the FuGENE 6 reagents. Then cells were immediately treated with IFN-alpha 2b (1000 IU/ml). After 24 hours cells were examined for the expression of GFP under a fluorescence microscope. Interferon treatment inhibits HCV IRES in sensitive cells but not in resistant cell line (See color insert.)

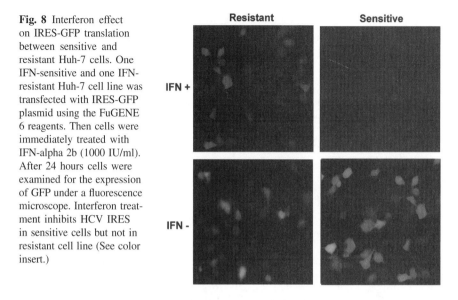

clearly indicate that GFP translation can be inhibited only in cells that are sensitive to interferon, but not in cells that are resistant to interferon. These results also suggest that the inhibition of viral IRES by interferon is a specific phenomenon. We have developed replicon cell clones that have altered Jak-Stat signaling. These cell clones can now be used to understand the role of viral and host factor involvement in the mechanisms of interferon resistance against HCV. These preliminary studies have now provided new evidence that an altered cellular response can make HCV replication resistant to interferon.

In summary, the results of all these studies indicate that there is a complex interaction between the virus and host factors that can interfere both with the intracellular production and with the cell's response to exogenous interferon. It is vital that the significance of viral and host factors regulation in the IFN action be further examined using HCV sequences and hepatic cell clones derived from patients who remain resistant to interferon therapy.

Conclusions

We discussed the progress made in our understanding of the mechanisms of interferon action and interferon resistance from basic research on HCV. Chronic hepatitis C virus infection is the major cause of liver cirrhosis and liver cancer in the United States and in many developed nations. The most effective way of preventing HCV-associated liver cancer is to eradicate chronic hepatitis C virus infection from the human population by developing effective antiviral strategies. Interferon therapy is a highly effective first line of treatment available against hepatitis C virus infection. The most challenging task is to cure those chronically infected patients not responding to interferon-based therapy. Research in this area will increase our understanding of the mechanisms of interferon resistance and develop alternative strategies to treat chronic HCV infections that are IFN nonresponders.

Acknowledgments This review was made possible through a series of investigations carried out in our laboratory over several years due to financial support from the National Institute of Health, CA54576, CA89121 and in part by the funds received from the Tulane Cancer Center. The authors are unable to cite many important contributions of other investigators due to the page limitation. The authors wish to acknowledge Jeanne Frois for critically reading the article.

References

Akira, S., Uematsu, S., Takeucji, O. (2006). Pathogen recognition and innate immunity. *Cell*, 124: 783–801.

Bach, E.A., Aguet, M., Schreiber, R.D. (1997). The IFN gamma receptor: a paradigm for cytokine receptor signaling. *Annual Review of Immunology*, 15: 571–591.

Basu, A., Meyer, K., Ray, R.B., Ray, R. (2001). Hepatitis C virus core protein modulates the interferon-induced transacting factors of JAK/STAT signaling pathways but does not affect the activation of downstream IRF-1 or 561 gene. *Virology*, 288: 379–390.

Bekisz, J., Schmeisser, H., Hernandez, J., Goldman, N.D., Zoon, K.C. (2004). Human interferons alpha, beta and omega. *Growth Factors*, 22: 243–251.

Blindenbacher, A., Duong, F.H.T., Hunziker, L., Stutvoet, S.T.D., Wang, X., Terracciano, L., Moradpour, D., Blum, H.E., Alonzi, T., Tripodi, M., Monika, N.L., Heim, M.H. (2003). Expression of hepatitis C virus proteins inhibits interferon alfa signaling in the liver of transgenic mice. *Gastroenterology*, 124: 1465–1475.

Blatt, L.M., Davis, J.M., Klein, S.B., Taylor, M.W. (1996). The biological activity and molecular characterization of a novel synthetic interferon alpha species, consensus interferon. *Journal of Interferon and Cytokine Research*, 16: 489–499.

Bosch, F.X., Ribes, J., Diaz, M., Cleries, R. (2004). Primary liver cancer: worldwide incidence and trends. *Gastroenterology*, 127: S5–S16.

Cheney, I.W., Lai, V.C.H., Zhong, W., Brodhag, T., Dempsey, S., Lim, C., Hong, Z., Lau, J.Y.N., Tam, R.C. (2002). Comparative analysis of anti-hepatitis C virus activity and gene expression mediated by alpha, beta, and gamma interferons. *Journal of Virolology*, 76: 11148–11154.

Choo, Q.L., Kuo, G., Weiner, A.J., Overby, L.R., Bradley, D.W., Houghton, M. (1989). Isolation of a cDNA clone derived from a blood-borne non-A, non-B viral hepatitis genome. *Science*, 244: 359–362.

Chung, R.T., He, W., Saquib, A., Contreras, A.M., Xavier, R.J., Chawla, A., Wang, T.C., Schmidt, E.V. (2001). Hepatitis C virus replication is directly inhibited by IFN-alfa in a full-length binary expression system. *Proceedings of National Academy of Sciences (USA)*, 98: 9847–9852.

Colamonici, O.R., Uyttendale, H., Domanski, P., Yan, H., Krolewski, J.J. (1994). P135 tyk2 an interferon-dependent tyrosine kinase, is physically associated with an interferon receptor. *Journal of Biological Chemistry*, 269: 3518–3522.

Darnell, J.E. Jr. (1998). Studies of interferon–induced transcriptional activation uncover the Jak–STAT pathway. *Journal of Interferon and Cytokines*, 18: 549–554.

Dash, S., Hiramatsu, N., Gerber M.A. (1997). Transfection of HepG2 cells with infectious hepatitis C virus RNA genome. *American Journal of Pathology*, 151: 363–373.

Dash, S., Kalkeri, G., McClure, H.M., Garry, R.F., Clejan, S., Thung, S.N., Murthy, K.K. (2001). Transmission of HCV to a chimpanzee using virus particles produced in an RNA transfected hepG2 cell culture. *Journal of Medical Virology*, 65: 276–281.

Dash, S., Prabhu, R., Hazari, S., Bastian, F., Garry, R.F., Zou, W., Haque, S., Joshi V., Regenstein, F.G., Thung S.N. (2005). Interferons alpha, beta, gamma each inhibits hepatitis C virus replication at the level internal ribosome entry site mediated translation. *Liver International*, 25: 1–15.

Davis, G., Lau, J. (1997). Factors predictive of a beneficial response to therapy of hepatitis C. *Hepatology*, 26 (suppl. 1): 122S–127S.

Davis, G.L., Esteban-Mur, R., Rustgi, V., Hoef, J., Gordon, S.C., Trepo, C., Shiffman, M.L., Zeuzem, S., Craxi, A., Ling, M.H., Albrecht, J. (1998). Interferon alpha-2b alone or in combination with ribavirin for the treatment of relapse of chronic hepatitis C. International Hepatitis International therapy group. *New England Journal of Medicine*, 339: 1493–1499.

Diaz, M.O., Pomykala, H.M., Bohlander, S.K., Maltepe, E., Malik, K., Brownstein, B. Olopade, O.I. (1994). Structure of the human type 1 interferon gene cluster determined from a YAC clone. *Genomics*, 22: 540–542.

Domanski, P., Colamonici, O.R. (1996). The type-1 interferon receptor. The long and short of it. Cytokine Growth Factor Review, 7: 143–151.

Domanski, P., Witte, M., Kellum, M., Runinstein, M., Hackett, R., Pitha, P., Colamonici O.R. (1995). Cloning and expression of a long form of beta subunit of the interferon alpha/beta receptor that is required for signaling. *Journal of Biological Chemistry*, 270: 21606–21611.

El-Serag, H.B. (2004). Hepatocellular carcinoma: recent trends in the United States. *Gastroenterology*, 127: S27–34.

Enomoto, N., Sakuma, I., Asahina, Y., Kurosaki, M., Murakami, T., Yamamoto, C., Izumi, N., Marumo, F., Sato, C. (1995). Comparison of full-length sequences of interferon-sensitive and resistant hepatitis C virus 1b. *Journal of Clinical Investigation*, 96: 224–230.

Enomoto, N., Sakuma, I., Asahina, Y., Kurasaki, M., Murakami, T., Yamamoto, C., Ogura, Y., Izumi, N., Murumo, F., Sato, C. (1996). Mutation in the nonstructural protein 5A gene and response to interferon in patients with chronic hepatitis C virus 1b infection. *New England Journal of Medicine*, 334: 77–81.

Feld, J.J., Hoofnagle, J.H. (2005). Mechanism of action of interferon and ribavirin treatment of hepatitis C. *Nature*, 436: 967–972.

Fenteany, G., Schreiber, S.L. (1998). Lactacystin, proteasome function and cell fate. *Journal of Biological Chemistry*, 273: 8545–8548.

Ferreon, J.C., Ferreon, A.C.M., Li, K., Lemon, S.M. (2005). Molecular determinants of TRIF proteolysis mediated by the hepatitis C virus NS3/4A protease. *Journal of Biological Chemistry*, 280: 20483–20492.

Francki, R.I.B., Fauquet, C.M., Knudson, D.L., Brown, F. (1991). Classification and nomenclature of viruses: fifth report of the international committee on taxonomy of viruses. *Archives Virology* (Suppl. 2), 223.

Francois, C., Duverlie, G., Rebouillat, D., Khorsi, H., Castelain, S., Blum, H.E., Gatignol, A., Wychowski, C., Moradpour, D., Meurs, E.F. (2000). Expression of hepatitis C virus proteins interferes with the antiviral action of interferon independently of PKR-mediated control of protein synthesis. *Journal of Virology*, 74: 5587–5596.

Frese, M., Schwarzle, V., Barth, K., Krieger N., Lohmann V., Mihm S., Haller O., Bartenschalager R. (2002). Interferon gamma inhibits replication of subgenomic and genomic hepatitis C virus RNAs. *Hepatology*, 35: 694–703.

Friebe, P., Bartenschlager, R. (2002). Genetic analysis of sequences in the $3'$ nontranslated region of hepatitis C virus that are important for RNA replication. *Journal of Virology*, 76: 5326–5338.

Friebe, P., Lohmann, V., Krieger, N., Bartenschalager, R. (2001). Sequences in the 5'UTR nontranslated region of hepatitis C virus required for RNA replication. *Journal of Virology*, 75: 12047–12057.

Foy, E., Li, K., Wang, C., Sumpter, R., Ikeda, M., Lemon, S.M., Gale, M. (2003). Regulation of interferon regulatory factor-3 by the hepatitis C virus serine protease. *Science*, 300: 1145–1148.

Foy, E., Li, K., Sumpter, R., Loo, Y.-M., Johnson, C.L., Wang, C., Fish, P.M., Yoneyama, M., Fujita, T., Lemon, S.M., Gale, M. (2005). Control of antiviral defense through hepatitis C virus disruption of retinoic acid-inducible gene-1 signaling. *Proceedings of National Academy of Sciences*, 102: 2986–2991.

Fried, M.W., Shiffman, M.L., Reddy, K.R., Smith, C., Marinos, G., Goncales, F.L. Jr, Haussinger, D., Diago, M., Carosi, G., Dhumeaux, D., Craxi A., Lin, A., Hoffman, J., Yu, J. (2002). Peginterferon alpha-2b plus ribavirin for chronic hepatitis C virus infection. *New England Journal of Medicine*, 347: 975–982.

Gale, M., Korth, M.J., Tang, N.M., Tan, S.L., Hopkins, D.A., Dever, T.E., Polyak, S.J., Gretch, D.R., Katze, M.G. (1997). Evidence that hepatitis C virus resistance to interferon is mediated through repression of the PKR protein kinase by the nonstructural 5A protein. *Virology*, 230: 217–227.

Gale, M., Blakely, C.M., Kwieciszewski, B., Tan, S.-L., Dossett, M., Tang, N.M., Korth, M.J., Polyak, S.J., Gretch, D.R., Katze, M.G. (1998). Control of PKR protein kinase by hepatitis C virus non-structural 5A protein: molecular mechanisms of kinase regulation. *Molecular and Cellular Biology*, 18: 5208–5218.

Garcia-Sastre, A., Biron, C.A. (2006). Type 1 interferons and the virus-host relationship: a lesson in détente. *Science* , 312: 879–882.

Goodbourn, S., Didcock, L., Randall, R.E. (2000). Interferons: cell signaling, immune modulation, antiviral responses and virus countermeasures. *Journal of General Virology*, 81: 2341–2364.

Guo, J., Bichko, V.V., Seeger, C. (2001). Effect of alpha interferon on the hepatitis C virus replicon. *Journal of Virology*, 75: 8516–8523.

Guo, J., Zhu, Q., Seeger, C. (2003). Cytopathic and noncytopathic interferon responses in cells expressing hepatitis C virus subgenomic replicons. *Journal of Virology*, 77: 10769–10779.

Guo, J., Sohn, J.A., Zhu, Q., Seeger, C. (2004). Mechanism of the interferon alpha response against hepatitis C virus replicons. *Virology*, 325: 71–81.

Hadziyannis, S.J., Sette, H. Jr., Morgan, T.R., Balan, V., Diago, M., Marcellin, P., Ramadori G., Bodenheimer H. Jr., Bernstein, D., Rizzetto, M., Zeuzem, S., Pockros, P.J., Lin, A., Ackrill, A.M., PEGASYS International Study Group. (2004). Peginterferon-alpha2a and ribavirin combination therapy in chronic hepatitis C. A randomized study of treatment duration and ribavirin dose. *Annals of Internal Medicine*, 140: 346–355.

Hazari, S., Bastian, F.O., Garry, R.F., Dash, S. (2004). Translational regulation of HCV-IRES mRNA in interferon treated Huh-7 cells. *Hepatology*, A-619, P-433A.

Hazari, S., Taylor, L., Garry, R.F., Loftig, R., Dash, S. (2005a). Lower activation of ISRE promoter among cell clones replicating HCV sub-genomic RNA leads to interferon resistance. *Hepatology*, 42: P-247A, A: 128.

Hazari, S., Patil, A., Joshi, V., Sullivan, D.E., Fermin, C.D., Garry, R.F., Elliott, R.M., Dash, S. (2005b). Alpha interferon inhibits translation mediated by the internal ribosome entry site of six different hepatitis C virus genotypes. *Journal of General Virology*, 86: 3047–3053.

He, Y., Yan, W., Coito, C., Li, Y., Gale, M. Jr., Katze, M.G. (2003). The regulation of hepatitis C virus (HCV) internal ribosome-entry site-mediated translation by HCV replicons and nonstructural proteins. *Journal of General Virology*, 84: 535–543.

Heathcote, J., Keeffe, E.B., Lee, S.S., Feinman, S.V., Tang, M.J., Reddy, K.R., Albert, D.G., Witt, K., Blatt, L.M. (1998). The consensus interferon study group. Retreatment of chronic hepatitis C with consensus interferon. *Hepatology* 27: 1136–1143.

Hemmi, S., Bohni, R., Stark, G., Di-Marco, F., Aguet, M. (1994). A novel member of the interferon receptor family complements functionally of the murine interferon gamma receptor in human cells. *Cell*, 76: 803–810.

Hoffmann, W.P., Zeuzem, S., Sarrazin, C. (2005). Hepatitis C virus-related resistance mechanisms to interferon alpha-based antiviral therapy. *Journal of Clinical Virology*, 32: 86–91.

Hoofnagle, J.H. (2004). Hepatocellular carcinoma: summary and recommendation. *Gastroenterology*, 127: S319–S323.

Hoofnagle, J.H. (2002). Course and outcome of hepatitis C. *Hepatology*, 36: S21–29.

Ji, H., Fraser, C.S., Yu, Y., Leary, J., Douda, J.A. (2004). Coordinated assembly of human translation initiation complexes by the hepatitis C virus internal ribosome entry site RNA. *Proceedings of National Academy of Sciences (USA)*, 101: 16990–16995.

Kalkeri, G., Khalap, N., Garry, R., Fermin, C., Dash, S. (2001). Hepatitis C viral protein affect cell viability and membrane permeability. *Experimental and Molecular Pathology*, 71: 194–208.

Karin, M., Lawrence, T., Nizet, V. (2006). Innate immunity gone awry: linking microbial infections to chronic inflammation and cancer. *Cell*, 124: 823–835.

Kato, T., Date, T., Miyamoto, M., Sugiyama, M., Tanaka, Y., Orito, E., Ohno, T., Sugihara, K., Hasegawa, I., Fujiwara, K., Ito, K., Ozasa, A., Mizokami, M., Wakita, T. (2005). Detection of anti-hepatitis C virus effect of interferon and ribavirin by a sensitive replicon system. *Journal of Clinical Microbiology*, 43: 5679–5684.

Kato, J., Kato, N., Moriyama, M., Goto, T., Taniguchi, H., Shiratori, Y., Omata, M. (2002). Interferons specifically suppress the translation from the internal ribosome entry site of hepatitis C virus through a double-stranded RNA- activated protein kinase-independent pathway. *Journal of Infectious Disease*, 186: 155–163.

Katze, M.G., He, Y., Gale, M. (2002). Viruses and interferon: a fight for supremacy. *Nature Review*, 675–687.

Kawai, T., Akira, S. (2006). Innate immune recognition of viral infection. *Nature Immunology*, 7: 131–137.

Keskinen, P., Melen, K., Julkunen, I (2002). Expression of HCV structural proteins impairs IFN-mediated antiviral response. *Virology*, 299: 164–171.

Khaber, K.S.A., Al-Zoghaibi, F., Al-Ahdal, M.N., Murayama, T., Dhalla, M., Mukaida, N., Taha, M., Al-Sedairy, S.T., Siddiqui, Y., Kessie, G., Matsushima, K. (1997). The alpha chemokine, interleukin 8 inhibits the antiviral action of interferon alpha. *Journal of Experimental Medicine*, 186: 1077–1085.

Kiyosawa, K., Umemura, T., Ichigo, T., Matsumoto, A., Yoshizawa, K., Tanaka, E. (2004). Hepatocellular carcinoma recent trends in Japan. *Gastroenterology*, 127 (Suppl.): 17–26.

Koev, G., Duncan, R.F., Lai, M.M. (2002). Hepatitis C virus IRES-dependent translation is insensitive to an eIF2alpha-independent mechanism of inhibition by interferon in hepatocyte cell lines. *Virology*, 297: 195–202.

Kolykhalov, A.A., Agapov, E.V., Blight, K.J., Mihalik, K., Feinstone, S.M., Rice, C.M. (1997). Transmission of hepatitis C by intrahepatic inoculation with transcribed RNA. *Science*, 277: 570–574.

Landolfo, S., Gribaudo, G., Angeretti, A., Gariglio, M. (1995). Mechanisms of viral inhibition by interferons. *Pharmacological Therapy*, 65: 415–442.

Langland, J.O., Cameron, J.M., Heck, M.C., Jancovich, J.K., Jacob, B.L. (2006). Inhibition of PKR by RNA and DNA viruses. *Virus Research*, 119: 100–110.

Li, K., Foy, E., Ferreon, J.C., Nakamura, M., Ferreon, A.C.M., Ikeda, M., Ray, S.C., Gale, M., Lemon, S.M. (2005). Immune evasion by hepatitis C virus NS3/4A protease-mediated cleavage of the toll-like receptor 3 adapter protein TRIF. *Proceedings of National Academy of Sciences (USA)*, 102: 2992–2997.

Lin, W., Choe, W.H., Hiasa, Y., Kamegaya, Y., Blackard, J.T., Schmidt, E.V., Chung, R.T. (2005). Hepatitis C virus expression suppresses interferon signaling by degrading STAT1. *Gastroenterology*, 128: 1034–1041.

Lindenbach, B.D., Evans, M.J., Syder, A.J., Wolk, B., Tellinghuisen, T.L., Liu, C.C., Maruyama, T., Hynes, R.O., Burton, D.R., McKeating, J.A., Rice, C.M. (2005). Complete replication of hepatitis C virus in cell culture. *Science*, 309: 623–626.

Lohmann, V., Korner, F., Koch, J.O., Herian, U., Theilmann, L., Bartenschlager, R. (1999). Replication of subgenomic hepatitis C virus RNAs in a hepatoma cell line. *Science*, 285: 110–113.

Lucas, S., Bartolome, J., Correno, V. (2005). Hepatitis C virus core protein down–regulates transcription of interferon-induced antiviral genes. *Journal of Infectious Disease*, 191: 93–99.

Manns, M.P., McHutchison, J.G., Gordon, S.C., Rustgi, V.K., Shiffman, M., Reindollar, R., Goodman, Z.D., Koury K, Ling, M., Albrecht, J.K., the International Hepatitis Interventional Therapy Group. (2001). Peginterferon alpha-2b plus ribavirin compared with interferon alpha 2b plus ribavirin for initial treatment of chronic hepatitis C. A randomized trial. *Lancet*, 358: 958–965.

Macquillan, G.C., Niu, X., Speers, D., English, S., Garas, G., Harnett, G.B., Reed, W.D., Allam, J.E., Jeffrey, G.P. (2004). Does sequencing the PKRBD of hepatitis C virus NS5A predict therapeutic response to combination therapy in an Australian population? *Journal of Gastroenterology Hepatology*, 19: 551–557.

Meng, L.R., Mohan, R., Kwok, B.H., Elofsson, M., Sin, N., Crew, C.M. (1999). Epoxomicin, a potent & selective proteosome inhibitor, exhibits in vivo antiinflammatory activity. *Proceedings of National Academy of Science (USA)*, 96: 10403–10408.

Miyamoto, M., Kato, T., Date, T., Mizokami, M., Wakita, T. (2006). Comparison between subgenomic replicons of hepatitis C virus genotypes 2a (JFH-1) and 1b (Con-1 NK 5.1). Intervirology, 49: 37–43.

Nousbaum, J., Polyak, S.J., Ray, S.C., Sullivan, D.G., Larson, A.M., Carithers, R.L., Gretch, D.R. (2000). Prospective characterization of full-length hepatitis C virus NS5A quasispecies during induction and combination antiviral therapy. *Journal of Virology*, 74: 9028–9038.

Novick, D., Cohen, B., Bubinstein, M. (1994). The human interferon alpha/beta receptor: characterization and molecular cloning. *Cell*, 77: 391–400.

Otto, G.A., Pulglisi, J.D. (2004). The pathway of HCV IRES-mediated translation initiation. *Cell*, 119: 369–380.

Pai, M., Prabhu, R., Panebra, A., Nangle, S., Haque, S., Bastian, F., Garry, R., Agrawal, K., Goodbourn, S., Dash, S. (2005). Activation of interferon stimulated response element (ISRE) in Huh-7 cells replicating HCV subgenomic RNA. *Intervirology*, 48: 301–311.

Paterson, M., Laxton, C.D., Thomas, H.C., Ackril, A.M., Foster, G.R. (1999). Hepatitis C virus NS5A protein inhibits interferon antiviral activity, but the effects do not correlate with clinical response. *Gastroenterology*, 117: 1187–1197.

Pawlotsky, J.M. (2000). Hepatitis C virus resistance to antiviral therapy. *Hepatology*, 32: 889–896.

Pawlotsky, J.M. (2003). The nature of interferon-alpha resistance in hepatitis C virus infection. *Current Opinion on Infectious Disease*, 16: 587–592.

Pesch, V., Lanaya, H., Renauld, J.C., Michiels, T. (2004). Characterization of the murine alpha interferon gene family. *Journal of Virology*, 78: 8219–8228.

Peterson, M., Laxton, C.D., Thomas, H.C., Ackrill, A.M., Foster, G.R. (1999). Hepatitis C virus NS5A protein inhibits interferon antiviral activity but the effects do not correlate with clinical response. *Gastroenterology*, 117: 1187–1197.

Platanias, L.C., Uddin, S., Domanski, P., Colamonici, O.R. (1996). Differences in interferon alfa and beta signalling. Interferon beta selectively induces the interaction of the alpha and beta L subunits of the type 1 interferon receptor. *Journal of Biological Chemistry*, 271: 23630–23633.

Polyak, S.J., Khabar, K.S.A., Rezeiq, M., Gretch, D.R. (2001). Elevated levels of interleukin-8 in serum are associated with hepatitis C virus infection and resistance to interferon therapy. *Journal of Virology*, 75: 6209–6211.

Polyak, S.J., Khabar, K.S.A., Paschal, D.M., Ezelle, H.J., Duverlie, G., Barber, G.N., Levy, D.E., Mukaida, N., Gretch, D.R. (2001). Hepatitis C virus nonstructural 5A protein induces interleukin-8 leading to partial inhibition of the interferon induced antiviral response. *Journal of Virology*, 75: 6095–6106.

Prabhu, R., Joshi, V., Garry, R.F., Bastian, F., Haque, S., Regenstein, F., Thung, S.N., Dash, S. (2004). Interferon alpha-2b inhibits negative strand RNA and protein expression from full-length HCV1a infectious clone. *Experimental and Molecular Pathology*, 76: 242–252.

Qi, Z., Kalkeri, G., Hanible, J., Prabhu, R., Bastian, F., Garry, R.F., Dash, S. (2003). Stem-loop structures (II-IV) of the 5′ untranslated sequences are required for the expression of the full-length hepatitis C virus genome. *Archives of Virology*, 148: 449–467.

Reed, K.E., Rice, C.M. (2000). Overview of hepatitis C virus genome structure, polyprotein processing and protein properties. *Current Topics of Microbiology & Immunology*, 242: 55–84.

Rice, C.M. (1996). Flaviviridae: the viruses and their replication. In: B.N. Fields, D.M. Knipe, P.M. Howley, et al. (eds). (pp. 936–959) *Fields Virology*, 3rd ed., Philadelphia: Lippincott-Raven.

Rivas-Estilla, A.M., Svitkin, Y., Lastra, M.L., Hatzoglou, M., Sherker, A., Koromilas, A.E. (2002). PKR-dependent mechanisms of gene expression from a subgenomic hepatitis C virus clone. *Journal of Virology*, 76: 10637–10653.

Robek, M.D., Boyd, B.S., Chisari, F.V. (2005). Lambda interferon inhibits hepatitis B and hepatitis C virus replication. *Journal of Virology*, 79: 3851–3854.

Schinkel, J., Spoon, W.J., Kroes, A.C. (2004). Meta-analysis of mutations in the NS5A gene and hepatitis C virus resistance to interferon therapy: uniting discordant conclusions. *Antiviral Therapy*, 9: 275–286.

Schroder, K., Hertzog, P.J., Ravasi, T., Hume, D.A. (2004). Interferon gamma: an overview of signals, mechanisms and functions. *Journal of Leukocyte Biology*, 75: 163–189.

Sen, G.C. (2001). Viruses and interferons. *Annual Revew of Microbiology*, 55: 255–281.

Shimazaki, T., Honda, M., Kaneko, S., Kobayashi, K. (2002). Inhibition of internal ribosomal entry site-directed translation of HCV by recombinant IFN-alpha correlates with a reduced La protein. *Hepatology*, 35: 199–208.

Shimizu, Y.K., Feinstone, S.M., Kohara, M., Purcell, R.H., Yoshikura, H. (1996). Hepatitis C virus: detection of intracellular virus particles by electron microscopy. *Hepatology*, 23: 205–209.

Shimizu, Y.K., Purcell, R.H., Yoshikura, H. (1993). Correlation between the infectivity of hepatitis C virus in vivo and its infectivity in vitro. *Proceedings of National Academy of Sciences (USA)*, 90: 6037–6041.

Shimizu, Y.K., Iwamoto, A., Hijikata, M., Purcell, R.H., Yoshikura, H. (1992). Evidence for in vitro replication of hepatitis C virus genome in a human T-cell line. *Proceedings of National Academy of Sciences (USA)*, 89: 5477–5481.

Shimizu, Y.K., Yoshikura, H. (1994). Multicycle infection of hepatitis C virus in cell culture and inhibition by alpha and beta interferons. *Journal of Virology*, 68: 8406–8408.

Simmonds, P. (2004). Genetic diversity and evolution of hepatitis C virus—15 years on. *Journal of General Virology*, 85: 3173–3188.

Stark, G.R., Kerr, I.M., William, B.R.G., Silverman, R.H., Schreiber, R.D. (1998). How cells respond to interferons. *Annual Review of Biochemistry*, 67: 227–264.

Strader, D.B., Wright, T., Thomas, D.L., Seeff, L.B. (2004). Diagnosis, management, and treatment of hepatitis C. *Hepatology*, 39: 1147–1171.

Sumpter, R., Loo, Y.-M., Foy, E., Li, K., Yoneyama, M., Fujita, T., Lemon, S.M., Gale, M. (2005). Regulating intracellular antiviral defense and permissiveness to hepatitis C virus RNA replication through a cellular RNA helicase, RIG-I. *Journal of Virology*, 79: 2689–2699.

Taylor, D.R., Shi, S.T., Romano, P.R., Barber, G.N., Lai, M.C. (1999). Inhibition of the interferon-inducible protein kinase R by HCV E2 protein. *Science*, 285: 107–109.

Uze, G., Lutfalla, G., Gresser, I. (1990). Genetic transfer of a functional human interferon alpha-receptor into mouse cells: cloning and expression of its cDNA. *Cell*, 60: 225–234.

Uze, G., Lutfalla, G., Mongensen, K.E. (1995). Alpha and beta interferons and their receptor and their friends and relations. *Journal of Interferon Research*, 15: 3–26.

Vrolijk, J.M., Kaul, A., Hanson, B.E., Lohmann, V., Haagmans, B.L., Schalm, S.W., Bartenschalager, R. (2003). A replicon-based bioassay for the measurement of interferons in patients with chronic hepatitis C. *Journal of Virological Methods*, 110: 201–209.

Wakita, T., Pietschmann, T., Kato, T., Date, T., Miyamoto, M., Zhao, Z., Murthy, K., Habermann, A., Krausslich, H.G., Mizokami, M., Bartenschlager, R., Liang, T.J. (2005). Production of infectious hepatitis C virus in tissue culture from cloned viral genome. *Nature Medicine*, 11: 791–796.

Wang, C., Pflugheber, J., Sumpter, R., Sodora, D.L., Hui, D., Sen, G., Gale, M. (2003). Alpha interferon induces distinct translational control programs to suppress hepatitis C virus RNA replication. *Journal of Virology*, 77: 3898–3912.

William, B.R.G. (1991). Transcriptional regulation of interferon stimulated genes. *European Journal of Biochemistry*, 200: 1–11.

Yanagi, M., Purcell, R.H., Emerson, S.U., Bukh, J. (1997). Transcripts from a single full-length cDNA clone of hepatitis C virus are infectious when directly transfected in to the livers of a chimpanzee. *Proceedings of National Academy of Sciences (USA)*, 94: 8738–8743.

Yi, M., Lemon, S.M. (2003). 3′ non-translated RNA signals required for replication of hepatitis C virus RNA. *Journal of Virology*, 77: 3557–3568.

Seroconversion in HCV occurs approximately 7 to 31 weeks after primary infection (Pawlotsky, 1999, 2004), and some HCV-specific antibodies are effective in blocking *in vitro* infection of target cells by HCV (Farci et al., 1994). However, in both chimpanzees and humans, naturally acquired anti-HCV antibodies generated during this infection do not seem to be protective upon secondary infection with HCV, indicating that these molecules play a limited role in preventing the spread of the virus (Farci et al., 1992; Lai et al., 1994). Moreover, studies in chimpanzees indicate that resolution of infection can occur without the development of detectable antibody responses (Cooper et al., 1999). In addition, recent studies using HCV pseudotyped particles indicate that neutralizing anti-HCV antibodies occur far more commonly in persistently infected individuals than in those who clear the virus (Bartosch et al., 2003; Logvinoff et al., 2004; Meunier et al., 2005). A recent report has suggested that antibodies might be essential for the control of non-cytopathic viruses and to maintain protective memory (Bachmann et al., 2004). It is therefore possible that antibodies play other unsuspected roles during HCV infections. These roles will certainly be explored in future studies.

Although the recent advent of HCV strains capable of *in vitro* replication and infectivity (Lindenbach et al., 2005; Wakita et al., 2005) is likely to enhance our understanding of the role of neutralizing antibodies in HCV infection, the role of T cells in viral control is better understood than the role of B cells. Thus, this review will focus on T cell-mediated immunity.

In contrast to immunoglobulins able to recognize soluble antigens, T lymphocytes express a T cell receptor that recognizes a degraded form of the antigen associated with a molecule encoded by the major histocompatibility complex (MHC). This peptide/MHC complex is formed, processed, and presented on the surface of an antigen-presenting cell (APC). Expression of CD4 or CD8 molecules distinguishes two distinct types of T lymphocytes exerting different functions: CD4+ T cells recognize MHC class II molecules, secrete cytokines, and are often referred to as T helper cells, while CD8+ T cells recognize MHC Class I molecules, secrete cytokines with antiviral properties, kill target cells, and are known as cytotoxic T cells (CTL).

Adaptive immune responses mediated by T cells are essential in the control of HCV and viral clearance. In both chimpanzees and humans, viral clearance is associated with sustained CD4+ and CD8+ T cells responses and an increase of IFN-γ expression in the liver (Thimme et al., 2002). Recent studies in which memory CD4+ and CD8+ T cells were depleted have confirmed the critical role of these cells in controlling HCV infection (Grakoui et al., 2003; Shoukry et al., 2003).

The chronological evolution of T cell immune responses in HCV infection can be classically divided into three phases (Bowen & Walker, 2005a) (Figure 1): the first phase corresponds to the first weeks following primary infection. Virus titers increase and become very high during this phase irrespective of future viral clearance or persistence. It is likely that CD4+ and CD8+ T cells specific for HCV are activated very early during this phase. However, one of the most remarkable observations that came out of studies investigating the kinetics of HCV infection in both chimpanzees and humans is that these responses are not detected in the blood before 1–3 months after initial infection (Bowen & Walker, 2005a; Cox et al.,

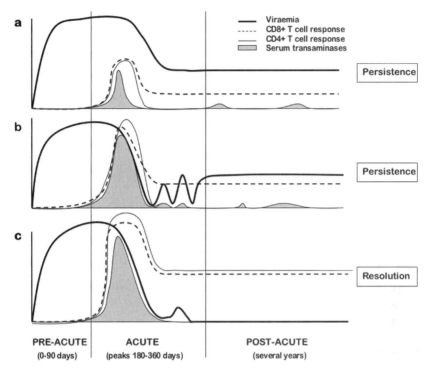

Fig. 1 Schematic representation of the three phases of HCV infection and of possible disease outcomes [modified version of a figure from Bowen & Walker (2005a)]. All HCV infections are characterized by three different phases: pre-acute, acute, and post-acute. The post-acute phase might lead to either viral persistence (a and b) or viral clearance (c). Three categories of infections can be distinguished: (a) infections in which viremia is never controlled, characterized by weak CD4+ and CD8+ T cell responses; (b) infections that seem to be initially controlled but rebound following loss of CD4+ T cells; (c) infections that are resolved and result in the generation of both memory CD4+ and CD8+ T cells

2005a). The median time for the development of a IFN-γ response is 33 days (Cox et al., 2005a). The reasons for this delay are not yet understood and will be discussed at the end of this chapter. The second phase of the disease is characterized by acute transient hepatitis that persists for a few weeks. Infected individuals may develop acute hepatitis irrespective of the outcome of infection. A rise in serum levels of alanine aminotransferase (ALT), a marker of hepatocyte damage, is associated with the emergence of detectable CD4+ and CD8+ T cell immune responses and a drop in viral titers. These responses have been shown to peak between 180–360 days after initial infection (Cox et al., 2005a). The last phase depends on the outcome of the disease; in 30% of infected individuals, virus is cleared and infection resolves. In these individuals, ALT levels normalize and $CD4^+$ and $CD8^+$ memory T cells persist. In approximately 70% of HCV infections, infection becomes chronic. This phase is usually characterized by an absent or almost undetectable HCV-specific $CD4^+$ T cell response and a poor or ineffective $CD8^+$ T cell response. In the following two sections we will review the characteristics of CD4 and CD8 T cells responses during acute and chronic HCV.

CD4+ T Cells During Acute and Chronic HCV Infection

CD4+ T Cells in Acute Infection

CD4+ T cells play a critical role in an immune response. Following contact with professional antigen-presenting cells [such as dendritic cells (DC)] in the lymph nodes (LNs), they trigger a signal that activates the maturation of the DC, a process known as licensing, secrete cytokines, and provide help to both antigen-specific CD8+ T cells and B cells. A recent report also suggests that IL-2 secreted by CD4+ T cells during the priming phase is critical for the programming and the maintenance of memory CD8+ T cells (Williams et al., 2006).

Most studies investigating CD4+ T cell responses in HCV have been performed using human peripheral blood mononuclear cells (PBMC). Early studies have suggested that most patients in whom the virus persists develop a defective blood CD4+ T cell response to the recombinant HCV proteins' core, nonstructural protein 3 (NS3), NS4, and NS5 during the acute phase of the disease (Diepolder et al., 1995) (Figure 2). In contrast, individuals who resolve infection exhibit robust HCV-specific CD4+ T cell proliferation (Figure 3). Although there are some exceptions (Thomson et al., 2003), these results have been confirmed in subsequent studies

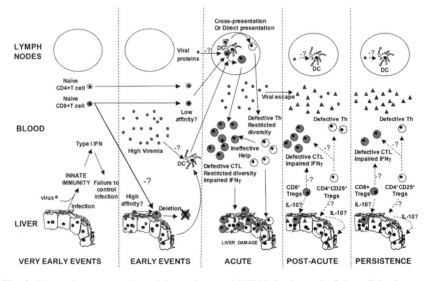

Fig. 2 Schematic representation of the evolution of HCV infection and of the cellular immune response in individuals developing persistent infection. Dashed lines and question marks represent hypothetical pathways of activation or regulation

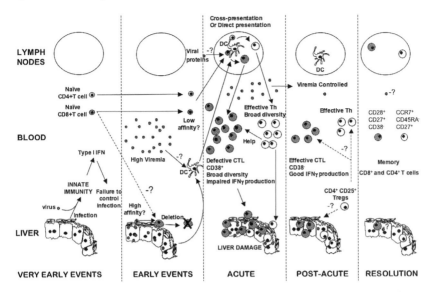

Fig. 3 Schematic representation of the HCV infection evolution and of the cellular immune response in individuals with self-limited disease. Dashed lines represent hypothetical pathways of activation or regulation. It is still uncertain whether the virus is completely cleared and whether the maintenance of memory T cells requires the presence of small amounts of virions subsiding in the host. This is represented by a question mark in the last panel of the figure

(Gerlach et al., 1999; Lechner et al., 2000; Takaki et al., 2000) and suggested that the vigor of the CD4+ T cell responses during the early stages of infection is a critical determinant of viral control and infection resolution.

The presence of a detectable CD4+ T cell response in the blood is not, however, an absolute predictor of resolution. Some patients initially displayed a strong CD4+ T cell response and clear the virus but, for reasons that remain unclear, lost this response, an event that was associated with HCV recurrence (Gerlach et al., 1999). Anti-HCV CD4+ T cell responses can be subdivided into three categories: (1) those that are strong and sustained, leading to viral clearance; (2) those that are weak and associated with the development of chronic disease; and (3) those that are initially strong and eliminate the virus but are subsequently lost, allowing HCV recurrence (Figure 1).

While detailed studies of intrahepatic human CD4+ T cells have not been performed during acute infection, analysis of T lymphocytes expanded from the liver of acutely infected chimpanzees has recapitulated classification into the three categories defined above (Thimme et al., 2002).

The reasons why CD4+ responses are absent or not sustained in some patients are still uncertain. It is now widely accepted that HCV infections that will eventually resolve are generally associated with a sustained CD4+ T cell response targeting multiple epitopes and most HCV proteins. This has been demonstrated in both humans and chimpanzees (Gerlach et al., 1999; Lechner et al., 2000; Takaki et al., 2000; Thimme et al., 2001, 2002; Day et al., 2002). Recent studies have shown that in addition to recognizing "promiscuous" HCV epitopes that are bound by multiple different HLA class II molecules, CD4+ T cells from individuals with resolved infection also recognized additional more heterogeneous and less frequently detected MHC class II-restricted epitopes (Schulze zur Wiesch et al., 2005). However, this observation is not universal, and it has been reported that some patients develop persistent infection despite the presence of a multispecific anti-HCV CD4+ T cell response during the acute phase of infection (Thimme et al., 2001). A study in chimpanzees has also highlighted the fact that immune responses to different epitopes can arise asynchronously during primary infection (Shoukry et al., 2004). In this study, an initial wave of CD4+ T cells targeting a set of dominant epitopes was followed by a second wave targeting subdominant epitopes several weeks later, after viral replication was largely contained. The hierarchy of the dominance was conserved seven years later during resolution of secondary infection with homologous virus. This report suggests that the hierarchical nature of the response could be another factor governing the outcome of HCV infection.

CD4+ T Cells in Chronic Infection

Chronic infections are characterized by a permanent, almost complete, loss of HCV-specific CD4+ T cells in the blood (Figure 2). PBMC harvested from patients infected for many years with HCV failed to proliferate and/or produce IFN-γ or exhibited oligoclonal diversity (Gerlach et al., 1999; Lechner et al., 2000; Schirren et al., 2000; Day et al., 2002; Rosen et al., 2002; Ulsenheimer et al., 2003; Wertheimer et al., 2003). The relative lack of circulating HCV-specific CD4+ T cells has recently been confirmed more directly using MHC class II tetramers (Day et al., 2003b). Although CD4+ T cells are hard to detect in the blood, they are not totally absent, however. Following HCV-specific stimulation, a very low percentage of CD4+ T cells that upregulated expression of the alpha chain of the IL-2R (CD25), an early marker of activation, was detected among PBMC from chronically infected subjects (Ulsenheimer et al., 2003). Furthermore, it has been possible to derive some CD4+ T cell lines specific for HCV from the blood of chronically infected patients by repeated stimulation with recombinant proteins, confirming the presence of such T cells (Minutello et al., 1993; Schirren et al., 2000; Penna et al., 2002). CD4+ T cells specific for the core antigen are more easily detected than those specific for the nonstructural proteins (MacDonald et al., 2002), indicating that the loss of CD4+ T cells in chronic HCV might depend on the nature of the epitope.

Loss of CD4+ T cells seems to be critical for the development of persistent viral infections (Day & Walker, 2003a) and might be responsible for the impaired function of CD8+ T cells during chronic HCV (see below). Some reports have shown that more CD4+ T cell lines were derived from the liver than the blood, suggesting that HCV-specific CD4+ T cells might be sequestered in the liver (Schirren et al., 2000; Penna et al., 2002). Future analysis of the intrahepatic lymphocyte repertoire of chronically infected patients using MHC class II tetramers will certainly shed more light on this important topic.

CD8+ T Cells in Acute and Chronic HCV Infection

CD8+ T Cells in Acute Infection

CD8+ T cells have been investigated extensively in HCV infections. This has been facilitated by the availability of MHC class I tetramers. Studies in infected chimpanzees and humans using both functional methods of CTL identification and MHC class I tetramers have demonstrated that increases in serum transaminase levels and clearance of the virus during the acute phase are generally associated with the emergence of a strong CTL response in the blood and the liver 1 to 3 months after infection (Cooper et al., 1999; Thimme et al., 2001, 2002; Shoukry et al., 2003; Cox et al., 2005a) (Figure 2). Up to 8% of the blood CD8+ T cells can be specific for a single HCV epitope at the peak of the acute response (Klenerman et al., 2002). Viral clearance follows the entry and accumulation of HCV-specific IFN-γ-producing T cells within the liver (Thimme et al., 2002). Consistent with this observation, a poor CD8+ T cell response is associated with viral persistence (Cooper et al., 1999) (Figure 2). The mechanism by which CTLs clear the virus during the acute phase is not entirely understood. Killing of infected cells by direct cytoxicity is a major mechanism by which CTLs eliminate viral infection. A role for this phenomenon in control of HCV replication is suggested by the temporal correlation among the detection of CTLs, a rise in transaminase levels, and a fall in viremia (Cooper et al., 1999; Thimme et al., 2001, 2002; Shoukry et al., 2003). However, resolution can sometimes occur without biochemical evidence of hepatitis (Thomson et al., 2003). It is therefore possible that CTLs eliminate the virus through a non-cytolyic mechanism involving cytokines such as IFN-γ (Thimme et al., 2001; Li et al., 2005).

Analysis of PBMCs from infected subjects indicates that CD8+ T cells specific for HCV acquire the phenotype of activated effector T cells (CD38+, CD69+ MHC Class II+, CD28+, CD27+) (Lechner et al., 2000; Gruener et al., 2001; Thimme et al., 2001; Appay et al., 2002). However, a striking feature of acute HCV infection is that despite acquisition of activation markers, these T cells are defective in IFN-γ/tumor necrosis factor (TNF)-α production and CTL activity during the early stages of infection when viremia is high. This early differentiation phenotype resembles anergy and has been designated by some investigators as "stunning" (Lechner et al., 2000) (Figure 2). It is not clear whether this phenotype reflects (1) arrested differentiation due to high antigen loads or inhibition by a viral factor or factors, (2) sequestration of effector cells in compartments other than the blood, (3) the early loss of mature cells, or (4) low-affinity cells remaining following deletion of high-affinity cells (see the upcoming section on mechanisms). In individuals who resolve infection, cytokine production is only restored several weeks later when HCV replication is controlled and when CD38- cells start to appear (Lechner et al.,

2000; Gruener et al., 2001; Thimme et al., 2001; Urbani et al., 2002; Shoukry et al., 2003). This unique behavior of CD8+ T cell responses during HCV infections seems to be independent of the outcome of the infection (resolved versus chronic) and appears to be a hallmark of this disease as it is not observed during infections with Epstein-Barr virus (EBV) or cytomegalovirus (CMV) (Gruener et al., 2001).

Similarly to CD4+ T cells, HCV infections that eventually resolve are generally associated with a sustained CD8+ T cell response targeting multiple epitopes and most HCV proteins in both humans and chimpanzees (Figure 3). A total of 80 HCV-CD8+ T cell clones generated from PBMC isolated during the acute phase of the response (week 0 to 24), were found to recognize 8 MHC class I epitopes in one patient who permanently cleared the virus (Lechner et al., 2000). Likewise, in one chimpanzee that resolved the disease, CTL isolated from the liver targeted at least nine epitopes that were restricted by all six MHC class I allotypes (Cooper et al., 1999). Other studies have confirmed the broad nature of the CD8+ T cell response by assessment of CTL function and/or IFN-γ production by these cells in humans and chimpanzees that resolve infection (Gruener et al., 2001; Thimme et al., 2002).

Conversely, the development of chronic HCV is often associated with a poor CD8+ T cell response or a response that targets fewer epitopes during acute infection (Lechner et al., 2000; Gruener et al., 2001) (Figure 2). However, this finding is not universal, and several studies have described chronic infections associated with multispecific CD8+ T cell responses (Koziel et al., 1992, 1993, 1995; Wong et al., 1998; Erickson et al., 2001).

CD8+ T Cells in Chronic Infection

Between the acute and chronic phases of HCV, the number of HCV-specific CD8+ T cells declines dramatically. Recent studies (Cox et al., 2005a) in which 23 infected patients were sampled monthly during this transition have revealed that the breadth of the response was set early in infection and that early responses became undetectable, while no new responses were formed. Despite early reports based on IFN-γ production or CTL activity suggesting that CD8+ T cells were not detected in the blood of chronically infected patients, CD8+ T cell clones specific for HCV can be derived from the liver of humans and chimpanzees in which the virus persists (Koziel et al., 1992, 1993, 1995; Kowalski et al., 1996; Nelson et al., 1997; Wong et al., 1998; Eckels et al., 1999). The presence of anti-HCV-specific CTLs have been confirmed using MHC class I tetramers. This powerful immunological tool has demonstrated that the liver contains a higher frequency of HCV-specific T cells than the blood (He et al., 1999; Grabowska et al., 2001), with HCV-specific T cells enriched within the liver up to 10- to 30-fold for CD8+ T cells, and 2-fold for CD4+ T cells (He et al., 1999; Schirren et al., 2000; Spangenberg et al. 2005). Furthermore, the intrahepatic HCV-specific CTL repertoire appears to be extremely stable over several years, at least in chimpanzees. In a chronically HCV-infected chimpanzee, intrahepatic CTLs of stable specificity were isolated over a 10-year period (Erickson et al., 2001).

It is notable that frequencies of HCV-specific CD8+ T cells found in the blood (0.0018–0.0660% of CD8+ T cells) are much lower (Rehermann et al., 1996b; He et al., 1999; Lechner et al., 2000; Chang et al., 2001; Barnes et al., 2004) than those specific for other persistent infections, such as human immunodeficiency virus (HIV), EBV, and CMV (He et al., 1999; Takaki et al., 2000; Chang et al., 2001). Repeated antigen stimulation of PBMC allows amplification of HCV-specific CD8+ T cells (Rehermann et al., 1996a, 1996b). Furthermore, as observed during acute infection, HCV-specific CD8+ T cells isolated from the liver and blood of chronically infected patients have been described as remaining persistently defective in cytotoxic function and IFN-γ production (Gruener et al., 2001; Spangenberg et al., 2005; Nisii et al., 2006). HCV-specific CD8+ T cells demonstrate a substantial enrichment of the early differentiation phenotype (CD62L^{low}, CCR7+, CD45Ro^{high}, CD69- MHC Class II^{low}, CD27+ CD28+ perforin^{low}) similar to the phenotype of CD8+ T cells detected in patients with self-limited infection (Lechner et al., 2000; Gruener et al., 2001; Appay et al., 2002; Spangenberg et al., 2005). The low frequency of CD8+ T cells in the blood combined with their apparent functional defects explains the difficulty experienced in early studies in detecting them using functional assays.

Memory T Cells in HCV Infections

Humans and chimpanzees resolving HCV infection develop acute hepatitis characterized by expansion of both CD4+ and CD8+ T cells and control of viremia. In these individuals, T cell numbers increase and then decrease as in other cellular immune responses. These individuals sometimes develop a resurgence of HCV replication that might reflect transient viral escape before the establishment of a stable memory CD4+ and CD8+ T cell pool (Bowen & Walker, 2005a) (Figure 3). Once established, T cell memory lasts for many years (Lechner et al., 2000; Takaki et al., 2000; Chang et al., 2001; Day et al., 2002, 2003b; Rosen et al., 2002; Wertheimer et al., 2003) even when antibodies to HCV can no longer be detected in serum. As it is the case for several viral infections, it is not clear whether memory T cells persist in the total absence of HCV or whether there is a reservoir of HCV that maintains the memory response. Memory T cells identified in the blood using MHC tetramers are present at extremely low frequencies [0.0009–0.083 % of the CD4+ T cells (Day et al., 2003b), 0.0018–0.0660% of the CD8+ T cells (Rehermann et al., 1996a; He et al., 1999; Lechner et al., 2000; Chang et al., 2001; Barnes et al., 2004)]. CD4+ T cells display a central memory T cell phenotype (CCR7+ CD45RA- CD27+) (Day et al., 2003b) indicative of a surveillance function for secondary lymphoid structures, and utilize a restricted Vβ repertoire suggesting that they had undergone significant *in vivo* selection. Memory CD8+ T cells in HCV infections exhibit a CD28+CD27+CD38- phenotype and restored expression of IFN-γ (Lechner et al., 2000; Thimme et al., 2001; Nascimbeni et al., 2003). T cells can also be identified in the liver of chimpanzees several years after spontaneous clearance, suggesting that memory T cells are also present intrahepatically (Cooper et al., 1999; Shoukry et al., 2003, 2004).

Although their efficacy may be limited due to viral diversity, T cell memory seems to be effective in conferring protective immunity in some individuals. Supporting this view, high-risk intravenous drug users who had been infected with HCV but resolved the infection were 12 times less likely to develop a chronic infection upon a second HCV infection (Mehta et al., 2002).

Protective immunity seems to be HCV-strain-specific and can be evaded by heterologous viruses (Sugimoto et al., 2005). The determinant role of memory CD4+ and CD8+ T cells in controlling HCV infection has been elegantly demonstrated in chimpanzees consecutively infected with the same HCV strain seven years later. The first infection was self-limited and led to the appearance of a CD8+ T cell pool in the blood when they were rechallenged with the virus. Upon reinfection, CD4+ and CD8+ T cell responses occurred within a shorter time frame than during the first infection (two weeks versus two months) and resulted in strong IFN-γ production, less liver damage, and rapid viral clearance. The same immunized chimpanzees were rechallenged for a third time with the virus. Antibody-depletion of memory CD8+ T cells before the third infection led to prolonged viral replication despite the presence of memory CD4+ T cells. Viremia was eventually terminated following the return of detectable intrahepatic HCV-specific CD8+ T cell responses (Shoukry et al., 2003), suggesting a critical role for CD8+ T cells in protection from viral persistence. Importantly, depletion of memory CD4+ T cells prior to homologous viral rechallenge in animals that had previously resolved infection resulted in viral persistence (Grakoui et al., 2003). Although small-scale studies, these important experiments suggest that protective immunity against chronic HCV requires the involvement of both memory CD4+ and CD8+ T cells.

Some Fundamental Aspects of HCV Infections

One of the most striking characteristics of the adaptive immune response to HCV is the inability of this response to mediate viral clearance in the majority of infected individuals, thus enabling viral persistence and ensuing chronic necroinflammatory liver disease. The outstanding question that arises from our current knowledge of the adaptive immune response to HCV could be summarized as follows: Why does infection persist in the majority of individuals despite a detectable immune response in many, while some individuals control viremia and resolve infection? Obviously, this is a very complex issue, with multiple mechanisms potentially involved in inhibition of the induction and maintenance of an effective anti-HCV immune response. However, any model of defective anti-HCV immunity needs to take into account the following observations:

1. *There is no general immunosuppression associated with HCV infections.* Critically, the inability to mount an effective immune response against HCV is antigen-specific, even at the site of infection. This has been recently illustrated by

examination of the function of intrahepatic CD8+ T cells specific for influenza virus in individuals chronically infected with HCV: only HCV-specific T cells exhibited impaired function (Spangenberg et al., 2005). Some HCV proteins have been reported to inhibit T cell immune responses in a non-antigen-specific manner *in vitro* , and it has been hypothesized that this mechanism could explain HCV persistence in the host. In particular, the core protein (nucleocapsid) of HCV has been shown to bind to the complement receptor gC1qR expressed by T cells and inhibits their proliferation by inducing expression of suppressor of cytokine signaling molecules (SOCS) (Kittlesen et al., 2000; Yao et al., 2001, 2004, 2005, 2006). Likewise, it has been suggested that DCs isolated from chronic HCV patients are impaired in their ability to present antigen and secrete IL-12, suggesting that the virus targets the antigen presentation ability of dendritic cells (Hiasa et al., 1998; Auffermann-Gretzinger et al., 2001; Bain et al., 2001). However, the relevance of these findings to *in vivo* HCV infections remains unclear as these mechanisms do not explain how non-HCV-specific T cells would be protected against the inhibitory effect of free core particles circulating within the bloodstream of HCV-infected individuals. Furthermore, recent studies have demonstrated no discernible differences in phenotype or function between DCs isolated from HCV-infected and uninfected chimpanzees (Larsson et al., 2004).

2. *Any proposed model should be able to explain why infection with the same HCV strain can lead to two different outcomes.* There are several theories about the liver being an immuno-privileged site in which activated T cells are killed in a non-antigen-specific manner (Huang et al., 1994; Crispe & Mehal, 1996; Mehal et al., 1999, 2001; O'Farrelly & Crispe, 1999; Crispe et al., 2000; Crispe, 2003). Although immune responses in the liver are often associated with immune tolerance, this model does not explain how effector/memory T cells generated during any immune response and recirculating through the liver would not be killed. It also does not explain how successful immune responses to liver-specific infections, such as self-limiting HCV, are generated in some individuals.

3. *T cells isolated from the blood might not be representative of intrahepatic lymphocytes.* It is likely that due to sequestration of HCV-specific T cells at the site of infection, the response to HCV is compartmentalized (Minutello et al., 1993; Schirren et al., 2000). As noted above, there is a higher frequency of HCV-specific T cells in the liver than in the blood. In addition, HCV-specific responses in the peripheral blood of chronically infected subjects are largely of Th2 (anti-inflammatory or pro-humoral response) and Th0 (undifferentiated) phenotype (Tsai et al., 1997; Woitas et al., 1997), while the intrahepatic milieu in chronic infection is largely dominated by Th1 (proinflammatory) cytokines (Napoli et al., 1996; Dumoulin et al., 1997; Penna et al., 2002). These findings indicate that results from several studies in which the phenotype and/or function of HCV-specific T cells derived from blood were analyzed need to be interpreted carefully as their conclusions might not reflect the intrahepatic HCV-specific immune response.

4. *The immune system is continuously stimulated by viral antigens.* The liver is able to regenerate very easily following injury. This extraordinary property places the

liver in a unique position among other solid organs. T cell-mediated damage in many tissues is irreversible. It has been shown that HCV kinetics are at a steady state during chronic infection and that 10^{12} HCV virions per day are produced and cleared by the host (Neumann et al., 1998). The immune system would have to deal with continuous presentation of antigen that might exhaust in the long term its extraordinary, but still limited, regenerative ability thus affecting the generation of memory. If HCV escapes the immune system for long enough, it might be difficult for memory to be established, and this might explain the establishment of persistent infection.

Considering the above, HCV persistence or clearance might be (1) genetically predetermined, (2) determined by early events of T cell activation occurring before the acute phase of the infection, (3) determined and/or maintained by mechanisms operating during the acute and chronic phases of the infection. It is important to emphasize that these mechanisms are not exclusive.

Role of MHC Haplotype in HCV Persistence

It is unlikely that the outcome of the immune response is entirely genetically determined. For CD4+ T cells, it is possible that the MHC class II haplotype and thus the range and type of epitopes that are targeted during the immune response play some role in HCV persistence. Some MHC class II allotypes, such as HLA-DRβ1*0701, have been associated with HCV persistence (Fanning et al., 2001), while other allotypes (DRβ1*0101, HLA-DRβ1*1101, and DQβ1*0301) are linked to sustained T helper response and/or a self-limited disease (Alric et al., 1997; Minton et al., 1998; Thursz et al., 1999; Harcourt et al., 2001). Likewise, the CD8+ T cell response during acute hepatitis does not seem to be preferentially biased toward any particular HCV protein or epitope. Some MHC class I allotypes, such as HLA-Cw*04, have been associated with HCV persistence (Thio et al., 2002) and HLA-B27 with protection (Neumann-Haefelin et al., 2006). However, some studies (Cooper et al., 1999; Lauer et al., 2002; Mizukoshi et al., 2002) have failed to identify any association between CTL specificity and the different MHC class I epitopes predicted by algorithms for binding to some common HLA molecules, such as HLA-A2.1 (Battegay et al., 1995; Cerny et al., 1995; Erickson et al., 2001).

Early Events Influencing the Outcome of HCV Infection

Several studies have characterized anti-HCV immune responses during the acute phase of infection when both CD4+ and CD8+ T cells are detectable, in an attempt to determine whether there are differences that distinguish self-limiting versus persisting infections. Although these studies have provided important information on the breadth and repertoire of these responses, it is tempting to speculate that differences in the adaptive immune response associated with infection outcome are determined before these responses are readily detectable. It is puzzling that despite the likelihood that the virus is recognized by the immune system very early after infection, CD4+ and CD8+ T cell responses remain undetectable for weeks. Why is the immune response so silent during this period? Is it possible that interactions between immune cells and the virus during this time determine T cell phenotype, function, and fate and hence the outcome of infection? In the following sections, we will speculate on the immunological mechanisms that might occur during the first weeks of infection and influence the outcome of the disease.

Role of Early Antigen Presentation in the Liver Leading to Deletion

Data are lacking regarding early events in HCV infection (days to a few weeks after infection), particularly those related to intrahepatic lymphocytes. Very few studies have investigated early events in HCV infection, as this requires sampling of individuals as soon as they are infected. In addition, the frequency of HCV-specific T cells is so low that it is extremely difficult to detect these T cells even using MHC tetramers. In the absence of data from very early cellular studies in HCV infection, we can, however, speculate on early events of activation following infection on the basis of early molecular studies and findings from other models.

Although non-cytopathic and hepatotropic, gene microarray analysis indicates that HCV is "seen" by the innate and adaptive immune system as early as day 2 post-infection in the chimpanzee model (Bigger et al., 2001; Su et al., 2002). Until 10 years ago, recognition of exogenous antigens by CD8+ T cells was not accepted as it was assumed that, unlike CD4+ T cells, CD8+ T cells were only able to recognize endogenously synthesized antigen on the surface of infected target cells. However, in recent years it has become apparent that some subsets of dendritic cells are able to take up soluble antigen released by dying cells, process it into the MHC class I pathway of presentation, and activate CD8+ T cells (Heath et al., 2004; Bevan, 2006). This process, known as cross-presentation, does not therefore require antigen expression by the antigen-presenting cells (APCs). In other words, dendritic cells do not need to be infected by the virus to present antigen. It is therefore likely that although HCV specifically targets hepatocytes, intracellular viral proteins would be cross-presented in the lymphoid tissues and activate CD4+ and CD8+ T cells specific for HCV intracellular proteins.

Some studies have suggested that cross-presentation is not an efficient process (Ochsenbein et al., 2001). Furthermore, it might require some time to build an antigen concentration high enough to activate T cells in LN. The delay between antigen production and cross-presentation might vary between DC subsets. Recent evidence suggests that it takes 24 hours for dermal DCs to migrate to LN, but this time is 72 hours for Langherans cells (Allan et al., 2006). The migration time for liver DCs is not known. Depending on the initial spread of the virus, it might thus take a few days before the adaptive immune response is activated in lymphoid tissues. This delay would probably be reinforced by the non-cytopathic nature of HCV infections. It is possible that the virus uses this window of opportunity to establish itself and spread within the host. The rapidity of viral spread at early stages of infection and the number of infected hepatocytes remain a matter of much debate. Viral titers within the serum have often been used as a proxy for viremia within the liver. Should this assumption be correct, data from some studies indicate that HCV spreads very rapidly. Viral particles are detected in the blood as soon as three days after infection of chimpanzees (Shimizu et al., 1990). Based on an exponential growth observed until 8 or 9 days after inoculation in the sera, the doubling time of HCV in the circulation was estimated at 6.3–8.6 hours and log time (time required to grow 10-fold) at 31.3–42.9 hours (Tanaka et al., 2005). Consistent with rapid spread of the virus in the liver, viral levels in 50% of HCV-infected patients receiving a liver transplantation increase to pre-transplantation levels in just 72 hours (Garcia-Retortillo et al., 2002). Some groups have hypothesized that the rapidity with which the virus spreads in the liver is a major difference between hepatitis B and C and is one of the parameters influencing the higher rate of chronic infections in HCV (Thimme et al., 2001). However, there is an alternative view that persisting infections are associated with a longer doubling time while resolving infections are associated with a shorter doubling time (Bocharov et al., 2003). This model also implies that the initial infecting dose of virions and initial CTL frequency might play a role in viral persistence (Ehl et al., 1998; Zinkernagel, 2000).

If HCV spreads intrahepatically at the beginning of infection in the absence of a detectable immune response in lymphoid tissues, it is likely that the liver will be the major site expressing viral antigens (Figures 2 and 3). In recent years, it has become apparent that the liver occupies a unique position among solid organs as it is able to very efficiently retain activated T cells (Ando et al., 1994a, 1994b; Mehal et al., 1999; Hamann, 2000) and to support activation of naive CD8+ T cells independently of lymphoid tissues (Bertolino et al., 2001, 2002; Bowen et al., 2002, 2004, 2005d). These findings contradict a prior immunological paradigm stating that naïve T cells can only be activated in secondary lymphoid organs. They also imply that the liver may play a more direct role in priming T cells than previously thought. However, unlike activation in the LN, recent data indicate that T cell activation in the liver induces deletion of antigen-specific activated T cells and thus tolerance (Bowen et al., 2004, 2005d). This property, which remains somewhat controversial (Klein & Crispe, 2006; Wuensch et al., 2006), has been evidenced using TCR transgenic mouse models and needs to be confirmed in a clinical setting. However, it might be critical as it would explain the ability of the liver to induce tolerance following transplantation and be exploited by virus, such as HCV, to persist in the host.

Although some reports indicate that HCV infects some dendritic cells (Hiasa et al., 1998; Auffermann-Gretzinger et al., 2001; Bain et al., 2001) or epithelial intestinal cells (Deforges et al., 2004), this virus is predominantly hepatotropic (Nouri-Aria et al., 1995). Hepatocytes would therefore represent the major APC at the beginning of HCV infection. Hepatocytes have been shown to be very efficient APCs *in vitro* (Bertolino et al., 1998, 1999). They are able to induce activation of naïve T cells, a property thought to be unique to DCs. However, unlike activation by DCs, T cells activated by hepatocytes die by neglect due to activation in the absence of co-stimulatory molecules (Bertolino et al., 1999). Using transgenic mouse models in which antigen expression is restricted to hepatocytes, the antigen-presenting capacity of this cell type has been confirmed *in vivo* (Bertolino et al., 2001; Bowen et al., 2004). Recent evidence has demonstrated that circulating T cells can directly contact hepatocytes and probe for MHC/peptide complexes through fenestrations present in liver sinusoidal endothelial cells (LSECs) (Warren et al., 2006). This contact is favored by polarized expression of MHC class I molecules and ICAM-1 on the basolateral membrane of hepatocytes.

If, in the absence of inflammation, intrahepatic activation induces tolerance, it is tempting to speculate that the virus has evolved to develop mechanisms aimed at favoring early antigen presentation in the liver (Figures 2 and 3). It is puzzling that some flaviviruses upregulate MHC class I molecules in infected cells independently of IFN-γ (King & Kesson, 1988; Lobigs et al., 1996). Normally, this upregulation should increase the visibility of the virus to the immune system. However, if such upregulation occurs early in the liver, it could lead to the deletion of any potential CTL and be beneficial to the virus. Other mechanisms that could be targeted by the virus are the migration of dendritic cells to lymphoid tissues and the initiation of the cross-presentation process. HCV has been shown to impair DC function *in vitro*, either indirectly via negative regulatory signals delivered to NK cells (Jinushi et al., 2004) or more directly by viral proteins signaling negative DC maturation (Kanto et al., 1999; Auffermann-Gretzinger et al., 2001; Bain et al., 2001). However, these effects remain controversial as the antigen-presenting function of DCs isolated from chronic infected individuals does not seem to be impaired. It is possible that these mechanisms affect the migration of DCs into the LN at the beginning of infection when priming of T cells in the LN is critical. Whether HCV affects the balance of presentation between the liver and the LN has yet to be demonstrated.

In summary, findings in transgenic mouse models indicate that in the absence of antigen presentation in LN, anti-HCV-specific T cells could be activated in the liver. As this type of activation induces deletional tolerance, it is possible that the virus uses this mechanism to eliminate a high proportion of antigen-specific T cells, to facilitate its spread, and to establish a chronic infection. This model might explain the remarkable delay observed before CD8+ T cell responses are detected (Bowen et al., 2005a). In addition, it is possible that only high-affinity T cells are deleted following activation in the liver, thus leaving only low-affinity T cells for activation in the LN. Low-affinity T cells have been described to have an anergic phenotype (Heath et al., 1992, 1995; Girgis et al., 1999) (Figure 4). This might explain the anergic or "stunned" phenotype of CD8+ T cells during acute and chronic HCV infection (Gruener et al., 2001; Urbani et al., 2002; Wedemeyer et al., 2002).

Role of a Defective Activation Leading to Anergy

An alternative explanation to the anergic phenotype of CD8+ T cells during HCV infection is that T cells are activated irrespective of their affinity but that this activation leads to defective signaling, resulting in anergy. It is difficult to distinguish

The more "classical" CD4+ CD25+ regulatory T cells have also been suggested to play a role in chronic HCV infections. Increased frequencies of this subset have been described in chronically HCV-infected individuals in comparison to subjects with resolved infection (Sugimoto et al., 2003; Cabrera et al., 2004; Boettler et al., 2005). Inhibition of HCV-specific responses by this subset in *in vitro* assays was dose-dependent, required direct cell-to-cell contact, and was independent of IL-10 and transforming growth factor beta (Boettler et al., 2005; Rushbrook et al., 2005). Interestingly, in some of these studies, regulatory T cell-mediated suppressive activity was not limited to HCV-specific CD8+ T cells: CD4+ CD25+ regulatory T cells also inhibited CD8+ T cells specific for EBV, CMV, and the influenza virus in chronically HCV-infected patients. Thus, the significance of these *in vitro* findings remains uncertain. In addition, a recent report (Manigold et al., 2006) indicates that CD4+ CD25+ FoxP3+ regulatory T cells can be isolated from HCV-infected chimpanzees irrespective of whether infection persists or is cleared.

While the presence of several subsets of regulatory T cells is not mutually exclusive, it is unclear whether these regulatory T cells play a role in predetermining the outcome at early stages of infection or appear at a later stage and possibly favor viral persistence. Recent studies demonstrated that early T cells responses generated during the acute phase of HCV infection decline dramatically during the chronic phase and that no new responses were formed (Cox et al., 2005a). Although loss of HCV-specific CD4+ T cell helper responses may be involved in the failure to generate new CTL helper responses to neo-epitopes generated following viral escape (see upcoming section), it is possible that the generation of regulatory T cells might also be involved in this phenomenon. In addition, the observation that infection persists in some individuals without viral escape (Urbani et al., 2005a) suggests that HCV-specific regulatory T cells may play a role in chronic infection.

Therefore, although regulatory T cells are generated during anti-HCV responses, further studies are required to determine whether they are really critical to HCV persistence and in which compartment they exert their regulatory function.

The Role of Viral Escape

It is now widely accepted that the immune system and the virus continuously challenge each other during the development of HCV infection. Evidence for escape mutation of MHC class II-restricted epitopes is still lacking. However, the development of mutations in class I-restricted epitopes that allow viral evasion of CD8+ T cell-mediated immune responses that were first demonstrated in chimpanzees (Erickson et al., 2001) have now been confirmed in humans (Timm et al., 2004; Cox et al., 2005b; Ray et al., 2005; Tester et al., 2005). Escape from CTL responses due to mutations in MHC class I-restricted epitopes can occur due to changes in epitopes that alter proteasomal processing, leading to epitope destruction, reduce MHC class I binding, or lead to alterations in CTL recognition (Bowen & Walker, 2005b).

Although CTL escape mutations appear early in infection and remain fixed in circulating quasispecies (Erickson et al., 2001) and the development of these mutations is associated with viral persistence, it is uncertain whether they represent a cause or a consequence of chronic infection (Cerny & Chisari, 1999; Bowen & Walker, 2005c). In particular, it has remained unclear how the virus can escape a multispecific CD8+ T cell response and why, despite this response, several HCV epitopes never mutate in persistent infections (Urbani et al., 2005a, 2005b; Komatsu et al., 2006). However, recent evidence indicates that several parameters may influence the development of escape mutations during the evolution of infection, and thus contribute to viral persistence. In particular, impaired HCV-specific CD4+ T cell responses may play a critical role in the development of CTL escape mutations: in a recent study in which CD4+ T cells were depleted in chimpanzees prior to homologous HCV rechallenge (Grakoui et al., 2003), evolution to persistent infection was associated with the appearance of multiple escape mutants, despite the presence of a memory CD8+ T cell response. In addition, recent evidence indicates that the development of CTL escape mutations is associated with epitope-specific immune responses of relatively narrow T cell receptor (TCR) diversity, whereas broader epitope-specific responses are associated with lack of viral escape (Meyer-Olson et al., 2004) Thus, where CD4+ T cell responses are absent or wane and where clonotypic CD8+ T cell responses are insufficient to allow recognition of viral variants, CTL escape mutations can occur and may perhaps contribute to viral persistence.

Conclusion

Our knowledge of HCV infections has increased considerably since the discovery of the virus in 1989. Several studies in both humans and chimpanzees have revealed that, as expected, several arms of the immune response are activated following infection. Current treatments use molecules, type I IFNs, which represent one of the first lines of defense against the virus. Although its role during infection needs to be better understood, innate immunity against HCV does not prevent the spread of the virus and is thought to be relatively inefficient. The development of more effective treatments and anti-HCV vaccines will be largely dependent on our understanding of adaptive immune responses during viral infection. It is now widely accepted that natural resolution of infection is critically dependent on the generation and the survival of a potent and broad CD4+ and CD8+ T cell response after the acute phase of the infection. Although CD8+ T cells are essential to clear the virus, CD4+ T cells also appear critical in the response and might be important for promoting CD8+ T cell survival and function. If this response fails, some CD8+ T cells survive but are functionally impaired, leading to viral persistence. Why CD4+ T cell responses wane in individuals with persistent infections remains unclear, and understanding of this phenomenon will certainly be central to the development of future HCV vaccines. An increased understanding of the very early stages of HCV infection before the adaptive immune response become readily detectable will also be critical: the delay observed before the development of both CD4+ and CD8+ T cells responses is puzzling, and it is possible that some crucial events predetermine the adaptive response during this period. Unfortunately, data on early events are lacking, and further experiments will be required to determine whether the virus is presented intrahepatically and uses this strategy to purge the repertoire of HCV-specific T cells and to establish itself in the host. Once established, other strategies might be used by the virus to persist, including regulatory T cells, anergy, and viral escape.

Acknowledgments This work was supported by a program grant of the National Health and Medical Research Council of Australia (NHMRC).

References

Accapezzato, D., Francavilla, V., Paroli, M., Casciaro, M., Chircu, L.V., Cividini, A., Abrignani, S., Mondelli, M.U., Barnaba, V. (2004). Hepatic expansion of a virus-specific regulatory CD8(+) T cell population in chronic hepatitis C virus infection. *Journal of Clinical Investigation* , 113: 963–972.

Allan, R., Waithman, J., Bedoui, S., Jones, C., Villadangos, J., Zhan, Y., Lew, A., Shortman, K., Heath, W., Carbone, F. (2006). Migratory dendritic cells transfer antigen to a lymph node-resident dendritic cell population for efficient CTL priming. *Immunity*, 25: 1–10.

Alric, L., Fort, M., Izopet, J., Vinel, J.P., Charlet, J.P., Selves, J., Puel, J., Pascal, J.P., Duffaut, M., Abbal, M. (1997). Genes of the major histocompatibility complex class II influence the outcome of hepatitis C virus infection. *Gastroenterology*, 113: 1675–1681.

Alter, H.J., Seeff, L.B. (2000). Recovery, persistence, and sequelae in hepatitis C virus infection: a perspective on long-term outcome. *Seminars in Liver Disease*, 20: 17–35.

Ando, K., Guidotti, L.G., Cerny, A., Ishikawa, T., Chisari, F.V. (1994a). CTL access to tissue antigen is restricted in vivo. *Journal of Immunology*, 153: 482–488.

Ando, K., Guidotti, L.G., Wirth, S., Ishikawa, T., Missale, G., Moriyama, T., Schreiber, R.D., Schlicht, H.J., Huang, S.N., Chisari, F.V. (1994b). Class I-restricted cytotoxic T lymphocytes are directly cytopathic for their target cells in vivo. *Journal of Immunology*, 152: 3245–3253.

Appay, V., Nixon, D.F., Donahoe, S.M., Gillespie, G.M., Dong, T., King, A., Ogg, G.S., Spiegel, H.M., Conlon, C., Spina, C.A., Havlir, D.V., Richman, D.D., Waters, A., Easterbrook, P., McMichael, A.J., Rowland-Jones, S.L. (2000). HIV-specific CD8(+) T cells produce antiviral cytokines but are impaired in cytolytic function. *Journal of Experimental Medicine*, 192: 63–75.

Appay, V., Dunbar, P.R., Callan, M., Klenerman, P., Gillespie, G.M., Papagno, L., Ogg, G.S., King, A., Lechner, F., Spina, C.A., Little, S., Havlir, D.V., Richman, D.D., Gruener, N., Pape, G., Waters, A., Easterbrook, P., Salio, M., Cerundolo, V., McMichael, A.J., Rowland-Jones, S.L. (2002). Memory CD8+ T cells vary in differentiation phenotype in different persistent virus infections. *Nature Medicine*, 8: 379–385.

Auffermann-Gretzinger, S., Keeffe, E.B., Levy, S. (2001). Impaired dendritic cell maturation in patients with chronic, but not resolved, hepatitis C virus infection. *Blood*, 97: 3171–3176.

Bachmann, M.F., Hunziker, L., Zinkernagel, R.M., Storni, T., Kopf, M. (2004). Maintenance of memory CTL responses by T helper cells and CD40-CD40 ligand: antibodies provide the key. *European Journal of Immunology*, 34: 317–326.

Bain, C., Fatmi, A., Zoulim, F., Zarski, J.P., Trepo, C., Inchauspe, G. (2001). Impaired allostimulatory function of dendritic cells in chronic hepatitis C infection. *Gastroenterology*, 120: 512–524.

Barnes, E., Ward, S.M., Kasprowicz, V.O., Dusheiko, G., Klenerman, P., Lucas, M. (2004). Ultra-sensitive class I tetramer analysis reveals previously undetectable populations of antiviral CD8+ T cells. *European Journal of Immunology*, 34: 1570–1577.

Bartosch, B., Bukh, J., Meunier, J.C., Granier, C., Engle, R.E., Blackwelder, W.C., Emerson, S. U., Cosset, F.L., Purcell, R.H. (2003). In vitro assay for neutralizing antibody to hepatitis C virus: evidence for broadly conserved neutralization epitopes. *Proceedings of the National Academy of Sciences USA*, 100: 14199–14204.

Battegay, M., Fikes, J., Di Bisceglie, A.M., Wentworth, P.A., Sette, A., Celis, E., Ching, W.M., Grakoui, A., Rice, C.M., Kurokohchi, K., et al. (1995). Patients with chronic hepatitis C have circulating cytotoxic T cells which recognize hepatitis C virus-encoded peptides binding to HLA-A2.1 molecules. *Journal of Virology*, 69: 2462–2470.

Bertolino, P., Trescol-Biemont, M.C., Rabourdin-Combe, C. (1998). Hepatocytes induce functional activation of naive CD8+ T lymphocytes but fail to promote survival. *European Journal of Immunology*, 28: 221–236.

Bertolino, P., Trescol-Biemont, M.C., Thomas, J., Fazekas de St. Groth, B., Pihlgren, M., Marvel, J., Rabourdin-Combe, C. (1999). Death by neglect as a deletional mechanism of peripheral tolerance. *International Immunology*, 11: 1225–1238.

Bertolino, P., Bowen, D.G., McCaughan, G.W., Fazekas De St. Groth, B. (2001). Antigen-specific primary activation of CD8+ T cells within the liver. *Journal of Immunology*, 166: 5430–5438.

Bertolino, P., McCaughan, G.W., Bowen, D.G. (2002). Role of primary intrahepatic T-cell activation in the 'liver tolerance effect'. *Immunology & Cell Biology*, 80: 84–92.

Bevan, M.J. (2006). Cross-priming. *Nature Immunology*, 7: 363–365.

Bigger, C.B., Brasky, K.M., Lanford, R.E. (2001). DNA microarray analysis of chimpanzee liver during acute resolving hepatitis C virus infection. *Journal of Virology*, 75: 7059–7066.

Bigger, C.B., Guerra, B., Brasky, K.M., Hubbard, G., Beard, M.R., Luxon, B.A., Lemon, S.M., Lanford, R.E. (2004). Intrahepatic gene expression during chronic hepatitis C virus infection in chimpanzees. *Journal of Virology*, 78: 13779–13792.

Blight, K.J., Kolykhalov, A.A., Rice, C.M. (2000). Efficient initiation of HCV RNA replication in cell culture. *Science*, 290: 1972–1974.

Blindenbacher, A., Duong, F.H., Hunziker, L., Stutvoet, S.T., Wang, X., Terracciano, L., Moradpour, D., Blum, H.E., Alonzi, T., Tripodi, M., La Monica, N., Heim, M.H. (2003). Expression of hepatitis C virus proteins inhibits interferon alpha signaling in the liver of transgenic mice. *Gastroenterology*, 124: 1465–1475.

Bocharov, G., Klenerman, P., Ehl, S. (2003). Modelling the dynamics of LCMV infection in mice: II. Compartmental structure and immunopathology. *Journal of Theoretical Biology*, 221: 349–378.

Boettler, T., Spangenberg, H.C., Neumann-Haefelin, C., Panther, E., Urbani, S., Ferrari, C., Blum, H.E., von Weizsacker, F., Thimme, R. (2005). T cells with a CD4+CD25+ regulatory phenotype suppress in vitro proliferation of virus-specific CD8+ T cells during chronic hepatitis C virus infection. *Journal of Virology*, 79: 7860–7867.

Bowen, D.G., Warren, A., Davis, T., Hoffmann, M.W., McCaughan, G.W., De St. Groth, B.F., Bertolino, P. (2002). Cytokine-dependent bystander hepatitis due to intrahepatic murine CD8+ T-cell activation by bone marrow-derived cells. *Gastroenterology*, 123: 1252–1264.

Bowen, D.G., Zen, M., Holz, L., Davis, T., McCaughan, G.W., Bertolino, P. (2004). The site of primary T cell activation is a determinant of the balance between intrahepatic tolerance and immunity. *Journal of Clinical Investigation*, 114: 701–712.

Bowen, D.G., Walker, C.M. (2005a). Adaptive immune responses in acute and chronic hepatitis C virus infection. *Nature*, 436: 946–952.

Bowen, D.G., Walker, C.M. (2005b). Mutational escape from CD8+ T cell immunity: HCV evolution, from chimpanzees to man. *Journal of Experimental Medicine*, 201: 1709–1714.

Bowen, D.G., Walker, C.M. (2005c). The origin of quasispecies: cause or consequence of chronic hepatitis C viral infection? *Journal of Hepatology* , 42: 408–417.

Bowen, D.G., McCaughan, G., Bertolino, P. (2005d). Intrahepatic immunity: a tale of two sites? *Trends in Immunology*, 26: 512–517.

Cabrera, R., Tu, Z., Xu, Y., Firpi, R.J., Rosen, H.R., Liu, C., Nelson, D.R. (2004). An immunomodulatory role for CD4(+)CD25(+) regulatory T lymphocytes in hepatitis C virus infection. *Hepatology*, 40: 1062–1071.

Cerny, A., Chisari, F.V. (1999). Pathogenesis of chronic hepatitis C: immunological features of hepatic injury and viral persistence. *Hepatology*, 30: 595–601.

Cerny, A., McHutchison, J.G., Pasquinelli, C., Brown, M.E., Brothers, M.A., Grabscheid, B., Fowler, P., Houghton, M., Chisari, F.V. (1995). Cytotoxic T lymphocyte response to hepatitis C virus-derived peptides containing the HLA A2.1 binding motif. *Journal of Clinical Investigation*, 95: 521–530.

Chan, C.W., Crafton, E., Fan, H.N., Flook, J., Yoshimura, K., Skarica, M., Brockstedt, D., Dubensky, T.W., Stins, M.F., Lanier, L.L., Pardoll, D.M., Housseau, F. (2006). Interferon-producing killer dendritic cells provide a link between innate and adaptive immunity. *Nature Medicine*, 12: 207–213.

Chang, K.M., Thimme, R., Melpolder, J.J., Oldach, D., Pemberton, J., Moorhead-Loudis, J., McHutchison, J.G., Alter, H.J., Chisari, F.V. (2001). Differential CD4(+) and CD8(+) T-cell responsiveness in hepatitis C virus infection. *Hepatology*, 33: 267–276.

Chen, C.M., You, L.R., Hwang, L.H., Lee, Y.H. (1997). Direct interaction of hepatitis C virus core protein with the cellular lymphotoxin-beta receptor modulates the signal pathway of the lymphotoxin-beta receptor. *Journal of Virology* , 71: 9417–9426.

Chisari, F.V. (2005). Unscrambling hepatitis C virus-host interactions. *Nature*, 436: 930–932.

Cooper, S., Erickson, A.L., Adams, E.J., Kansopon, J., Weiner, A.J., Chien, D.Y., Houghton, M., Parham, P., Walker, C.M. (1999). Analysis of a successful immune response against hepatitis C virus. *Immunity*, 10: 439–449.

Cox, A.L., Mosbruger, T., Lauer, G.M., Pardoll, D., Thomas, D.L., Ray, S.C. (2005a). Comprehensive analyses of CD8+ T cell responses during longitudinal study of acute human hepatitis C. *Hepatology*, 42: 104–112.

Cox, A.L., Mosbruger, T., Mao, Q., Liu, Z., Wang, X.H., Yang, H.C., Sidney, J., Sette, A., Pardoll, D., Thomas, D.L., Ray, S.C. (2005b). Cellular immune selection with hepatitis C virus persistence in humans. *Journal of Experimental Medicine*, 201: 1741–1752.

Crispe, I.N. (2003). Hepatic T cells and liver tolerance. *Nature Reviews in Immunology*, 3: 51–62.

Crispe, I.N., Mehal, W.Z. (1996). Strange brew: T cells in the liver. *Immunology Today*, 17: 522–525.

Crispe, I.N., Dao, T., Klugewitz, K., Mehal, W.Z., Metz, D.P. (2000). The liver as a site of T-cell apoptosis: graveyard, or killing field? *Immunological Reviews*, 174: 47–62.

Day, C.L., Lauer, G.M., Robbins, G.K., McGovern, B., Wurcel, A.G., Gandhi, R.T., Chung, R.T., Walker, B.D. (2002). Broad specificity of virus-specific CD4+ T-helper-cell responses in resolved hepatitis C virus infection. *Journal of Virology*, 76: 12584–12595.

Day, C.L., Walker, B.D. (2003a). Progress in defining CD4 helper cell responses in chronic viral infections. *Journal of Experimental Medicine*, 198: 1773–1777.

Day, C.L., Seth, N.P., Lucas, M., Appel, H., Gauthier, L., Lauer, G.M., Robbins, G.K., Szczepiorkowski, Z.M., Casson, D.R., Chung, R.T., Bell, S., Harcourt, G., Walker, B.D., Klenerman, P., Wucherpfennig, K.W. (2003b). Ex vivo analysis of human memory CD4 T cells specific for hepatitis C virus using MHC class II tetramers. *Journal of Clinical Investigation*, 112: 831–842.

Deforges, S., Evlashev, A., Perret, M., Sodoyer, M., Pouzol, S., Scoazec, J.Y., Bonnaud, B., Diaz, O., Paranhos-Baccala, G., Lotteau, V., Andre, P. (2004). Expression of hepatitis C virus proteins in epithelial intestinal cells in vivo. *Journal of General Virology*, 85: 2515–2523.

Diepolder, H.M., Zachoval, R., Hoffmann, R.M., Wierenga, E.A., Santantonio, T., Jung, M.C., Eichenlaub, D., Pape, G.R. (1995). Possible mechanism involving T-lymphocyte response to non-structural protein 3 in viral clearance in acute hepatitis C virus infection. *Lancet*, 346: 1006–1007.

Dumoulin, F.L., Bach, A., Leifeld, L., El-Bakri, M., Fischer, H.P., Sauerbruch, T., Spengler, U. (1997). Semiquantitative analysis of intrahepatic cytokine mRNAs in chronic hepatitis C. *Journal of Infectious Disease*, 175: 681–685.

Eckels, D.D., Zhou, H., Bian, T.H., Wang, H. (1999). Identification of antigenic escape variants in an immunodominant epitope of hepatitis C virus. *International Journal of Immunolology*, 11: 577–583.

Ehl, S., Klenerman, P., Zinkernagel, R.M., Bocharov, G. (1998). The impact of variation in the number of CD8(+) T-cell precursors on the outcome of virus infection. *Cellular Immunology*, 189: 67–73.

Erickson, A.L., Kimura, Y., Igarashi, S., Eichelberger, J., Houghton, M., Sidney, J., McKinney, D., Sette, A., Hughes, A.L., Walker, C.M. (2001). The outcome of hepatitis C virus infection is predicted by escape mutations in epitopes targeted by cytotoxic T lymphocytes. *Immunity*, 15: 883–895.

Fanning, L.J., Levis, J., Kenny-Walsh, E., Whelton, M., O'Sullivan, K., Shanahan, F. (2001). HLA class II genes determine the natural variance of hepatitis C viral load. *Hepatology*, 33: 224–230.

Farci, P., Alter, H.J., Govindarajan, S., Wong, D.C., Engle, R., Lesniewski, R.R., Mushahwar, I.K., Desai, S.M., Miller, R.H., Ogata, N., et al. (1992). Lack of protective immunity against reinfection with hepatitis C virus. *Science*, 258: 135–140.

Farci, P., Alter, H.J., Wong, D.C., Miller, R.H., Govindarajan, S., Engle, R., Shapiro, M., Purcell, R.H. (1994). Prevention of hepatitis C virus infection in chimpanzees after antibody-mediated in vitro neutralization. *Proceedings of the National Academy of Sciences USA*, 91: 7792–7796.

Foy, E., Li, K., Wang, C., Sumpter, R., Jr., Ikeda, M., Lemon, S.M., Gale, M., Jr. (2003). Regulation of interferon regulatory factor-3 by the hepatitis C virus serine protease. *Science*, 300: 1145–1148.

Foy, E., Li, K., Sumpter, R., Jr., Loo, Y.M., Johnson, C.L., Wang, C., Fish, P.M., Yoneyama, M., Fujita, T., Lemon, S.M., Gale, M., Jr. (2005). Control of antiviral defenses through hepatitis C virus disruption of retinoic acid-inducible gene-I signaling. *Proceedings of the National Academy of Sciences USA*, 102: 2986–2991.

Gale, M.J. Jr., Foy, E.M. (2005). Evasion of intracellular host defence by hepatitis C virus. *Nature*, 436: 939–945.

Gale, M.J. Jr., Korth, M.J., Tang, N.M., Tan, S.L., Hopkins, D.A., Dever, T.E., Polyak, S.J., Gretch, D.R., Katze, M.G. (1997). Evidence that hepatitis C virus resistance to interferon is mediated through repression of the PKR protein kinase by the nonstructural 5A protein. *Virology*, 230: 217–227.

Gale, M., Jr., Blakely, C.M., Kwieciszewski, B., Tan, S.L., Dossett, M., Tang, N.M., Korth, M.J., Polyak, S.J., Gretch, D.R., Katze, M.G. (1998). Control of PKR protein kinase by hepatitis C virus nonstructural 5A protein: molecular mechanisms of kinase regulation. *Molecular and Cellular Biology*, 18: 5208–5218.

Garcia-Retortillo, M., Forns, X., Feliu, A., Moitinho, E., Costa, J., Navasa, M., Rimola, A., Rodes, J. (2002). Hepatitis C virus kinetics during and immediately after liver transplantation. *Hepatology*, 35: 680–687.

Gerlach, J.T., Diepolder, H.M., Jung, M.C., Gruener, N.H., Schraut, W.W., Zachoval, R., Hoffmann, R., Schirren, C.A., Santantonio, T., Pape, G.R. (1999). Recurrence of hepatitis C virus after loss of virus-specific CD4(+) T-cell response in acute hepatitis C [see comments]. *Gastroenterology*, 117: 933–941.

Girgis, L., Davis, M.M., Fazekas de St. Groth, B. (1999). The avidity spectrum of T cell receptor interactions accounts for T cell anergy in a double transgenic model. *Journal of Experimental Medicine*, 189: 265–278.

Grabowska, A.M., Lechner, F., Klenerman, P., Tighe, P.J., Ryder, S., Ball, J.K., Thomson, B.J., Irving, W.L., Robins, R.A. (2001). Direct ex vivo comparison of the breadth and specificity of the T cells in the liver and peripheral blood of patients with chronic HCV infection. *European Journal of Immunology*, 31: 2388–2394.

Grakoui, A., Shoukry, N.H., Woollard, D.J., Han, J.H., Hanson, H.L., Ghrayeb, J., Murthy, K.K., Rice, C.M., Walker, C.M. (2003). HCV persistence and immune evasion in the absence of memory T cell help. *Science*, 302: 659–662.

Gruener, N.H., Lechner, F., Jung, M.C., Diepolder, H., Gerlach, T., Lauer, G., Walker, B., Sullivan, J., Phillips, R., Pape, G.R., Klenerman, P. (2001). Sustained dysfunction of antiviral CD8+ T lymphocytes after infection with hepatitis C virus. *Journal of Virology*, 75: 5550–5558.

Guo, J.T., Bichko, V.V., Seeger, C. (2001). Effect of alpha interferon on the hepatitis C virus replicon. *Journal of Virology*, 75: 8516–8523.

Hamann, A., Klugewitz, K., Austrup, F., Jablonski-Westrich, D. (2000). Activation induces rapid and profound alterations in the trafficking of T cells. *European Journal of Immunology*, 30: 3207–3218.

Harcourt, G., Hellier, S., Bunce, M., Satsangi, J., Collier, J., Chapman, R., Phillips, R., Klenerman, P. (2001). Effect of HLA class II genotype on T helper lymphocyte responses and viral control in hepatitis C virus infection. *Journal of Viral Hepatitis*, 8: 174–179.

He, X.S., Rehermann, B., Lopez-Labrador, F.X., Boisvert, J., Cheung, R., Mumm, J., Wedemeyer, H., Berenguer, M., Wright, T.L., Davis, M.M., Greenberg, H.B. (1999). Quantitative analysis of hepatitis C virus-specific CD8(+) T cells in peripheral blood and liver using peptide-MHC tetramers. *Proceedings of the National Academy of Sciences USA*, 96: 5692–5697.

Heath, W.R., Allison, J., Hoffmann, M.W., Schonrich, G., Arnold, B., Miller, J.F. (1992). Autoimmune diabetes as a consequence of locally produced interleukin-2. *Nature*, 359: 547–549.

Heath, W.R., Karamalis, F., Donoghue, J., Miller, J.F. (1995). Autoimmunity caused by ignorant CD8+ T cells is transient and depends on avidity. *Journal of Immunology*, 155: 2339–2349.

Heath, W.R., Belz, G.T., Behrens, G.M., Smith, C.M., Forehan, S.P., Parish, I.A., Davey, G.M., Wilson, N.S., Carbone, F.R., Villadangos, J.A. (2004). Cross-presentation, dendritic cell subsets, and the generation of immunity to cellular antigens. *Immunological Reviews*, 199: 9–26.

Hiasa, Y., Horiike, N., Akbar, S.M., Saito, I., Miyamura, T., Matsuura, Y., Onji, M. (1998). Low stimulatory capacity of lymphoid dendritic cells expressing hepatitis C virus genes. *Biochemical and Biophysical Research Communications*, 249: 90–95.

Huang, L., Soldevila, G., Leeker, M., Flavell, R., Crispe, I.N. (1994). The liver eliminates T cells undergoing antigen-triggered apoptosis in vivo. *Immunity*, 1: 741–749.

Jaeckel, E.C., Raja, S., Tan, J., Das, S.K., Dey, S.K., Girod, D.A., Tsue, T.T., Sanford, T.R. (2001). Correlation of expression of cyclooxygenase-2, vascular endothelial growth factor, and peroxisome proliferator-activated receptor delta with head and neck squamous cell carcinoma. *Archives of Otolaryngology—Head and Neck Surgery*, 127: 1253–1259.

Jinushi, M., Takehara, T., Tatsumi, T., Kanto, T., Miyagi, T., Suzuki, T., Kanazawa, Y., Hiramatsu, N., Hayashi, N. (2004). Negative regulation of NK cell activities by inhibitory receptor CD94/NKG2A leads to altered NK cell-induced modulation of dendritic cell functions in chronic hepatitis C virus infection. *Journal of Immunology*, 173: 6072–6081.

Kanto, T., Hayashi, N., Takehara, T., Tatsumi, T., Kuzushita, N., Ito, A., Sasaki, Y., Kasahara, A., Hori, M. (1999). Impaired allostimulatory capacity of peripheral blood dendritic cells recovered from hepatitis C virus-infected individuals. *Journal of Immunology*, 162: 5584–5591.

Katze, M.G., He, Y., Gale, M., Jr. (2002). Viruses and interferon: a fight for supremacy. *Nature Reviews in Immunology*, 2: 675–687.

King, N.J., Kesson, A.M. (1988). Interferon-independent increases in class I major histocompatibility complex antigen expression follow flavivirus infection. *Journal of General Virology*, 69 (Pt. 10): 2535–2543.

Kittlesen, D.J., Chianese-Bullock, K.A., Yao, Z.Q., Braciale, T.J., Hahn, Y.S. (2000). Interaction between complement receptor gC1qR and hepatitis C virus core protein inhibits T-lymphocyte proliferation. *Journal of Clinical Investigation*, 106: 1239–1249.

Klein, I., Crispe, I.N. (2006). Complete differentiation of CD8+ T cells activated locally within the transplanted liver. *Journal of Experimental Medicine*, 203: 437–447.

Klenerman, P., Lucas, M., Barnes, E., Harcourt, G. (2002). Immunity to hepatitis C virus: stunned but not defeated. *Microbes and Infection*, 4: 57–65.

Komatsu, H., Lauer, G., Pybus, O.G., Ouchi, K., Wong, D., Ward, S., Walker, B., Klenerman, P. (2006). Do antiviral CD8+ T cells select hepatitis C virus escape mutants? Analysis in diverse epitopes targeted by human intrahepatic CD8+ T lymphocytes. *Journal of Viral Hepatitis*, 13: 121–130.

Konan, K.V., Giddings, T.H., Jr., Ikeda, M., Li, K., Lemon, S.M., Kirkegaard, K. (2003). Nonstructural protein precursor NS4A/B from hepatitis C virus alters function and ultrastructure of host secretory apparatus. *Journal of Virology*, 77: 7843–7855.

Kowalski, H., Erickson, A.L., Cooper, S., Domena, J.D., Parham, P., Walker, C.M. (1996). Patr-A and B, the orthologues of HLA-A and B, present hepatitis C virus epitopes to CD8+ cytotoxic T cells from two chronically infected chimpanzees. *Journal of Experimental Medicine*, 183: 1761–1775.

Koziel, M.J., Dudley, D., Wong, J.T., Dienstag, J., Houghton, M., Ralston, R., Walker, B.D. (1992). Intrahepatic cytotoxic T lymphocytes specific for hepatitis C virus in persons with chronic hepatitis. *Journal of Immunology*, 149: 3339–3344.

Koziel, M.J., Dudley, D., Afdhal, N., Choo, Q.L., Houghton, M., Ralston, R., Walker, B.D. (1993). Hepatitis C virus (HCV)-specific cytotoxic T lymphocytes recognize epitopes in the core and envelope proteins of HCV. *Journal of Virology*, 67: 7522–7532.

Koziel, M.J., Dudley, D., Afdhal, N., Grakoui, A., Rice, C.M., Choo, Q.L., Houghton, M., Walker, B.D. (1995). HLA class I-restricted cytotoxic T lymphocytes specific for hepatitis C virus. Identification of multiple epitopes and characterization of patterns of cytokine release. *Journal of Clinical Investigation*, 96: 2311–2321.

Lai, M.E., Mazzoleni, A.P., Argiolu, F., De Virgilis, S., Balestrieri, A., Purcell, R.H., Cao, A., Farci, P. (1994). Hepatitis C in multiple episodes of acute hepatitis in polytransfused thalassaemic children. *Lancet*, 343: 388–390.

Lanford, R.E., Guerra, B., Lee, H., Averett, D.R., Pfeiffer, B., Chavez, D., Notvall, L., Bigger, C. (2003). Antiviral effect and virus-host interactions in response to alpha interferon, gamma interferon, poly(i)-poly(c), tumor necrosis factor alpha, and ribavirin in hepatitis C virus subgenomic replicons. *Journal of Virology*, 77: 1092–1104.

Larsson, M., Babcock, E., Grakoui, A., Shoukry, N., Lauer, G., Rice, C., Walker, C., Bhardwaj, N. (2004). Lack of phenotypic and functional impairment in dendritic cells from chimpanzees chronically infected with hepatitis C virus. *Journal of Virology*, 78: 6151–6161.

Lauer, G.M., Ouchi, K., Chung, R.T., Nguyen, T.N., Day, C.L., Purkis, D.R., Reiser, M., Kim, A.Y., Lucas, M., Klenerman, P., Walker, B.D. (2002). Comprehensive analysis of CD8(+)-T-cell responses against hepatitis C virus reveals multiple unpredicted specificities. *Journal of Virology*, 76: 6104–6113.

Lechner, F., Wong, D.K., Dunbar, P.R., Chapman, R., Chung, R.T., Dohrenwend, P., Robbins, G., Phillips, R., Klenerman, P., Walker, B.D. (2000). Analysis of successful immune responses in persons infected with hepatitis C virus. *Journal of Experimental Medicine*, 191: 1499–1512.

Li, K., Foy, E., Ferreon, J.C., Nakamura, M., Ferreon, A.C., Ikeda, M., Ray, S.C., Gale, M., Jr., Lemon, S.M. (2005). Immune evasion by hepatitis C virus NS3/4A protease-mediated cleavage of the Toll-like receptor 3 adaptor protein TRIF. *Proceedings of the National Academy of Sciences USA*, 102: 2992–2997.

Li, Y., Wang, X., Douglas, S.D., Metzger, D.S., Woody, G., Zhang, T., Song, L., Ho, W.Z. (2005). CD8+ T cell depletion amplifies hepatitis C virus replication in peripheral blood mononuclear cells. *Journal of Infectious Diseases*, 192: 1093–1101.

Lindenbach, B.D., Evans, M.J., Syder, A.J., Wolk, B., Tellinghuisen, T.L., Liu, C.C., Maruyama, T., Hynes, R.O., Burton, D.R., McKeating, J.A., Rice, C.M. (2005). Complete replication of hepatitis C virus in cell culture. *Science*, 309: 623–626.

Lobigs, M., Blanden, R.V., Mullbacher, A. (1996). Flavivirus-induced up-regulation of MHC class I antigens; implications for the induction of CD8+ T-cell-mediated autoimmunity. *Immunological Reviews*, 152: 5–19.

Logvinoff, C., Major, M.E., Oldach, D., Heyward, S., Talal, A., Balfe, P., Feinstone, S.M., Alter, H., Rice, C.M., McKeating, J.A. (2004). Neutralizing antibody response during acute and chronic hepatitis C virus infection. *Proceedings of the National Academy of Sciences USA*, 101: 10149–10154.

Lucas, M., Vargas-Cuero, A.L., Lauer, G.M., Barnes, E., Willberg, C.B., Semmo, N., Walker, B.D., Phillips, R., Klenerman, P. (2004). Pervasive influence of hepatitis C virus on the phenotype of antiviral CD8+ T cells. *Journal of Immunology*, 172: 1744–1753.

MacDonald, A.J., Duffy, M., Brady, M.T., McKiernan, S., Hall, W., Hegarty, J., Curry, M., Mills, K.H. (2002). CD4 T helper type 1 and regulatory T cells induced against the same epitopes on the core protein in hepatitis C virus-infected persons. *Journal of Infectious Diseases*, 185: 720–727.

Manigold, T., Shin, E.C., Mizukoshi, E., Mihalik, K., Murthy, K.K., Rice, C.M., Piccirillo, C.A., Rehermann, B. (2006). Foxp3+CD4+CD25+ T cells control virus-specific memory T cells in chimpanzees that recovered from hepatitis C. *Blood*, 107: 4424–4432.

Manns, M.P., McHutchison, J.G., Gordon, S.C., Rustgi, V.K., Shiffman, M., Reindollar, R., Goodman, Z.D., Koury, K., Ling, M., Albrecht, J.K. (2001). Peginterferon alfa-2b plus ribavirin compared with interferon alfa-2b plus ribavirin for initial treatment of chronic hepatitis C: a randomised trial. *Lancet*, 358: 958–965.

Matsumoto, M., Hsieh, T.Y., Zhu, N., VanArsdale, T., Hwang, S.B., Jeng, K.S., Gorbalenya, A.E., Lo, S.Y., Ou, J.H., Ware, C.F., Lai, M.M. (1997). Hepatitis C virus core protein interacts with the cytoplasmic tail of lymphotoxin-beta receptor. *Journal of Virology*, 71: 1301–1309.

Mehal, W.Z., Juedes, A.E., Crispe, I.N. (1999). Selective retention of activated CD8+ T cells by the normal liver. *Journal of Immunology*, 163: 3202–3210.

Mehal, W.Z., Azzaroli, F., Crispe, I.N. (2001). Antigen presentation by liver cells controls intrahepatic T cell trapping, whereas bone marrow-derived cells preferentially promote intrahepatic T cell apoptosis. *Journal of Immunology*, 167: 667–673.

Mehta, S.H., Cox, A., Hoover, D.R., Wang, X.H., Mao, Q., Ray, S., Strathdee, S.A., Vlahov, D., Thomas, D.L. (2002). Protection against persistence of hepatitis C. *Lancet*, 359: 1478–1483.

Meunier, J.C., Engle, R.E., Faulk, K., Zhao, M., Bartosch, B., Alter, H., Emerson, S.U., Cosset, F.L., Purcell, R.H., Bukh, J. (2005). Evidence for cross-genotype neutralization of hepatitis C virus pseudo-particles and enhancement of infectivity by apolipoprotein C1. *Proceedings of the National Academy of Sciences USA*, 102: 4560–4565.

Meyer-Olson, D., Shoukry, N.H., Brady, K.W., Kim, H., Olson, D.P., Hartman, K., Shintani, A.K., Walker, C.M., Kalams, S.A. (2004). Limited T cell receptor diversity of HCV-specific T cell responses is associated with CTL escape. *Journal of Experimental Medicine*, 200: 307–319.

Minton, E.J., Smillie, D., Neal, K.R., Irving, W.L., Underwood, J.C., James, V. (1998). Association between MHC class II alleles and clearance of circulating hepatitis C virus. Members of the Trent Hepatitis C Virus Study Group. *Journal of Infectious Diseases*, 178: 39–44.

Minutello, M.A., Pileri, P., Unutmaz, D., Censini, S., Kuo, G., Houghton, M., Brunetto, M.R., Bonino, F., Abrignani, S. (1993). Compartmentalization of T lymphocytes to the site of disease: intrahepatic CD4+ T cells specific for the protein NS4 of hepatitis C virus in patients with chronic hepatitis C. *Journal of Experimental Medicine*, 178: 17–25.

Mizukoshi, E., Nascimbeni, M., Blaustein, J.B., Mihalik, K., Rice, C.M., Liang, T.J., Feinstone, S.M., Rehermann, B. (2002). Molecular and immunological significance of chimpanzee major histocompatibility complex haplotypes for hepatitis C virus immune response and vaccination studies. *Journal of Virology*, 76: 6093–6103.

Napoli, J., Bishop, G.A., McGuinness, P.H., Painter, D.M., McCaughan, G.W. (1996). Progressive liver injury in chronic hepatitis C infection correlates with increased intrahepatic expression of Th1-associated cytokines. *Hepatology*, 24: 759–765.

Nascimbeni, M., Mizukoshi, E., Bosmann, M., Major, M.E., Mihalik, K., Rice, C.M., Feinstone, S.M., Rehermann, B. (2003). Kinetics of CD4+ and CD8+ memory T-cell responses during hepatitis C virus rechallenge of previously recovered chimpanzees. *Journal of Virology*, 77: 4781–4793.

Nelson, D.R., Marousis, C.G., Davis, G.L., Rice, C.M., Wong, J., Houghton, M., Lau, J.Y. (1997). The role of hepatitis C virus-specific cytotoxic T lymphocytes in chronic hepatitis C. *Journal of Immunology*, 158: 1473–1481.

Neumann, A.U., Lam, N.P., Dahari, H., Gretch, D.R., Wiley, T.E., Layden, T.J., Perelson, A.S. (1998). Hepatitis C viral dynamics in vivo and the antiviral efficacy of interferon-alpha therapy. *Science*, 282: 103–107.

Neumann-Haefelin, C., McKiernan, S., Ward, S., Viazov, S., Spangenberg, H.C., Killinger, T., Baumert, T.F., Nazarova, N., Sheridan, I., Pybus, O., von Weizsacker, F., Roggendorf, M.,

Kelleher, D., Klenerman, P., Blum, H.E., Thimme, R. (2006). Dominant influence of an HLA-B27 restricted CD8+ T cell response in mediating HCV clearance and evolution. *Hepatology*, 43: 563–572.

Nisii, C., Tempeshill, M., Agrati, C., Poccia, F., Tocci, G., Longo, M.A., D'offici, G., Tersigni, R., Lo Lacono, O., Antonucci, G., Oliva, A. (2006). Accummulation of dysfunctional effcion CD8+ T cells in the lives of patients with chronic HCV infection, *Journal of Hepatology*, 44: 475–483.

Nouri-Aria, K.T., Sallie, R., Mizokami, M., Portmann, B.C., Williams, R. (1995). Intrahepatic expression of hepatitis C virus antigens in chronic liver disease. *Journal of Pathology*, 175: 77–83.

O'Farrelly, C., Crispe, I.N. (1999). Prometheus through the looking glass: reflections on the hepatic immune system. *Immunology Today*, 20: 394–398.

Ochsenbein, A.F., Sierro, S., Odermatt, B., Pericin, M., Karrer, U., Hermans, J., Hemmi, S., Hengartner, H., Zinkernagel, R.M. (2001). Roles of tumour localization, second signals and cross priming in cytotoxic T-cell induction. *Nature*, 411: 1058–1064.

Pavio, N., Taylor, D.R., Lai, M.M. (2002). Detection of a novel unglycosylated form of hepatitis C virus E2 envelope protein that is located in the cytosol and interacts with PKR. *Journal of Virology* , 76: 1265–1272.

Pawlotsky, J.M. (1999). Diagnostic tests for hepatitis C. *Journal of Hepatology* , 31 Suppl. 1: 71–79.

Pawlotsky, J.M. (2004). Pathophysiology of hepatitis C virus infection and related liver disease. *Trends in Microbiology*, 12: 96–102.

Pearlman, B.L. (2004). Hepatitis C treatment update. *American Journal of Medicine*, 117: 344–352.

Penna, A., Missale, G., Lamonaca, V., Pilli, M., Mori, C., Zanelli, P., Cavalli, A., Elia, G., Ferrari, C. (2002). Intrahepatic and circulating HLA class II-restricted, hepatitis C virus-specific T cells: functional characterization in patients with chronic hepatitis C. *Hepatology*, 35: 1225–1236.

Ray, S.C., Fanning, L., Wang, X.H., Netski, D.M., Kenny-Walsh, E., Thomas, D.L. (2005). Divergent and convergent evolution after a common-source outbreak of hepatitis C virus. *Journal of Experimental Medicine* , 201: 1753–1759.

Rehermann, B., Nascimbeni, M. (2005). Immunology of hepatitis B virus and hepatitis C virus infection. *Nature Reviews in Immunology*, 5: 215–229.

Rehermann, B., Chang, K.M., McHutchinson, J., Kokka, R., Houghton, M., Rice, C.M., Chisari, F.V. (1996a). Differential cytotoxic T-lymphocyte responsiveness to the hepatitis B and C viruses in chronically infected patients. *Journal of Virology*, 70: 7092–7102.

Rehermann, B., Chang, K.M., McHutchison, J.G., Kokka, R., Houghton, M., Chisari, F.V. (1996b). Quantitative analysis of the peripheral blood cytotoxic T lymphocyte response in patients with chronic hepatitis C virus infection. *Journal of Clinical Investigation*, 98: 1432–1440.

Rosen, H.R., Miner, C., Sasaki, A.W., Lewinsohn, D.M., Conrad, A.J., Bakke, A., Bouwer, H.G., Hinrichs, D.J. (2002). Frequencies of HCV-specific effector CD4+ T cells by flow cytometry: correlation with clinical disease stages. *Hepatology*, 35: 190–198.

Rushbrook, S.M., Ward, S.M., Unitt, E., Vowler, S.L., Lucas, M., Klenerman, P., Alexander, G.J. (2005). Regulatory T cells suppress in vitro proliferation of virus-specific CD8+ T cells during persistent hepatitis C virus infection. *Journal of Virology* , 79: 7852–7859.

Santantonio, T., Fasano, M., Sinisi, E., Guastadisegni, A., Casalino, C., Mazzola, M., Francavilla, R., Pastore, G. (2005). Efficacy of a 24-week course of PEG-interferon alpha-2b monotherapy in patients with acute hepatitis C after failure of spontaneous clearance. *Journal of Hepatology*, 42: 329–333.

Schirren, C.A., Jung, M.C., Gerlach, J.T., Worzfeld, T., Baretton, G., Mamin, M., Hubert Gruener, N., Houghton, M., Pape, G.R. (2000). Liver-derived hepatitis C virus (HCV)-specific CD4(+) T cells recognize multiple HCV epitopes and produce interferon gamma. *Hepatology*, 32: 597–603.

Schulze zur Wiesch, J., Lauer, G.M., Day, C.L., Kim, A.Y., Ouchi, K., Duncan, J.E., Wurcel,

A.G., Timm, J., Jones, A.M., Mothe, B., Allen, T.M., McGovern, B., Lewis-Ximenez, L., Sidney, J., Sette, A., Chung, R.T., Walker, B.D. (2005). Broad repertoire of the CD4+ Th cell response in spontaneously controlled hepatitis C virus infection includes dominant and highly promiscuous epitopes. *Journal of Immunology* , 175: 3603–3613.

Shimizu, Y.K., Weiner, A.J., Rosenblatt, J., Wong, D.C., Shapiro, M., Popkin, T., Houghton, M., Alter, H.J., Purcell, R.H. (1990). Early events in hepatitis C virus infection of chimpanzees. *Proceedings of the National Academy of Sciences USA*, 87: 6441–6444.

Shoukry, N.H., Grakoui, A., Houghton, M., Chien, D.Y., Ghrayeb, J., Reimann, K.A., Walker, C.M. (2003). Memory CD8+ T cells are required for protection from persistent hepatitis C virus infection. *Journal of Experimental Medicine* , 197: 1645–1655.

Shoukry, N.H., Sidney, J., Sette, A., Walker, C.M. (2004). Conserved hierarchy of helper T cell responses in a chimpanzee during primary and secondary hepatitis C virus infections. *Journal of Immunology* , 172: 483–492.

Spangenberg, H.C., Viazov, S., Kersting, N., Neumann-Haefelin, C., McKinney, D., Roggendorf, M., von Weizsacker, F., Blum, H.E., Thimme, R. (2005). Intrahepatic CD8+ T-cell failure during chronic hepatitis C virus infection. *Hepatology*, 42: 828–837.

Su, A.I., Pezacki, J.P., Wodicka, L., Brideau, A.D., Supekova, L., Thimme, R., Wieland, S., Bukh, J., Purcell, R.H., Schultz, P.G., Chisari, F.V. (2002). Genomic analysis of the host response to hepatitis C virus infection. *Proceedings of the National Academy of Sciences USA*, 99: 15669–15674.

Sugimoto, K., Ikeda, F., Stadanlick, J., Nunes, F.A., Alter, H.J., Chang, K.M. (2003). Suppression of HCV-specific T cells without differential hierarchy demonstrated ex vivo in persistent HCV infection. *Hepatology*, 38: 1437–1448.

Sugimoto, K., Kaplan, D.E., Ikeda, F., Ding, J., Schwartz, J., Nunes, F.A., Alter, H.J., Chang, K.M. (2005). Strain-specific T-cell suppression and protective immunity in patients with chronic hepatitis C virus infection. *Journal of Virology* , 79: 6976–6983.

Takaki, A., Wiese, M., Maertens, G., Depla, E., Seifert, U., Liebetrau, A., Miller, J.L., Manns, M.P., Rehermann, B. (2000). Cellular immune responses persist and humoral responses decrease two decades after recovery from a single-source outbreak of hepatitis C. *Nature Medicine*, 6: 578–582.

Tanaka, J., Katayama, K., Kumagai, J., Komiya, Y., Yugi, H., Kishimoto, S., Mizui, M., Tomoguri, T., Miyakawa, Y., Yoshizawa, H. (2005). Early dynamics of hepatitis C virus in the circulation of chimpanzees with experimental infection. *Intervirology*, 48: 120–123.

Tardif, K.D., Siddiqui, A. (2003). Cell surface expression of major histocompatibility complex class I molecules is reduced in hepatitis C virus subgenomic replicon-expressing cells. *Journal of Virology* , 77: 11644–11650.

Tester, I., Smyk-Pearson, S., Wang, P., Wertheimer, A., Yao, E., Lewinsohn, D.M., Tavis, J.E., Rosen, H.R. (2005). Immune evasion versus recovery after acute hepatitis C virus infection from a shared source. *Journal of Experimental Medicine* , 201: 1725–1731.

Thimme, R., Oldach, D., Chang, K.M., Steiger, C., Ray, S.C., Chisari, F.V. (2001). Determinants of viral clearance and persistence during acute hepatitis C virus infection. *Journal of Experimental Medicine* , 194: 1395–1406.

Thimme, R., Bukh, J., Spangenberg, H.C., Wieland, S., Pemberton, J., Steiger, C., Govindarajan, S., Purcell, R.H., Chisari, F.V. (2002). Viral and immunological determinants of hepatitis C virus clearance, persistence, and disease. *Proceedings of the National Academy of Sciences USA*, 99: 15661–15668.

Thio, C.L., Gao, X., Goedert, J.J., Vlahov, D., Nelson, K.E., Hilgartner, M.W., O'Brien, S.J., Karacki, P., Astemborski, J., Carrington, M., Thomas, D.L. (2002). HLA-Cw*04 and hepatitis C virus persistence. *Journal of Virology* , 76: 4792–4797.

Thomson, M., Nascimbeni, M., Havert, M.B., Major, M., Gonzales, S., Alter, H., Feinstone, S.M., Murthy, K.K., Rehermann, B., Liang, T.J. (2003). The clearance of hepatitis C virus infection in chimpanzees may not necessarily correlate with the appearance of acquired immunity. *Journal of Virology* , 77: 862–870.

Thursz, M., Yallop, R., Goldin, R., Trepo, C., Thomas, H.C. (1999). Influence of MHC class II genotype on outcome of infection with hepatitis C virus. The HENCORE group. Hepatitis C

European Network for Cooperative Research. *Lancet*, 354: 2119–2124.

Timm, J., Lauer, G.M., Kavanagh, D.G., Sheridan, I., Kim, A.Y., Lucas, M., Pillay, T., Ouchi, K., Reyor, L.L., Schulze zur Wiesch, J., Gandhi, R.T., Chung, R.T., Bhardwaj, N., Klenerman, P., Walker, B.D., Allen, T.M. (2004). CD8 epitope escape and reversion in acute HCV infection. *Journal of Experimental Medicine* , 200: 1593–1604.

Trobonjaca, Z., Leithauser, F., Moller, P., Schirmbeck, R., Reimann, J. (2001). Activating immunity in the liver. I. Liver dendritic cells (but not hepatocytes) are potent activators of IFN-gamma release by liver NKT cells. *Journal of Immunology*, 167: 1413–1422.

Tsai, S.L., Liaw, Y.F., Chen, M.H., Huang, C.Y., Kuo, G.C. (1997). Detection of type 2-like T-helper cells in hepatitis C virus infection: implications for hepatitis C virus chronicity. *Hepatology*, 25: 449–458.

Ulsenheimer, A., Gerlach, J.T., Gruener, N.H., Jung, M.C., Schirren, C.A., Schraut, W., Zachoval, R., Pape, G.R., Diepolder, H.M. (2003). Detection of functionally altered hepatitis C virus-specific CD4 T cells in acute and chronic hepatitis C. *Hepatology*, 37: 1189–1198.

Urbani, S., Boni, C., Missale, G., Elia, G., Cavallo, C., Massari, M., Raimondo, G., Ferrari, C. (2002). Virus-specific CD8+ lymphocytes share the same effector-memory phenotype but exhibit functional differences in acute hepatitis B and C. *Journal of Virology* , 76: 12423–12434.

Urbani, S., Amadei, B., Cariani, E., Fisicaro, P., Orlandini, A., Missale, G., Ferrari, C. (2005a). The impairment of CD8 responses limits the selection of escape mutations in acute hepatitis C virus infection. *Journal of Immunology* , 175: 7519–7529.

Urbani, S., Amadei, B., Fisicaro, P., Pilli, M., Missale, G., Bertoletti, A., Ferrari, C. (2005b). Heterologous T cell immunity in severe hepatitis C virus infection. *Journal of Experimental Medicine* , 201: 675–680.

Wakita, T., Pietschmann, T., Kato, T., Date, T., Miyamoto, M., Zhao, Z., Murthy, K., Habermann, A., Krausslich, H.G., Mizokami, M., Bartenschlager, R., Liang, T.J. (2005). Production of infectious hepatitis C virus in tissue culture from a cloned viral genome. *Nature Medicine*, 11: 791–796.

Warren, A., Le Couteur, D., Fraser, R., Bowen, D.G., G.W[KN3]., M., Bertolino, P. (2006). T lymphocytes interact with hepatocytes through fenestrations in liver sinusoidal endothelial cells. *Hepatology*, *in press.*

Wedemeyer, H., He, X.S., Nascimbeni, M., Davis, A.R., Greenberg, H.B., Hoofnagle, J.H., Liang, T.J., Alter, H., Rehermann, B. (2002). Impaired effector function of hepatitis C virus-specific CD8+ T cells in chronic hepatitis C virus infection. *Journal of Immunology*, 169: 3447–3458.

Wertheimer, A.M., Miner, C., Lewinsohn, D.M., Sasaki, A.W., Kaufman, E., Rosen, H.R. (2003). Novel CD4+ and CD8+ T-cell determinants within the NS3 protein in subjects with spontaneously resolved HCV infection. *Hepatology*, 37: 577–589.

Williams, M.A., Tyznik, A.J., Bevan, M.J. (2006). Interleukin-2 signals during priming are required for secondary expansion of CD8+ memory T cells. *Nature*, 441: 890–893.

Woitas, R.P., Lechmann, M., Jung, G., Kaiser, R., Sauerbruch, T., Spengler, U. (1997). CD30 induction and cytokine profiles in hepatitis C virus core-specific peripheral blood T lymphocytes. *Journal of Immunology* , 159: 1012–1018.

Wong, D.K., Dudley, D.D., Afdhal, N.H., Dienstag, J., Rice, C.M., Wang, L., Houghton, M., Walker, P.D., Koziel, M.J. (1998). Liver-derived CTL in hepatitis C virus infection: breadth and specificity of responses in a cohort of persons with chronic infection. *Journal of Immunology*, 160: 1479–1488.

Wuensch, S.A., Pierce, R.H., Crispe, I.N. (2006). Local intrahepatic CD8+ T cell activation by a non-self-antigen results in full functional differentiation. *Journal of Immunology* , 177: 16891–16897.

Yao, Z.Q., Nguyen, D.T., Hiotellis, A.I., Hahn, Y.S. (2001). Hepatitis C virus core protein inhibits human T lymphocyte responses by a complement-dependent regulatory pathway. *Journal of Immunology* , 167: 5264–5272.

Yao, Z.Q., Eisen-Vandervelde, A., Waggoner, S.N., Cale, E.M., Hahn, Y.S. (2004). Direct binding of hepatitis C virus core to gC1qR on CD4+ and CD8+ T cells leads to impaired activation of

Lck and Akt. *Journal of Virology* , 78: 6409–6419.

Yao, Z.Q., Waggoner, S.N., Cruise, M.W., Hall, C., Xie, X., Oldach, D.W., Hahn, Y.S. (2005). SOCS1 and SOCS3 are targeted by hepatitis C virus core/gC1qR ligation to inhibit T-cell function. *Journal of Virology* , 79: 15417–15429.

Yao, Z.Q., Shata, M.T., Tricoche, N., Shan, M.M., Brotman, B., Pfahler, W., Hahn, Y.S., Prince, A.M. (2006). gC1qR expression in chimpanzees with resolved and chronic infection: potential role of HCV core/gC1qR-mediated T cell suppression in the outcome of HCV infection. *Virology*, 346: 324–337.

You, L.R., Chen, C.M., Lee, Y.H. (1999). Hepatitis C virus core protein enhances NF-kappaB signal pathway triggering by lymphotoxin-beta receptor ligand and tumor necrosis factor alpha. *Journal of Virology* , 73: 1672–1681.

Zhu, N., Khoshnan, A., Schneider, R., Matsumoto, M., Dennert, G., Ware, C., Lai, M.M. (1998). Hepatitis C virus core protein binds to the cytoplasmic domain of tumor necrosis factor (TNF) receptor 1 and enhances TNF-induced apoptosis. *Journal of Virology*, 72: 3691–3697.

Zinkernagel, R.M. (2000). Localization dose and time of antigens determine immune reactivity. *Seminars in Immunology*, 12: 163–171; discussion, 257–344.

Immune Responses Against the Hepatitis C Virus and the Outcome of Therapy

Paul Klenerman* and **Eleanor Barnes**

Introduction

The hepatitis C virus (HCV) sets up persistence in the majority of those infected. In doing so, it evades both innate and adaptive immune responses. However, in a reasonable fraction of patients (20–50%), there is long-term control of viremia through some effective combination of host responses. It is generally considered that cellular immune responses—mediated by CD4+ and CD8+ T cells—play a major role in determining this successful outcome, although they do so in concert with many other cellular and humoral mediators (Ward et al., 2002).

During acute HCV infection, cellular immune responses are generated in most patients against a diverse set of viral epitopes (Lechner et al., 2000b; Cox et al., 2005b). Typically, in those with persistent infection, once this has been established, the host cellular immune responses seem to be diminished. Although T cell responses may be detected in the liver (He et al., 1999; Grabowska et al., 2001), the number of antiviral T cells found in the blood is very low and the function of antiviral T cells may be impaired (Gruener et al., 2001; Wedemeyer et al., 2002). It is not known whether this is a cause or a consequence of infection. Three major pathways are thought to lead to the downregulation of these responses—immune escape through viral variation, cellular exhaustion, and induction of T cell regulatory subsets (Klenerman & Hill, 2005).

As discussed elsewhere in this volume, interferon/ribavirin therapy for the hepatitis C virus is variably successful in chronically infected patients. This depends on a number of host factors but critically on the viral genotype. The same or even more limited therapy is much more successful in the treatment of those who have been recently infected. It is well known that such therapies may have important effects on a number of cellular components of the immune system. Therefore, it has been generally considered that the immune system might play some role in the overall therapeutic effect of combination therapy. This could happen through three main mechanisms: as a direct consequence of interferon/ribavirin; as an indirect action as

---

* Address Correspondence: Nuffield Department of Medicine, Peter Medawar Building for Pathogen Research, University of Oxford, South Parks Road, Oxford OX1 3SY UK, Tel 0044 1865 281885, Fax 0044 01865 2818236, E-mail: paul.klenerman@ndm.ox.ac.uk

et al., 1999; Grabowska et al., 2001). Interestingly, a proportion of these cells secrete IL-10 in response to antigenic stimulation, as opposed to IFNg. It is suggested that these play a regulatory role (Accapezzato et al., 2004).

Overall these data suggest a picture where robust CD4+ and CD8+ T cell responses directed against a broad range of targets and sustained over time are associated with good control of virus, while weaker, narrower, or less well-sustained populations are associated with persistence. These associations, though now found consistently between groups, leave open the question of cause and effect. Some longitudinal data suggest that failure of CD4+ T cell responses to proliferate *in vitro* in response to antigen, or deletion of such responses, precedes the rise in viral load associated with early failure of control (possibly after a period of RNA negativity in blood) (Gerlach, 1999). Further important evidence pointing to the role of CD4+ and CD8+ T cell responses as mediators of control rather than markers of control comes from depletion experiments in the chimpanzee model. Here spontaneously resolved infection readily occurs, but depletion of either subset has a marked effect on the ability of a subsequently rechallenged animal to resist infection (Grakoui et al., 2003; Shoukry et al., 2003). These data strongly suggest that both sets of cells, working in parallel, are crucial in natural host control over the virus. Before we discuss their importance in therapy-mediated control, it is important, however, to consider why such T cells may fail in persistent infection.

Mechanisms of T Cell Failure

Immune Escape Through Mutation

Probably the most important cause of virus persistence in HCV is the capacity of this RNA virus to generate variants during the early course of infection (Bowen & Walker, 2005). It is well recognized that such mutations, which are very readily observed in the envelope hypervariable regions, may allow evasion of envelope-directed antibody and thus resist neutralization (Farci et al., 2000). However, it is increasingly clear that the generation of CD8+ T cell escape mutations in particular plays a critical role in allowing the virus to resist CTL attack. This was first shown in HCV infection of the chimpanzee, following on from a wealth of data in human and simian immunodeficiency viruses (HIV and SIV), as well as the murine model lymphocytic choriomeningitis virus (LCMV) (Erickson et al., 2001). The advantage of this model is that the input virus is well characterized, a feature unusual in human infection. However, a number of careful studies have now revealed immune escape in a number of key epitopes in acute infection (Timm et al., 2004; Cox et al., 2005b). This is a particularly important point in the discussion of the impact of therapy as T cells that target a virus that has mutated to evade recognition are effectively of no value *in vivo*, even though they can be readily measured.

Immune escape is readily demonstrable but by no means universal. A number of key epitopes are not commonly mutated, and for these no evidence of selection is seen either within individuals or at a population level (Cox et al., 2005b; Gaudieri et al., 2006). What is not yet clear is whether such responses are still exerting an important antiviral effect, but the virus is unable to generate escape mutations due to fitness constraints, or whether such responses are simply "marker" or "passenger" T cells that do not have a major impact on viral load (Zafiropoulos et al., 2004).

Immune Exhaustion

Whether T cell epitopes undergo escape mutation or not—and this is particularly likely to be the case for CD4+ T cells, where such mutation has been recognized but is considered rare—T cell responses do dwindle in chronic infection. This process has been recognized in murine models of chronic viral infection—notably LCMV—and has been termed "T cell exhaustion"(Moskophidis, 1993). Such a process is associated with high levels of viral replication and thus antigen exposure and promoted by lack of T cell help. Exhausted T cells ultimately are deleted, but prior to this populations can be detected with various degrees of loss of function (including loss of cytokine secretion, killing, and proliferation, all of which have been described in chronic HCV infection) (Wherry & Ahmed, 2004). Lowering the viral load in the murine model, with drugs or the addition of antiviral effectors, can reverse this process, although for a long period the mechanism behind it remained obscure. Recent studies in the LCMV model and supported by work in HIV suggest that expression of the molecule PD-1 (programmed death 1) on T cells plays a key role in this process. Blockade of the interaction using an antibody to PDL-1 (PD-1 Ligand) can *in vivo* restore the function of the exhausted T cells and leads to regain of control over this chronic infection (Barber et al., 2006; Vertuani et al., 2002). This remarkable result has been repeated *in vitro* in HIV and does suggest that overactivation of the PD-1 pathway can contribute to the failure of T cells to control the virus (Day et al., 2006). More recent work using blockade of IL-10 receptors also indicates that this cytokine may play a role (Brooks et al., 2006). Both these pathways might be relevant to HCV and also to the impact of therapy.

Immune Regulation

Much recent focus in the pathogenesis of auto-immunity has been on recently described subsets of CD4+ T cells described as regulatory T cells, or Tregs. Such cells express high levels of the IL-2 receptor chain CD25 and are characterized by expression of the forkhead transcription factor FOXP3 (Fehervari & Sakaguchi, 2004). More recently it has emerged that such cells may play a key role in the pathogenesis of chronic infections, including HCV (Sugimoto et al., 2003; Boettler et al., 2005; Rushbrook et al., 2005). A number of groups have found elevated regulatory activity in chronic infection, although whether this is due to elevated frequencies of Treg populations or increased functionality is not yet known. Certainly,

Acute Disease

Since acute infection is associated with strong CD4+ and CD8+ T cell responses, and treatment of acute disease is associated with an excellent clinical outcome, it was at first considered that therapy at this stage of disease was successful due to effects on T cell responses. However, the conclusions from the most recent studies do not necessarily support this idea. One study in this area, from Kamal and colleagues again, of 40 patients, shows an association between reconstitution of strong CD4 responses and a good clinical outcome (Kamal et al., 2004). This trial was aimed largely at analysis of CD4 responses.

In contrast, two detailed studies of both CD4 and CD8 responses did not report an obvious association. In one study, by Lauer and colleagues, the impact of interferon-alpha was largely to suppress existing CD8+ T cell responses, and overall no association was found between the treatment outcome and the size or quality of CD8+ T cell response (Lauer et al., 2005). Similar data applied to the CD4 T cell response. A study by Rahman found a very similar result, with no enhancement of responses and no clear association with outcome (Rahman et al., 2004).

Since these latter two studies were performed in great detail, including in the case of the Lauer analysis (Lauer et al., 2005), the mapping of each individual T cell response across the entire genome, it does seem that the conclusion from the studies of acute disease to date is that a close relationship between enhancement of T cell responses during therapy and the outcome of treatment does not exist. The Lauer study also presented a particularly interesting case where, following successful treatment, therapy with a T cell-depleting antibody OKT3 was used. This removed the HCV-specific T cell response entirely, but no recrudescence of virus was seen. This is a unique case, but here it suggests additionally that the presence of T cell responses is not required for the maintenance of the RNA-ve state following a successful end-of-treatment response.

Mechanisms of Action of the Adaptive Immune System During Combination Therapy

Having understood how T cell responses may control HCV naturally, or fail to do so, and also the impact of therapy upon such responses, we now consider how these responses may act during therapy. Broadly we can consider the main models:

1. Cellular immune responses modulated as a direct consequence of interferon/ribavirin (stimulation model);
2. Cellular immune responses modulated an indirect action as a result of the antiviral effect of combination therapy (indirect model);
3. A synergistic action between the immune system and interferon/ribavirin (synergy model).

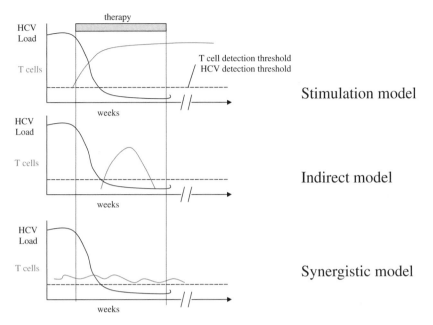

Fig. 1 Models of T cell responses during combination therapy. The upper panel shows successful therapy accompanied by boosting of T cell responses (stimulation model). This could accelerate clearance in the first phase, viral decay in later phases, or recurrence in the post-treatment phase. The middle panel shows the indirect model, where responses re-emerge on therapy, irrelevant to outcome, as a consequence of viral control by drugs. The lower panel (synergy model) indicates no specific influence of therapy on T cells (or even a decay is possible), but nevertheless the presence of T cells remains—acting as a "third drug." The fourth (irrelevant) model is not illustrated, but any of the above changes in T cells could be seen, or no relevant T cell activity could be detectable

We will discuss these models and the evidence in favor of each (Figure 1). There is also a fourth model to consider, that T cell responses are not involved at all in the treatment response (irrelevant model), which will be discussed at the end of the chapter. This may be most relevant in the context of newly developed potent antivirals.

Stimulation Model

Both interferon and ribavirin are well recognized as immune modulators, although the molecular evidence in favor of the former is much more substantial. Interferon-alpha acts to modulate—largely upregulate, although there is a significant dose dependence—most of the key steps in generation of an antiviral T cell response. Importantly, interferon-alpha is a major stimulator of dendritic cells, providing a stimulus that leads to maturation, along similar lines to conventional stimuli such as LPS (Luft et al., 1998; Santini et al., 2000). Mature dendritic cells are especially potent and induce CD4+ and CD8+ T cell responses through appropriate antigen stimulation, co-stimulation (through CD80 and CD86, among others), as well as cytokine secretion (most importantly IL-12). The transition of mature dendritic cells to the appropriate lymphoid environment under inflammatory stimuli also serves to maximize their potency. The action of interferon-alpha on dendritic cells includes an effect on "cross-presentation"—the ability of such cells to take up soluble antigen and present it through the Class I pathway (Le Bon et al., 2003). This could be very important in generating CD8+ T cell responses against HCV, where the virus is not thought to efficiently infect these cells.

In addition to these effects, interferon-alpha has a positive stimulatory effect on NK cells and T cells directly, improving cytokine secretion, although conflicting effects on proliferation have been reported (Brinkmann et al., 1993). The impact of interferons on regulatory T cells requires further investigation, although it has been suggested that such regulatory activity, at least in the short term, may be inhibited (Rushbrook et al., 2005).

Overall, therefore, a positive impact on T cell responses of interferon alone is to be expected. However, it should be noted that priming and restimulation of such antiviral T cell populations are antigen-dependent. Therefore, a potent antiviral effect, which may restore "exhausted" T cells, may paradoxically remove antigens necessary for priming and restimulation. This issue is further explored below.

While a considerable literature exists on the activity of interferon on the immune response, the data on ribavirin are much more limited. Most data are *in vitro* and rely on cultures of mixed cell populations. It is evident that ribavirin can act as an immunomodulator, with actions on cellular proliferation most obvious, although these are clearly dose-dependent, and can modulate cytokine secretion patterns, with changes described as switching toward a T helper 1 (Th1) profile (Martin et al., 1998; Tam et al., 1999). An analysis of dendritic cell function in the presence of ribavirin suggests that inhibition of both IL-12 and IL-10 occurs when used alone. However, the combination of interferon and ribavirin leads to a relative increase of IL-12 over IL-10 (Barnes et al., 2004). Since IL-10 is the major immunosuppressive cytokine in this interaction, the net result is likely to improve the immunostimulatory capacity of such dendritic cell populations—although this has not been proven *in vivo*.

Thus, the model here is that combination therapy will serve largely to boost T cell responses, numerically and/or functionally. This is backed up by some evidence from the clinical trials, but such evidence is far from uniform. Early responses, within the first 3 to 6 months, might be boosted in some cases, but similar therapies in acute infection do little to boost responses in carefully followed patients. Thus, although this model is popular, it is not well supported by the evidence.

Indirect Model

The impact of therapy through lowering of the viral load may have a profound effect on the quality of T cell responses observed, through a reduction in "exhaustion" (Wherry et al., 2004). A parallel situation is likely to occur in superinfection, where one virus is effectively "cleared" by another. T cell responses specific for the original virus can recover—indeed a substantial proportion of the responses mapped down to individual peptides in chronically infected patients is directed against such "cleared" epitopes (Lauer et al., 2004). If not deleted through exhaustion, such cells may recover numerically and functionally once the virus is controlled.

An alternative indirect consequence of the antiviral effects of combination therapy may be through modulation of regulatory subsets. Tregs may be stimulated by chronic inflammatory stimuli, and interestingly, interferon can induce IL-15 production, which appears to impair the inhibitory capacity of Tregs (Rushbrook et al., 2005). Inhibition of Tregs is an attractive hypothesis to explain the strong association of interferon therapy with auto-immune thyroid disease in HCV.

Thus, this model would propose that what is being observed in the clinical studies, where there is typically some boosting of T cell responses in chronic infection, is that there is a reduction in either the degree of "exhaustion" of the T cell populations or of the "regulation" of such cells. This would release populations for cytokine release or proliferation in *in vitro* assays and potentially with effector function *in vivo*. However, such reactivity could be regarded as a consequence rather than a cause of viral suppression. Again, since this feature is apparently not observed in acute infection, either the status of the T cells or that of the regulatory cells may be different—induction of both exhaustion and regulation may require long periods of antigen exposure.

Synergy Model

The clinical evidence does not strongly support a major role for induction of effective T cell immunity in the treatment effect. However, does this mean T cells play no role at all? One possibility that may fit the evidence best is that T cells may play a role, but this is independent of a boosting effect, rather just a synergistic action. In this model T cell responses are present in the chronic phase as well as in the acute phase of infection, especially within liver tissue, but are insufficient, in the face of high viral loads, to exert substantial control over viral replication. However, at low viral loads, once the interferon-ribavirin has been efficacious over a period of time, the same T cell responses (boosted or not according to the models above) may be relatively more effective at clearing residual nests of virally infected tissue. In other words, at very low viral loads, the additional effect of T cell responses might be sufficient to drive the virus below some critical threshold for long-term survival.

Evidence in favor of a requirement for T cells in successful treatment of chronic HCV comes from an immunogenetic study, which showed a beneficial influence of HLA DR11, the same molecule/haplotype that is protective in acute infection (Thursz et al., 1999). It is thought that DR11 serves to direct T cell responses to conserved or high-avidity epitopes (Godkin, 2001). Another line of evidence comes from the link between pre-existing CD8+ T cells found pre-treatment in the liver and the treatment response. Additionally, there is indirect evidence from HIV/HCV co-infection. In this setting, viral loads are higher and the treatment response is lower (Plosker & Keating, 2004). The cause of the higher viral loads is not fully understood, but there are lower levels of both CD8+ and CD4+ T cells, which might well be contributory (Kim et al., 2005; Harcourt et al., 2006). An attractive hypothesis, therefore, is that the lack of effective T cell responses might influence the overall treatment response in the co-infected cohort, through failure of synergy at low-level virus loads. (It is alternatively possible that the effect of HIV co-infection is mediated through high pre-treatment viral load alone, leading to a reduced drug-mediated clearance.) What is not known is to what extent treatment of HIV with highly active antiretroviral therapy (HAART) fully restores the T cell responses against HCV, which might potentially influence both the viral load and the response to therapy.

Finally, the synergy model would explain the difference between the treatment responses in acute and chronic infection. In acute infection, the T cell responses are of higher magnitude and functionality—the three crucial influences of escape, exhaustion, and regulation are yet to become fully active, and thus the impact of such a synergistic activity may be more evident.

Overall, therefore, the synergy model does have some evidence in its favor. In order to test its validity further, it would be necessary to correlate the pre-treatment T cell responses (accounting for the individual's viral sequence variability), ideally within the liver, with the outcome of treatment. This is actually quite a technical effort, and it will be difficult to truly segregate groups with and without detectable T cell responses in which to compare outcomes. Comparisons cross-genotype may also be of value in defining the validity of this model.

A Fourth Model: No Role for T Cells

One final model to consider, in opposition somewhat to all of these three, is that T cells play no role at all in combination therapy, which is acting purely as an antiviral ("irrelevant" model). In the absence of really strong evidence in favor of one of the above three models, this is still a possibility. One possible setting would be if T cells that are still detectable are no longer able to recognize the virus, which has evolved within the patient. This would render them blind to current infection. Another would be if all such T cells were fully exhausted. This may occur, although typically some weak specificities are detectable *ex vivo* in about half the patients studied overall. The experiment of nature where such T cells are depleted in HIV co-infection, and where both viral load increases and treatment response decreases, is one piece of evidence that however weak these T cells are (in blood), they are still capable of influencing host virus interactions. However, the null hypothesis, that T cells play no part at all in a successful outcome, still needs to be rejected.

Conclusions

Treatment of HCV influences the adaptive immune response, but the question remains as to what extent the adaptive immune response influences the outcome of treatment. So far the data are scant and contradictory, but careful studies have largely come to the conclusion that successful treatment is not dependent on boosting T cells, even if T cell augmentation can be observed. However, this does not rule out a role for T cells, and certainly such responses could still be involved under the "synergy" model. If such T cell populations can contribute to the enhanced treatment response seen in acute therapy or with Genotype 3 infection, it leaves the door open for more direct modulation of T cell responses through immunotherapy, to improve treatment outcome. Indeed, if treatment only weakly or inconsistently boosts T cell responses currently, this suggests that this avenue could be exploited—but only if escape, exhaustion, and regulation are overcome. By inducing responses at a time when virus load is low or undetectable and liver inflammation has normalized, exhaustion and regulation may be minimized. However, to what extent the virus has evolved to evade the key HLA-determined T cell responses in any given patient is an important—as yet unanswered—question and ultimately the one on which the success of any such exercise will hinge.

Acknowledgments PK is funded by the Wellcome Trust and EB by the Medical Research Council (UK). The group also receives funding from the EU, the James Martin School of the 21st Century (Oxford). We thank Nasser Semmo, Gillian Harcourt, Isabelle Sheridan, Alison Turner, Isla Humphreys, and Rodney Phillips for their long-term input into the HCV research at the Peter Medawar Building.

References

Accapezzato, D., Francavilla, V., Paroli, M., Casciaro, M., Chircu, L.V., Cividini, A., Abrignani, S., Mondelli M.U., Barnaba, V. (2004). Hepatic expansion of a virus-specific regulatory CD8(+) T cell population in chronic hepatitis C virus infection. *Journal of Clinical Investigation*, 113: 963–972.

Barber, D.L., Wherry, E.J., Masopust, D., Zhu, B., Allison, J.P., Sharpe, A.H.,. Freeman, G.H., Ahmed, R. (2006). Restoring function in exhausted CD8 T cells during chronic viral infection. *Nature*, 439: 682–687.

Barnes, E., Harcourt, G., Brown, O., Lucas, M., Phillips, R., Dusheiko, G., Klenerman, P. (2002). The dynamics of T-lymphocyte responses during combination therapy for chronic hepatitis C virus infection. *Hepatology*, 36: 743–754.

Barnes, E., Salio, M., Cerundolo, V., Medlin, J., Murphy, S., Dusheiko, G., Klenerman, P. (2004). Impact of alpha interferon and ribavirin on the function of maturing dendritic cells. *Antimicrobial Agents & Chemotherapy*, 48: 3382–3389.

Boettler, T., Spangenberg, H.C., Neumann-Haefelin, C., Panther, E., Urbani, S.,. Ferrari, C., Blum, H.E., von Weizsacker, F., Thimme, R. (2005). T cells with a CD4+CD25+ regulatory phenotype suppress in vitro proliferation of virus-specific CD8+ T cells during chronic hepatitis C virus infection. *Journal of Virology*, 79: 7860–7867.

Bowen, D.G., Walker, C.M. (2005). Mutational escape from CD8+ T cell immunity: HCV evolution, from chimpanzees to man. *Journal of Experimental Medicine*, 201: 1709–1714.

Brinkmann, V., Geiger, T., Alkan, S., Heusser, C.H. (1993). Interferon alpha increases the frequency of interferon gamma-producing human CD4+ T cells. *Journal of Experimental Medicine*, 178: 1655–1663.

Brooks, D.G., Trifilo, M.J., Edelmann, K.H., Teyton, L., McGavern, L.B., Oldstone, M.B. (2006). Interleukin-10 determines viral clearance or persistence in vivo. *Nature Medicine*, 12: 1301–1309.

Cooper, S., Erickson, A.L., Adams, E.J., Kansopon, J., Weiner, A.J., Chien, D.Y., Houghton, M., Parham, P., Walker, C.M. (1999). Analysis of a successful immune response against hepatitis C virus. *Immunity*, 10: 439–449.

Cox, A.L., Mosbruger, T., Lauer, G.M., Pardoll, D., Thomas, D.L., Ray, S.C. (2005a). Comprehensive analyses of CD8+ T cell responses during longitudinal study of acute human hepatitis C. *Hepatology*, 42: 104–412.

Cox, A.L., Mosbruger, T., Mao, Q., Liu, Z., Wang, X.H., Yang, H.C., Sidney, J., Sette, A., Pardoll, D., Thomas, D.L., Ray, S.C. (2005b). Cellular immune selection with hepatitis C virus persistence in humans. *Journal of Experimental Medicine*, 201: 1741–1752.

Cramp, M.E., Rossol, S., Chokshi, S., Carucci, P., Williams, R., Naoumov, N.V. (2000). Hepatitis C virus-specific T-cell reactivity during interferon and ribavirin treatment in chronic hepatitis C. *Gastroenterology*, 118: 346–355.

Day, C.L., Kaufmann, D.E., Kiepiela, P., Brown, J.A., Moodley, E.S., Reddy, S., Mackey, E.W., Miller, J.D., Leslie, A.J., DePierres, C., Mncube, Z., Duraiswamy, J.,. Zhu, B., Eichbaum, Q., Altfeld, M., Wherry, E.J., Coovadia, H.M., Goulder, P.J.,. Klenerman, P., Ahmed, R., Freeman, G.J., Walker, B.D. (2006). PD-1 expression on HIV-specific T cells is associated with T-cell exhaustion and disease progression. *Nature*, 443: 350–354.

Day, C.L., Lauer, G.M., Robbins, G.K., McGovern, B., Wurcel, A.G., Gandhi, R.T., Chung, R.T., Walker, B.D. (2002). Broad specificity of virus-specific CD4+ T-helper-cell responses in resolved hepatitis C virus infection. *Journal of Virology*, 76: 12584–12595.

Diepolder, H., Zachoval, R., Hoffman, R., Wierenga, E., Santorino, T., Jung, M., Eichenlaub, D., Pape, G. (1995). Possible mechanism involving T lymphocyte response to NS3 in viral clearance in acute HCV infection. *Lancet*, 346: 1006–1007.

Erickson, A.L., Kimura, Y., Igarashi, S., Eichelberger, J., Houghton, M., Sidney, J., McKinney, D., Sette, A., Hughes, A.L., Walker, C.M. (2001). The outcome of hepatitis C virus infection is predicted by escape mutations in epitopes targeted by cytotoxic T lymphocytes. *Immunity*, 15: 883–895.

Farci, P., Shimoda, A., Coiana, A., Diaz, G., Peddis, G., Melpolder, J.C., Strazzera, A., Chien, A.Y., Munoz, S.J., Balestrieri, A., Purcell, R.H., Alter, H.J. (2000). The outcome of acute hepatitis C predicted by the evolution of the viral quasispecies. *Science*, 288: 339–344.

Fehervari, Z., Sakaguchi, S. (2004). Development and function of CD25+CD4+ regulatory T cells. *Current Opinion in Immunology*, 16: 203–208.

Gaudieri, S., Rauch, A., Park, L.P., Freitas, E., Herrmann, S., Jeffrey, G., Cheng, W.,. Pfafferott, K., Naidoo, K., Chapman, R., Battegay, M., Weber, R., Telenti, A., Furrer, H., James, I., Lucas, M., Mallal, M.S. (2006). Evidence of viral adaptation to HLA class I-restricted immune pressure in chronic hepatitis C virus infection. *Journal of Virology*, 80: 11094–11104.

Gerlach, J., Diepolder, H., Jung, M.-C., Gruener, N., Schraut, W., Zachoval, R., Hoffman, R., Schirren, C., Santantonio, T., Pape, G. (1999). Recurrence of HCV after loss of virus specific CD4+ T cell response in acute hepatitis C. *Gastroenterology*, 117: 933–941.

Godkin, A. (2001). Characterization of novel HLA-DR11-restricted HCV epitopes reveals both qualitative and quantitative differences in HCV-specific CD4+ T cell responses in chronically infected and non-viremic patients. *European Journal of Immunology*, 31: 1438–1446.

Grabowska, A.M., Lechner, F., Klenerman, P., Tighe, P.J., Ryder, S., Ball, J.K., Thomson, J.K., Irving, W.L., Robins, R.A. (2001). Direct ex vivo comparison of the breadth and specificity of the T cells in the liver and peripheral blood of patients with chronic HCV infection. *European Journal of Immunology*, 31: 2388–2394.

Grakoui, A., Shoukry, N.H., Woollard, D.J., Han, J.H., Hanson, H.L., Ghrayeb, J., Murthy, K.K., Rice, C.M.,Walker, C.M. (2003). HCV persistence and immune evasion in the absence of memory T cell help. *Science*, 302: 659–662.

Gruener, N.H., Lechner, F., Jung, M.C., Diepolder, H., Gerlach, T., Lauer, G., Walker, B., Sullivan, J., Phillips, R., Pape, G.R., Klenerman, P. (2001). Sustained dysfunction of antiviral CD8+ T lymphocytes after infection with hepatitis C virus. *Journal of Virology*, 75: 5550–5558.

Harcourt, G., Gomperts, E., Donfield, S., Klenerman, P. (2006). Diminished frequency of hepatitis C virus specific interferon {gamma} secreting CD4+ T cells in human immunodeficiency virus/hepatitis C virus coinfected patients. *Gut*, 55: 1484–1487.

Harcourt, G., Hellier, S., Bunce, M., Satsangi, J., Collier, J., Chapman, R., Phillips, R., Klenerman, P. (2001). Effect of HLA class II genotype on T helper lymphocyte responses and viral control in hepatitis C virus infection. *Journal of Viral Hepatitis*, 8: 174–179.

Harcourt, G.C., Lucas, M., Godkin, A.J., Kantzanou, M., Phillips, R.E., Klenerman, P. (2003). Evidence for lack of cross-genotype protection of CD4+ T cell responses during chronic hepatitis C virus infection. *Clinical & Experimental Immunology*, 131: 122–129.

Harcourt, G.C., Lucas, M., Sheridan, I., Barnes, E., Phillips, R., Klenerman, P. (2004). Longitudinal mapping of protective CD4+ T cell responses against HCV: analysis of fluctuating dominant and subdominant HLA-DR11 restricted epitopes. *Journal of Viral Hepatitis*, 11: 324–331.

He, X.-S., Rehermann, B., Lopez-Labrador, F., Boisvert, J., Cheung, R., Mumm, J., Wedermeyer, H., Berenguer, M., Wright, T., Davis, M. (1999). Quantitative analysis of HCV-specific CD8+ T cells in peripheral blood and liver using peptide-MHC tetramers. *Proceedings of the National Academy of Sciences (USA)*, 96: 5692–5697.

Kamal, S.M., Fehr, J., Roesler, B., Peters, T.,Rasenack, J.W. (2002). Peginterferon alone or with ribavirin enhances HCV-specific CD4 T-helper 1 responses in patients with chronic hepatitis C. *Gastroenterology*, 123: 1070–1083.

Kamal, S.M., Ismail, A., Graham, C.S., He, W., Rasenack, J.W., Peters, T., Tawil, A.A., Fehr, J.J., Khalifa Kel, S., Madwar, M.M., Koziel, M.J. (2004). Pegylated interferon alpha therapy in acute hepatitis C: relation to hepatitis C virus-specific T cell response kinetics. *Hepatology*, 39: 1721–1731.

Kim, A.Y., Lauer, G.M., Ouchi, K., Addo, M.M., Lucas, M., Schulze Zur Wiesch, J.,. Timm, J., Boczanowski, M., Duncan, J.E., Wurcel, A.G., Casson, D., Chung, R.T., Draenert, R., Klenerman, P., Walker, B.D. (2005). The magnitude and breadth of hepatitis C virus-specific CD8+ T cells depend on absolute CD4+ T-cell count in individuals coinfected with HIV-1. *Blood*, 105: 1170–1178.

Klenerman, P., Hill, A. (2005). T cells and viral persistence: lessons from diverse infections. *Nature Immunology*, 6: 873–879.

Koziel, M., Walker, B. (1997). Characteristics of the intrahepatic CTL response in chronic HCV infection. *Springer Seminars in Immunopathology*, 19: 69–83.

Lauer, G., Barnes, E., Lucas, M., Timm, J., Ouchi, K., Kim, A., Day, C., Robbins, G., Casson, D., Reiser, M., Dusheiko, G., Allen, T., Chung, R., Walker, B., Klenerman, P. (2004). High resolution analysis of cellular immune responses in resolved and persistent hepatitis C virus infection. *Gastroenterology*, 127: 924–936.

Lauer, G. M., Lucas, M., Timm, J., Ouchi, K., Kim, A.Y., Day, C.L., Schulze Zur Wiesch, J., Paranhos-Baccala, G., Sheridan, I., Casson, D.R., Reiser, M., Gandhi, R.T., Li, B., Allen, T.M., Chung, R.T., Klenerman, P., Walker, B.D. (2005). Full-breadth analysis of CD8+ T-cell responses in acute hepatitis C virus infection and early therapy. *Journal of Virology*, 79: 12979–12988.

Le Bon, A., Etchart, N., Rossmann, C., Ashton, M., Hou, S., Gewert, D., Borrow, P.,. Tough, D.F. (2003). Cross-priming of CD8(+) T cells stimulated by virus-induced type I interferon. *Nature Immunology* 4: 1009–1015.

Lechner, F., Gruener, N., Urbani, S., Uggeri, J., Santantonio, T., Cerny, A., Phillips, R., Ferrari, C., Pape, G., Klenerman, P. (2000a). CTL responses are induced during acute HCV infection but are not sustained. *European Journal of Immunology*, 30: 2479–2487.

Lechner, F., Wong, D.K., Dunbar, P.R., Chapman, R., Chung, R.T., Dohrenwend, P.,. Robbins, G., Phillips, R., Klenerman, P., Walker, B.D. (2000b). Analysis of successful immune responses in persons infected with hepatitis C virus. *Journal of Experimental Medicine*, 191: 1499–1512.

Luft, T., Pang, K.C., Thomas, E., Hertzog, P., Hart, D.N., Trapani, J., Cebon, J. (1998). Type I IFNs enhance the terminal differentiation of dendritic cells. *Journal of Immunology*, 161: 1947–19453.

Marinho, R.T., Pinto, R., Santos, M.L., Lobos, I.V., Moura, M.C. (2004). Effects of interferon and ribavirin combination therapy on CD4+ proliferation, lymphocyte activation, and Th1 and Th2 cytokine profiles in chronic hepatitis C. *Journal of Viral Hepatitis*, 11: 206–216.

Martin, J., Navas, S., Quiroga, J.A., Pardo, M., Carreno, V. (1998). Effects of the ribavirin-interferon alpha combination on cultured peripheral blood mononuclear cells from chronic hepatitis C patients. *Cytokine*, 10: 635–644.

Missale, G., Bertoni, R., Lamonaca, V., Valli, A., Massari, M., Mori, C., Rumi, M., Houghton, M., Fiaccadori, F., Ferrari, C. (1996). Different clinical behaviours of acute HCV infection are associated with different vigor of the anti-viral T cell response. *Journal of Clinical Investigation*, 98: 706–714.

Moskophidis, D., Laine, E., Zinkernagel, R.M. (1993). Peripheral clonal deletion of antiviral memory CD8+ T cells. *European Journal of Immunology*, 23: 3306–3311.

Plosker, G.L., Keating, G.M. (2004). Peginterferon-alpha-2a (40kD) plus ribavirin: a review of its use in hepatitis C virus and HIV co-infection. *Drugs*, 64: 2823–2843.

Rahman, F., Heller, T., Sobao, Y., Mizukoshi, E., Nascimbeni, M., Alter, H.,. Herrine, S.,.Hoofnagle, J., Liang, T.J., Rehermann, B. (2004). Effects of antiviral therapy on the cellular immune response in acute hepatitis C. *Hepatology*, 40: 87–97.

Rushbrook, S.M., Ward, S.M., Unitt, E., Vowler, S.L., Lucas, M., Klenerman, P.,. Alexander, G.J. (2005). Regulatory T cells suppress in vitro proliferation of virus-specific CD8+ T cells during persistent hepatitis C virus infection. *Journal of Virology*, 79: 7852–7859.

Santini, S.M., Lapenta, C., Logozzi, M., Parlato, S., Spada, M., Di Pucchio, T., Belardelli, F. (2000). Type I interferon as a powerful adjuvant for monocyte-derived dendritic cell development and activity in vitro and in Hu-PBL-SCID mice. *Journal of Experimental Medicine*, 191: 1777–1788.

Semmo, N., Day, C.L., Ward, S.M., Lucas, M., Harcourt, G., Loughry, A., Klenerman, P. (2005). Preferential loss of IL-2-secreting CD4+ T helper cells in chronic HCV infection. *Hepatology*, 41: 1019–1028.

Shoukry, N.H., Grakoui, A., Houghton, M., Chien, D.Y., Ghrayeb, J., Reimann, K.A., Walker, C.M. (2003). Memory CD8+ T cells are required for protection from persistent hepatitis C virus infection. *Journal of Experimental Medicine*, 197: 1645–1655.

Sugimoto, K., Ikeda, F., Stadanlick, J., Nunes, F.A., Alter, H.J., Chang, K.M. (2003). Suppression of HCV-specific T cells without differential hierarchy demonstrated ex vivo in persistent HCV infection. *Hepatology*, 38: 1437–1448.

Tam, R.C., Pai, B., Bard, J., Lim, C., Averett, D.R., Phan, U.T., Milovanovic, T. (1999). Ribavirin polarizes human T cell responses towards a type 1 cytokine profile. *Journal of Hepatology*, 30: 376–382.

Tang, K.H., Herrmann, E., Cooksley, H., Tatman, N., Chokshi, S., Williams, R., Zeuzem, S., Naoumov, N.V. (2005). Relationship between early HCV kinetics and T-cell reactivity in chronic hepatitis C genotype 1 during peginterferon and ribavirin therapy. *Journal of Hepatology*, 43: 776–782.

Thursz, M., Yallop, R., Goldin, R., Trepo, C., Thomas, H.C. (1999). Influence of MHC class II genotype on outcome of infection with hepatitis C virus. The HENCORE group. Hepatitis C European Network for Cooperative Research. *Lancet*, 354: 2119–2124.

Timm, J., Lauer, G.M., Kavanagh, D.G., Sheridan, I., Kim, A.Y., Lucas, M., Pillay, T., Ouchi, K., Reyor, L.L., Zur Wiesch, J.S., Gandhi, R.T., Chung, R.T., Bhardwaj, N., Klenerman, P., Walker, B.D., Allen, B.M. (2004). CD8 epitope escape and reversion in acute HCV infection. *Journal of Experimental Medicine*, 200: 1593–1604.

Vertuani, S., Bazzaro, M., Gualandi, G., Micheletti, F., Marastoni, M., Fortini, C., Canella, A., Marino, M., Tomatis, R., Traniello, S., Gavioli, R. (2002). Effect of interferon-alpha therapy on epitope-specific cytotoxic T lymphocyte responses in hepatitis C virus-infected individuals. *European Journal of Immunology*, 32: 144–54.

Ward, S., Lauer, G., Isba, R., Walker, B., Klenerman, P. (2002). Cellular immune responses against hepatitis C virus: the evidence base. *Clinical & Experimental Immunology* 128: 195–203.

Wedemeyer, H., He, X.S., Nascimbeni, M., Davis, A.R., Greenberg, H.B., Hoofnagle, J.H., Liang, T.J., Alter, H., Rehermann, B. (2002). Impaired effector function of hepatitis C virus-specific CD8+ T cells in chronic hepatitis C virus infection. *Journal of Immunology*, 169: 3447–3458.

Wherry, E.J., Ahmed, R. (2004). Memory CD8 T-cell differentiation during viral infection. *Journal of Virology*, 78: 5535–5545.

Zafiropoulos, A., Barnes, E., Piggott, C., Klenerman, P. (2004). Analysis of "driver" and "passenger" CD8+ T-cell responses against variable viruses. *Proceedings of the Royal Society B: Biological Sciences* 271 Suppl. 3: S53–536.

Other Microbial Components Associated with Hepatitis C Virus Infection: Their Effects on Interferon-α/Ribavirin Treatment

Luigi Amati, Vittorio Pugliese, and Emilio Jirillo

Introduction

The hepatic immune response in patients with hepatitis C virus (HCV) infection has been the object of intensive studies. In fact, HCV is a lymphotropic virus and CD81 acts as its co-receptor on CD4+, CD8+, CD19+, and CD56+ lymphocytes, respectively (Kronenberger et al., 2006). Generally, antigen (AG)-specific CD8+ cell cytotoxic response is very effective in viral clearance, being that this function is sustained by the intervention of CD4+ cells via release of interferon (IFN)-γ and interleukin (IL)-2, which expands the pool of T cytolitic cells (Jellison et al., 2005). Additionally, activated AG-specific CD8+ cells generate a pool of memory cells, which protect the host against subsequent infections (Jabbari & Harty, 2006). However, this seems not to be the case in HCV infection because of mutations within immunodominant CD8 epitopes, which allow HCV to escape from immunosurveillance (Chang, 1998; Urbani et al., 2005). On the other hand, early defects of CD8+ cell cytolytic function have been reported in HCV infection, even including the capacity of these cells to exert antiviral activity (Rehermann & Nascimbeni, 2005; Bowen & Walker, 2005). In fact, in this precocious phase, CD8+ T cells interacting with high avidity with AG-presenting cells (APCs) may undergo deletion or poor activation (Bowen et al., 2005). On the other hand, later in infection CD8+ cells activated by HCV in the lymph nodes (LN) may interact with lower avidity with APCs, thus leading to an ineffective viral clearance (Snyder et al., 2003). Taken together, these events seem to contribute to HCV disease chronicity. Conversely, CD4+ cells seem to be very active against some conserved protein epitopes of the HCV, and their response correlates with viral clearance (Harcourt et al., 2004). Either in HCV-infected patients or in HCV-infected chimpanzees (Shoukry et al., 2004), production of IFN-γ from T helper(h)-1 cells is associated with a rapid clearance of circulating HCV.

Consequently, development of chronicity in HCV-infected patients may rely on a defective HCV AG presentation to T cells and, therefore, on a failure to maintain and sustain a robust Th1 response against immunodominant proteins (Neumann-Haefelin et al., 2005).

In HCV infection, AG presentation is strongly supported by dendritic cells (DCs), and *in vitro* studies have clearly demonstrated that human DCs expressing

HCV core and NS3 AGs were able to activate T cells to proliferate and release cytokines (Li et al., 2006).

Of note, early AG presentation in the LN during HCV infection abolishes intra-hepatic tolerance, leading to a more efficacious cytotoxic immune response. In this framework, it is worth mentioning the role played by NKT cells, a T cell subset specific for CD1d, predominantly present in mouse liver and also in human liver chronically infected with HCV (Sandberg & Liunggren, 2005). Quite interestingly, NKT cells arise in the thymus and are positively selected via interaction with MHC class I molecules on double-positive CD4+ CD8+ thymocytes (Sandberg & Ljung-gren, 2005). As far as the trafficking pattern of these cells is concerned, NKT cells express a chemokine receptor profile very similar to that of Th1 inflammatory cells (Godfrey & Kronenberg, 2004). Functionally, the interaction between liver DCs and NKT cells leads to a dramatic release of IFN-γ and IL-4 in the latter (Trobonjaca et al., 2001). This event might suggest the role played by NKT cells in the initiation and regulation of the immune response.

Just recently, the role of the chemokine receptor CCR5 has been emphasized as an important modulator of the inflammatory response in the course of HCV infection (Ajuebor et al., 2006). In fact, interaction of CCR5 with its intrahepatic ligands favors the recruitment of Th1 cells into the liver, thus promoting clearance of HCV during acute infection (Boisvert et al., 2003). Quite interestingly, it seems that IFN-α possesses the ability to increase expression of CCR5 on T cells during HCV infection, thus contributing to viral clearance (Yang et al., 2001).

However, further data are required to confirm whether downregulation of CCR5 expression on T cells renders individuals more susceptible to HCV infection.

In HCV infection, B lymphocytes are present intrahepatically and harbor HCV (Sansonno et al., 1998). Under HCV antigenic pressure, B cells undergo clonal expansion often associated with generation of rheumatoid factor, which, in turn, interacts with human MHC class I AGs, thus interfering with peptide recognition by T cells (Williams et al., 1994). With special reference to auto-antibodies, antilactoferrin (LF) auto-antibodies have been detected in HCV+ patients and, mostly, in nonresponders to IFN-α/ribavirin treatment (Amati et al., 2004). In particular, LF is an iron binding protein endowed with antiviral activity and able to bind to the lipid A moiety of bacterial endotoxins (Caccavo et al., 1999, 2002). Therefore, production of anti-LF auto-antibodies in HCV disease may aggravate its clinical course by depotentiating the anti-inflammatory activities of LF.

Finally, the lack of correlation between a strong Th response and a corresponding robust antibody response in HCV-infected patients may depend on the attitude of CD4+ cells to promote CD8+-dependent cytotoxicity rather than immunoglobulin production (Napoli et al., 1996; Cacciarelli et al., 1996).

As far as hepatic innate immunity is concerned, Kupffer cells (KCs) represent the largest contigent of resident macrophages (M$\emptyset$) present in the body (Fax et al., 1989). KCs are able to capture and present AGs and express the co-stimulatory CD80 and CD86 molecules (Burgio et al., 1998). In contrast to DCs, KCs do not migrate out of the liver and present AGs locally (MacPhee et al., 1992). MHC-I-positive hepatocytes, despite the lack of co-stimulatory molecules (Ni et al., 1999), can directly present AGs to uncommitted T cells via ICAM-1 (Bertolino et al., 1998). KCs are able to recognize AGs via Toll-like receptors (TLRs) with a subsequent release of pro-inflammatory cytokines and oxygen free radicals (Liu et al., 1998). Therefore, in the course of HCV infection, hyperactivation of KCs can cause further liver damage through the release of the above-cited mediators (Fearns et al., 1995). However, a mechanism of hepatic immune tolerance has been established by liver sinusoidal endothelial cells (LSECs) (Knolle & Limmer, 2001). They behave as APCs and resemble immature DCs, which are resistant to maturation even under tumor necrosis factor (TNF)-α and endotoxin stimulation. Functionally, LSECs attenuate Th1-mediated responses, thus facilitating antibody response. Furthermore, as far as intrahepatic tolerance is concerned, DCs express IL-10 in the liver, thus rendering APCs tolerogenic (Goddard et al., 2004). On the other hand, the role of NKT cells in intrahepatic tolerance is still debated (Godfrey & Kronenberg, 2004).

Quite interestingly, polymorphonuclear cells (PMN) represent 1–2% of the total nonparenchymal cells found in normal mouse liver (Gregory et al., 1996). In *Listeria*-infected mice, a massive infiltration of immigrating PMN and their co-localization with KCs into the liver have been reported (Gregory et al., 2002). *Listeriae* organisms were phagocytosed by PMN and subsequently found within KCs. PMN and KCs interacted via adhesion molecules [CD11b/CD18(MAC-1) and CD54(ICAM-1)], and, finally, adherent PMN were ingested and lysed by KCs. This mechanism can also be interpreted as an attempt by KCs to decrease the release of inflammatory mediators by activated PMN. The intrahepatic role of PMN has also been demonstrated by experiments of neutrophil depletion in mice that led to an accumulation of various bacteria previously inoculated intravenously (IV) (Verdrengh & Tarkowski, 1997; Van Andel et al., 1997; Conlan, 1997).

Concurrent Effects of Bacterial Endotoxins in the Course of HCV Infection

Bacterial endotoxins or lipopolysaccharides (LPS) from the outer cell membrane of gram-negative bacteria are deeply involved in the pathogenesis of sepsis (Opal et al., 1999). For cell activation to occur, LPS interact with CD14-bearing inflammatory cells [monocytes (MO)-MØ, PMN, and endothelial cells] and through TLR-4 lead to the release of a plethora of pro-inflammatory cytokines, free radicals, platelet activating factor, complement components, tissue factor, and various noxious mediators (Wright et al., 1990; Poltorak et al., 1998).

Evidence has been provided that TLRs require accessory molecules for microbial recognition. In the case of LPS binding to TLR-4, LPS binding protein (LBP), CD14, and MD-2 play specific roles in that LBP and CD14 determine the magnitude of LPS responses and type I IFN production (Miyake, 2006). On the other hand, MD-2 is responsible for ligand binding and receptor activation (Miyake, 2006). Also, in the case of TLR-2, intervention of similar accessory molecules is needed.

All these harmful substances participate in the generation of the systemic inflammatory response syndrome, and the liver is the principal organ devoted to LPS
detoxification (Jirillo et al., 2002). Experimentally, IV-injected LPS are taken up by
hepatocytes, which seem to be involved in the clearance of these bacterial products
by virtue of CD14 and TLR-4 expressed on their membrane (Vodovotz et al., 2001).
Moreover, as a result of LPS injection into rats, endotoxins have been found in the
bile and then excreted into the gut (Maitra et al., 1981).

In experimental hepatitis the major mediator involved in the liver damage is represented by tumor necrosis factor (TNF)-α, as demonstrated in mice treated with
D-galactosamine (D-GalN) (Galanos et al., 1979). D-GalN increases the sensitivity of mice to LPS, and mortality occurs via a massive apoptosis of hepatocytes
(Mignon et al., 1999). In this regard, the role of Fas ligand (FasL) in the induction
of TNF-α-mediated liver apoptosis is quite controversial. In fact, in the LPS-D-
GalN model, a defective Fas or a lack of functional Fas could not prevent mortality
and liver damage (Tannahil et al., 1999). Instead, this was the case in the model
of *Corynebacterium parvum* LPS, where blockade of FasL with soluble Fas fusion
protein was protective in mice (Kondo et al., 1997).

Another mediator able to cause liver injury in response to LPS is IL-18 released
by KCs. In the murine model, *Propionibacterium acnes*/LPS-induced liver injury,
IL-18 induced Fas-dependent hepatocyte apoptosis via natural killer (NK) cell-
induced increase of FasL (Tsutsui et al., 2000). This described model of acute liver
damage is similar to that of fulminant hepatitis in mice with a gene transfection of
FasL (Li et al., 2001). Conclusively, this pathogenic mechanism may serve to elucidate some aspects of the fulminant hepatitis described in the human HCV infection
(Vento, 2000).

Taken together, the above data indicate that in experimental hepatitis, liver damage occurs through apoptosis and caspases seem to mediate cell death (Van Molle
et al., 1999). Treatment with α1-antitrypsin may represent an anti-apoptotic mechanism, thus indicating that acute-phase proteins are able to prevent caspase activation
(Van Molle et al., 1999).

Finally, a role has been attributed to LPS in ethanol-induced liver injury. There
is evidence that ethanol increases the intestinal permeability to LPS (Enomoto
et al., 1998) and, at the same time, antibiotics and lactobacilli treatments mit-
igate ethanol-dependent hepatic damage by reducing gram-negative intestinal
flora (Adachi et al., 1995; Nanji et al., 1994). In addition, ethanol chronically
administered to CD14 knockout mice generates less damage than that observed
in the wild counterpart (Adachi et al., 1994). On the other hand, ablation of
KCs by gadolinium chloride attenuates ethanol-mediated liver injury since these
cells are the source of pro-inflammatory cytokines in response to LPS (Adachi
et al., 1994). Finally, anti-TNF-α monoclonal antibody treatment (Iimuro et al.,
1997) or the use of receptor-1 knockout mice (Yin et al., 2001) prevents hepatic damage caused by ethanol, thus further supporting the role of LPS in this
pathology.

In humans, the presence of endotoxins in liver disease has been documented
in many reports. Several authors have found endotoxemia cirrhotic patients, and
it seems that amounts of LPS progressively augment as liver function deteriorates

(Nolan, 1975; Liehz et al., 1976; Prytz et al., 1976). This last finding may be predictive of short-term survival in cirrhosis (Chan et al., 1997). Furthermore, the evidence for a reduced phagocytic activity of KCs in cirrhosis may be the cause of endotoxemia in this clinical condition (Kuratsune et al., 1983). In this framework, it should be mentioned that postoperative hepatic failure in cirrhotic patients seems to be the result of an exaggerated release of pro-inflammatory cytokines by primed MØ activated by LPS spilling over in the blood during hepatic resection (Sato et al., 1997).

In other related studies conducted in HCV patients, a correlation was found between endotoxemia and elevated levels of serum CD14 (Jirillo et al., 1998). Similar findings were also reported in the case of alcoholic cirrhosis and HBV infection (Oesterreicher et al., 1995).

Over recent years, a number of investigations have attempted to clarify the origin of endotoxemia in the course of HCV infection. Quite interestingly, in HCV infection a defect of innate immunity has been described in terms of the reduced ability of phagocytosis and killing exerted by PMN and MO (Jirillo et al., 1995, 1996). Moreover, T-cell-mediated antibacterial activity was also impaired in these patients as a further demonstration of natural immunity depression (Jirillo et al., 1995, 1996). Therefore, gram-negative bacteria can gain easier access into the HCV host with subsequent liberation of endotoxins at systemic or tissue levels. At the same time, the altered architecture of the liver can reduce its detoxifying capacity, thus leading to the accumulation of various toxic products into the host, even including LPS (Jirillo et al., 2000). In relevance to this event, bacterial toxins of intestinal derivation, accumulated in the HCV-infected liver because of a putative altered intestinal permeability, may further aggravate the hepatic damage (Jirillo et al., 2000).

These data are in agreement with a recent view according to which activation of the innate immune system in the liver abrogates APC tolerance, thus avoiding T cell apoptosis (Bowen et al., 2005). Consequentially, increased survival of CD8+ cells gives rise to a more efficient intrahepatic cytotoxicity.

On these grounds, our group has conducted a series of investigations on the putative effects of endotoxins in HCV patients receiving 6 months' treatment with IFN-α/ribavirin (RIB) (Amati et al., 2002; Caradonna et al., 2002). Before therapy (T0), HCV individuals were subdivided into two groups—endotoxemic and nonendotoxemic—in order to evaluate the influence of LPS on their immune status. Thus, at T0, in endotoxemic HCV+ patients, absolute numbers of CD3+, CD4+, CD14+, and CD19+ cells were higher than those observed in the non-endotoxemic HCV+ counterpart.

Additionally, MO intracellular content of TNF-α and IL-1β was more elevated in endotoxemic patients than in non-endotoxemic ones under resting conditions. Following in vitro LPS stimulation of MO, in endotoxemic individuals values of these cytokines were even higher than in non-endotoxemic ones. The same group of patients, divided into responders and non-responders at the end of IFN-α/RIB treatment over a period of six months (T6), was immunologically re-evaluated. In responders, endotoxemia present at T0 was no longer detectable, while non-responders were still endotoxemic.-

Quite interestingly, in responders there was a parallel increase of serum levels of IFN-γ and IL-10, while in non-responders the increase in IFN-γ was not paralleled by an equivalent increase in IL-10.

Consequently, in non-responders the MO intracellular content of IL-1β and TNF-α was more elevated than in responders. Taken together, all these data suggest that in responders to IFN-α/RIB treatment a re-equilibrium between Th1 (inflammatory) and Th2 (anti-inflammatory) cytokines occurs. As a result of this balance, bacterial AGs, even including LPS, can be neutralized in a more efficient way by the effects of intestinal and hepatic phagocytes as well as of epithelial and liver endothelial cells. Binding and/or de-activation of LPS or enhanced phagocytosis of opsonized microorganisms seem to be the major immune mechanisms elicited in response to a successful treatment with IFN-α/RIB. In relevance to the above-described mechanisms, it has been hypothesized that in patients who resolve HCV infection depression of the innate immune response might not occur, thus leading to a breaking of intrahepatic tolerance (Rehermann & Nascimbeni, 2005). Resulting inflammatory status may be beneficial to the host in terms of HCV eradication.

In non-responders, the lack of anti-inflammatory activity exerted by IL-10 seems to be responsible for MO hyperactivation, as evidenced by the more elevated content of pro-inflammatory cytokines (IL-1β and TNF-α) in these cells. Similar findings have been reported in alcoholic cirrhosis, where a decreased release of IL-10 from MO in response to LPS has been discovered (Le Moine et al., 1995). Consequently, exaggerated secretion of TNF-α in alcoholic cirrhosis may be attributed to the observed lack of IL-10 production.

In HCV+ non-responder patients, hepatic and/or systemic oversecretion of pro-inflammatory cytokines aggravates liver damage. In particular, IL-1β *in vivo* is able to inhibit IFN-α/β-induced Stat1 tyrosine phosphorylation, thus hampering IFN-mediated antiviral activity (Tian et al., 2000). Conversely, in Stat1 knockout mice, IFN-dependent signaling pathways are absent, thus provoking a reduced antimicrobial immune response. According to these data, IL-1β may account for refractoriness to IFN-α therapy in HCV disease, thus representing a putative therapeutic target in this pathology (Diehl, 1999).

The Role of β-Glucans in the Course of HCV Infection

β-glucans (BG) are natural polysaccharides that represent normal components of the cell wall of fungi and bacteria, as well as of oats, barley, and yeast (Williams et al., 1998). BG are ubiquitously distributed in the environment and, therefore, living organisms possess pattern recognition molecules able to interact with these polysaccharides (Amati et al., 2005b). In fungi, in addition to polysaccharides, glycoproteins (mannoproteins) are also present, and, in particular, mannose residues can elicit a robust immune response into a susceptible host (Williams et al., 1998; Fraser et al., 2006). Here, emphasis will be placed on BG, whose ability to regulate immune response will be illustrated below.

As far as the interaction of BG with phagocytes and, in particular, with MØ is concerned, besides the MØ mannose receptor and the complement receptor 3, dectin-1 has recently been considered as the major MØ receptor for these molecules (Brown et al., 2002). LPS and BG activate MØ, both leading to an increase in NFKB (Underhill & Olinsky, 2002; Gantner et al., 2003). However, LPS utilize TLR-4 on MØ (Underhill & Olinsky, 2002), while BG activate MØ in TLR-4-deficient mice for the production of TNF-α (Kataoka et al., 2002). By contrast, in mice with a defect in the adapter protein MyD88, the BG-induced MØ response is lower than that observed in the wild-type murine counterpart (Marr et al., 2003). These data indicate that LPS and BG share some common postreceptorial pathways.

According to current literature, BG seem to be protective toward LPS-mediated toxic effects. In this respect, mice administered with BG before the induction of sepsis (cecal ligation and puncture) underwent less mortality than untreated animals (Williams et al., 1999). Actually, preadministration of BG correlates with less expression of TLR-2 and TLR-4 mRNA and less concentration of serum TLR-4. This mechanism of downregulation of LPS-induced NFKB activation could depend on BG-mediated inhibition of IKKβ kinase activity and altered phosporylation and degradation of IKβ-α (Williams et al., 2002).

In vitro studies with BG are quite controversial. In fact, it has been demonstrated that peripheral blood mononuclear cells and murine MØ, stimulated with BG, produced IL-1 and TNF-α (Abel & Czop, 1992; Seljelial et al., 1989). However, murine MØ pretreated with BG and then stimulated with LPS produced higher amounts of IL-6, while TNF-α production was suppressed (Soltys & Quinn, 1999). In general terms, according to studies in an *in vitro* human model, BG seem to promote production of IL-8 and IL-10 and to suppress IL-2 and IFN-γ release from Th-1 cells in response to endotoxins (Nakagawa et al., 2003). Furthermore, it is worth mentioning that other investigations have emphasized the role of BG in the release of TNF-α from zymosan-activated MØ or in the upregulation of immunocompetent cell response, by their own or in synergy with LPS (Engstad et al., 2002). However, all the above discrepancies can be explained by the concentration of BG present in a given host. For instance, Hoffman et al. (1993) found that concentrations of BG less than 500 µg suppressed TNF-α release from rat alveolar MØ, while concentrations greater than 500 µg enhanced production of this cytokine in response to LPS.

In order to evaluate the effects of BG on the immune response in HCV patients undergoing IFN-α/RIB therapy, endotoxemia and β-glucanemia were measured in their blood at T0 and T6, respectively, as previously described in this chapter. Patients were subdivided into two subsets, LPS+/BG+ and LPS-/BG+, respectively, and then immune parameters were determined (Amati et al., 2005a). When serum levels of BG and plasma endotoxins were evaluated, endotoxemia, at T0, was detected in 22 of 46 patients, while BG were present in the sera of 44 of 46 patients. At T6, among 41 patients evaluated, endotoxemia was detected in 20 of them, while β-glucanemia was present in 38 individuals.

In terms of absolute numbers and percentage of lymphocyte phenotypes, no significant differences were observed between patients (at T0 and at T6) and normal donors. Quite interestingly, when patients were divided into two subsets, namely

LPS+/BG+ and LPS-/BG+ subjects, some interesting findings emerged. In particular, at T6 vs. T0, in the LPS-/BG+ subset there was an increase of CD3-CD8+
cells (a subset of NK cells) and of CD71+ cells (Figure 1), while memory cells
(CD45RO+ cells) decreased (Figure 2). In the LPS+/BG+ counterpart, similar findings were not detected.

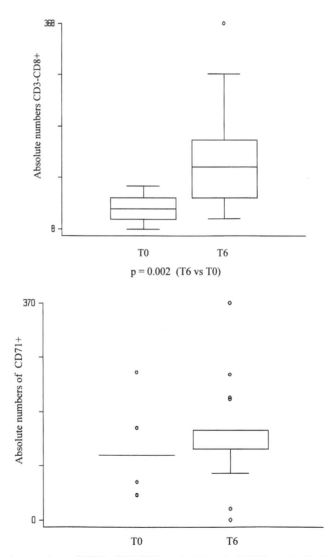

Fig. 1 Absolute numbers of HCV+ CD3-CD8+ cells (A) and of CD71+ cells (B). Samples from
HCV+ LPS-/BG+ patients were analyzed at T0 and at T6, respectively, on a FACSCalibur [(Bec-
ton Dickinson Immunocytometry System, San Josè, CA (BDIS)] by cell surface staining with
FITC/PE/FITC-conjugated monoclonal antibodies to CD3, CD8, and CD71 antigens, respectively

p< 0.001 (T6 vs T0)

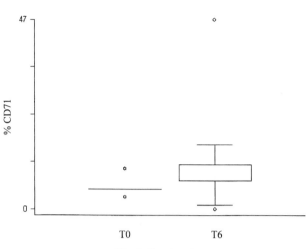

Fig. 1 *Continued*

Fig. 2 Absolute numbers of HCV+ CD45RO+ cells. Samples from HCV+ LPS-/BG+ patients were analyzed at T0 and at T6, respectively, on a FACSCalibur (Becton Dickinson) by cell surface staining with FITC-conjugated monoclonal antibodies to CD45RO antigen

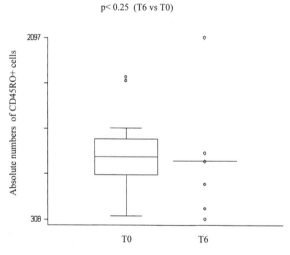

p< 0.25 (T6 vs T0)

In another set of experiments, at T0, a correlation between LPS/BG levels and immune/enzymatic parameters was performed in LPS+ BG+ HCV+ patients. A negative correlation was found with CD25+ cells, gamma glutamyl transpeptidase (γ-GT) values, total bilirubin, and direct bilirubin. On the other hand, no significant correlations were found in the case of LPS-/BG+ patients. Furthermore, at T6, in the LPS+/BG+ patients a positive correlation was detected with CD3+ and CD4+ cells, glutamic-ossalacetic transaminase (GOT) and glutamil-piruvic transaminase (GPT) and direct bilirubin. Quite interestingly, at T6, in the LPS-/BG+ counterpart a positive correlation was determined with CD25+ (Figure 3) and CD95+ cells

p=0.06 (T6)

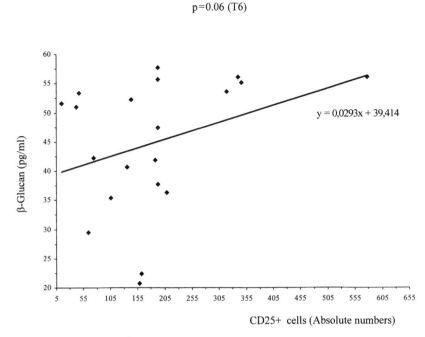

Fig. 3 Correlation between β-glucan serum concentration and absolute numbers of CD25+ cells, at T6, in LPS-/BG+ HCV+ patients. Spearman's rho = 0.42; p = 0.06

p=0.03 (T6)

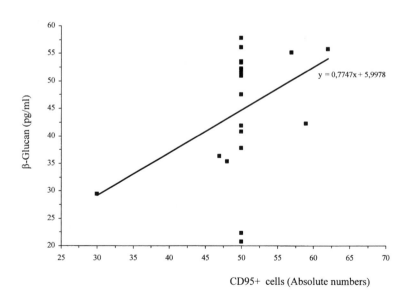

Fig. 4 Correlation between β-glucan serum concentration and absolute numbers CD95+ cells, at T6, in LPS-/BG+ HCV+ patients. Spearman's rho = 0.48; p = 0.03

p=0.04 (T6)

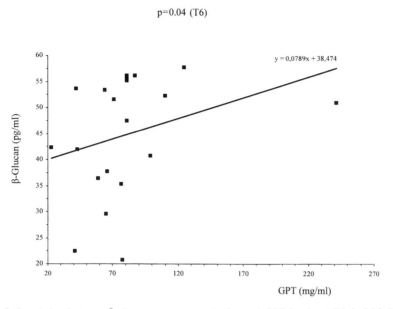

Fig. 5 Correlation between β-glucan serum concentration and GPT levels, at T6, in LPS-/BG+ HCV+ patients. Spearman's rho = 0.45; p = 0.04

(Figure 4) and GPT concentration (Figure 5), respectively. Conversely, a negative correlation was observed in the case of total bilirubin and direct bilirubin.

Conclusions

Taken together, the bulk of data reported in the previous sections clearly indicates the presence of both circulating endotoxins and BG in HCV+ patients before and after IFN-α/RIB therapy.

In the case of LPS, in responders endotoxemia is no longer detectable, while the status of a non-responder coincides with the presence of plasma LPS. On the other hand, at T6, levels of β-glucanemia are present in a percentage similar to that observed at T0. This suggests that therapy with IFN-α /RIB does not influence BG concentration.

The pathogenic mechanism accounting for the presence of circulating BG in HCV+ patients could be the same as that invoked in the case of endotoxemia. In fact, the impaired natural immunity in HCV disease, the reduced hepatic clearance exerted by KCs toward fungi and/or their components, and the increased intestinal permeability may represent important cofactors in the generation of glucanemia in an elevated percentage of patients (Amati et al., 2002; Jirillo et al., 1998).

As far as the role of LPS is concerned in HCV disease, our previous data have pointed out that, after IFN-α/RIB treatment in HCV+ patients, non-responders were still endotoxemic while responders were no longer endotoxemic (Amati et al., 2002; Caradonna et al., 2002). Endotoxemia was associated with an increased content of IL-1β and TNF-α in MO and with an exaggerated production of NO (Caradonna et al., 2002). On the contrary, in responders the increased release of IL-10 led to an anti-inflammatory response that neutralized the production of pro-inflammatory cytokines (Caradonna et al., 2002). Conclusively, in non-responders the uncontrolled inflammatory process could aggravate liver damage (Amati et al., 2002; Caradonna et al., 2002).

To the best of our knowledge, here we have provided the first evidence on the relationship between β-glucanemia and liver function in HCV disease. In fact, at T6, in HCV+ patients who were LPS+/BG+, we have determined a positive correlation with GOT, GPT, and direct bilirubin serum levels, respectively. By the way, in the LPS+/BG+ patients, at T0, a negative correlation was found with levels of γ-GT and direct and total bilirubin. On the other hand, in the LPS-/BG+ subjects at T6, a negative correlation was detected with total and direct bilirubin, respectively. In addition, no correlation was found with GOT, while a positive correlation was determined with GPT only. Taken together, these data suggest that BG contribute to a lesser extent to the hepatic damage in the course of HCV disease, while LPS seem to exert more noxious effects on the liver (Jirillo et al., 2002). In this respect, evidence has been provided that intrahepatic DCs and LSECs are refractory to LPS effects (De Creus et al., 2005; Uhrig et al., 2005). Therefore, abrogation of LPS tolerance in HCV+ non-responder patients might contribute to disease progression. In addition, at T6, the presence of circulating BG, in the absence of LPS, is associated with an increase in CD3-CD8+ cells, a subset of NK cells, in CD71+ cells, and with a decrease in CD45RO+ cells, while positively correlating with CD25+ and CD95+ cells. By contrast, at T0, in LPS+/BG+ patients a negative correlation was found with CD25+ cells. Collectively, these findings allow us to formulate the following hypothesis. In HCV disease, BG seem to expand the pool of CD25+ cells and likely of CD4+CD25+ T regulatory (TREG) cells. TREG cells could exert a potent anti-inflammatory activity via production of IL-10 and Transforming Growth Factor β. In support of this view, we provided clear-cut evidence that serum levels of IL-10 are increased in HCV+ patients who terminated IFN-α/RIB treatment free of circulating endotoxins (Caradonna et al., 2002). In a recent paper by Finkelman et al. (2006), evidence has been provided that allergen extracts contaminated with the highest content of BG are those endowed with the most successful therapeutic properties in allergic diseases. In particular, an effective allergen immunotherapy is associated with an increase in circulating IL10+ CD4+ CD25+ T cells and in mucosal IFN-γ-secreting T cells (Francis et al., 2003).

Just recently, according to a review by Sutmuller et al. (2006), *Candida albicans* infection leads to an immunosuppressive pathway mediated through TLR-2 and subsequent generation of TREG cells. At the same time, *Candida glucans* via Dectin-1/TLR-2 on MØ mediates an IL-10-dependent immunosuppression. Collectively, these data indicate that BG-mediated anti-inflammatory and immunosuppressive activities could reduce the hepatic damage in HCV infection.

Another important finding is represented by the increase of CD3- CD8+ cells in LPS-/BG+ patients. In general terms, NK cells are able to either exert a direct antimicrobial effect or modulate the innate or adaptive immune response via production of IFN-γ (Sher et al., 1993). At the moment, in our group of patients the expansion of this subset of NK cells is difficult to interpret; however, it may contribute to the observed increase of serum IFN-γ in responders to IFN-γ/RIB therapy (Amati et al., 2002; Caradonna et al., 2002).

With regard to other T cell surface markers in the LPS-/BG+ subset, the increase in CD71+ cells might be related to the expansion of CD25+ cells and NK cells. On the other hand, the increase of CD95+ cells could imply an apoptosis of CD45+RO cells, whose number is decreased, as previously reported in this chapter. These T memory cells likely comprise CD4+ and CD8+ cells specific for viral epitopes and, therefore, actively involved in the hepatolysis. In this respect, evidence has been provided that *in vivo* soluble BG could enhance spontaneous lymphocyte apoptosis, thus contributing to the multiple anti-inflammatory activities of the entire molecule (Abel & Czop, 2002*).* Conversely, in the LPS+/BG+ subset, a positive correlation was found with CD3+ and CD4+ cells, thus suggesting a continuous proliferation of T effector cells capable of maintaining the inflammatory status.

Quite interestingly, in mice BG abrogate induction of endotoxin tolerance leads to an increased expression of IFN-γ in response to IL-12 and IL-18 (Sherwood et al., 2001). The ability of BG to augment the expression of IFN-γ in LPS-tolerant mice suggests their potential use in the recovery of trauma and sepsis-induced immunosuppression. Therefore, also in HCV+ patients with endotoxemia, BG could contribute to the host protection by enhancing antimicrobial immunity, also attenuating the noxious effect of LPS. At the same time, BG in the absence of LPS may express with higher potency their beneficial role for the host, thus contributing to HCV eradication.

In summary, the HCV+ host is under multiple antigenic challenges (e.g., bacteria and fungi), and the mutual balance between LPS- and BG-induced regulation of the immune system may have important clinical reflections in terms of response or refractoriness to IFN-α/RIB therapy. This fact may correlate with the hypothesis according to which activation of the innate immune system abrogates APC tolerance in the liver, thus rendering CD8+ cells more efficient in their cytotoxic response (Bowen et al., 2005). On the other hand, Wuensch et al. (2006) have demonstrated that infecting hepatocytes with an adeno-associated virus vector, T cell activation is exclusively intrahepatic and does not lead to liver tolerance. In this case local CD8+ cell activation seems to bypass the need for CD4+ T cell help, thus indicating that the liver immune response and tolerance also depend on the type of antigenic challenge involved.

Conclusively, these findings suggest that BG may represent potential new drugs for mitigating the exaggerated hepatic inflammation induced by LPS or other microbial AGs that have entered the HCV+ host. Therefore, calibration of intrahepatic activation of the immune system in HCV disease seems essential for eradicating the virus on the one hand and for avoiding liver damage on the other hand.

References

Abel, G., Czop, J.K. (1992). Stimulation of human monocyte Beta-Glucan receptor by glucan particles production of TNF-alfa and IL-1 beta. *International Journal of Immunopharmacology,* 14: 1363–1373.

Adachi, Y., Moore, L.E., Bradford, B.V., Gao, W., Bojes, H.K., Thurman, R.G. (1995). Antibiotics prevent liver injury in rats following long-term exposure to ethanol. *Gastroenterology,* 108: 218–224.

Adachi, Y., Bradford, V., Gao, W., Bojes, H.K., Thurman, R.G. (1994). Inactivation of Kupffer cells prevents early alcohol-induced liver injury. *Hepatology,* 20: 453–460.

Ajuebor, M.N., Carey, J.A., Swain, M.G. (2006). CCR5 in T-cell mediated liver disease: what's going on? *Journal of Immunology,* 177: 2039–2045.

Amati, L., Caradonna, L., Magrone, T., Mastronardi, M.L., Cuppone, R., Cozzolongo, R., Manghisi, O.G., Caccavo, D., Amoroso, A., Jirillo, E. (2002). Modifications of the immune responsiveness in patients with hepatitis C virus infection following treatment with IFN-α/Ribavirin. *Current Pharmaceutical Design,* 8: 981–993.

Amati, L., Cozzolongo, R., Manghisi, O.G., Cuppone, R., Pellegrino, N.M., Caccavo, D., Jirillo, E. (2004). The immune responsiveness in hepatitis C virus infected patients: effects of interferon-alpha/ribavirin combined treatment on the lymphocyte response with special reference to B cells. *Current Pharmaceutical Design,* 10: 2093–2100.

Amati, L., Leogrande, D., Finkelman, M.A., Tamura, H., Jirillo, E. (2005a). Peripheral immunity in patients with hepatitis C virus (HCV) infection: different roles of endotoxins (LPS) and β-glucans. *Journal of Endotoxin Research,* 10: 362 [abstract].

Amati, L., Leogrande, D., Passeri, M.E., Mastronardi, M.L., Passantino, L., Venezia, P., Jirillo, E. (2005). β-glucans: old molecules with newly discovered immunological activities. *Drug Design Reviews-Online,* 2: 251–258.

Bertolino, P., Trescol-Biemont, M.C., Rabourdin-Combe, C. (1998). Hepatocytes induce functional activation of naive CD8+ T lymphocytes but fail to promote survival. *European Journal of Immunology,* 28: 221–236.

Bowen, D.G., Walker, C.M. (2005). Adaptive immune response in acute and chronic hepatitis C virus infection. *Nature,* 436: 946–952.

Bowen, D.G., McCaughan, G.W., Bertolino, P. (2005). Intrahepatic immunity: a tale of two sites? *Trends in Immunology,* 26: 512–517.

Brown, G.D., Taylor, P.R., Reid, D.M., Willment, J.A., Williams, D.L., Martinez-Pomares, L., Wong, S.Y.C., Gordon, S. (2002). Dectin-1 is a major β-glucan receptor on macrophages. *The Journal of Experimental Medicine,* 196: 407–412.

Burgio, V.L., Ballardini, G., Artini, M., Caratozzolo, M., Bianchi, F.B., Levriero, M. (1998). Expression of co-stimulatory molecules by Kupffer cells in chronic hepatitis of hepatitis C virus etiology. *Hepatology,* 27: 1600–1606.

Caccavo, D., Afeltra, A., Pece, S., Giuliani, G., Freudenberg, M., Galanos, C., Jirillo, E. (1999). Lactoferrin-lipid A-lipopolysaccharide interaction: inhibition by anti-human lactoferrin monoclonal antibody AMG 10.14. *Infection and Immunity,* 67: 4668–4672.

Caccavo, D., Pellegrino, N.M., Altamura, M., Rigon, A., Amati, L., Amoroso, A., Jirillo, E. (2002). Antimicrobial and immunoregulatory functions of lactoferrin and its potential therapeutic application. *Journal of Endotoxin Research,* 8: 403–417.

Cacciarelli, T.V., Martinez, O.M., Gish, R.G., Villanueva, J.C., Krams, S.M. (1996). Immunoregulatory cytokines in chronic hepatitis C virus infection: pre- and posttreatment with interferon alpha. *Hepatology,* 24: 6–9.

Caradonna, L., Mastronardi, M.L., Magrone, T., Cozzolongo, R., Cuppone, R., Manghisi, O.G., Caccavo, D., Pellegrino, N.M., Amoroso, A., Jirillo, E., Amati, L. (2002). Biological and clinical significance of endotoxemia in the corse of hepatitis C virus infection. *Current Pharmaceutical Design,* 8: 995–1005.

Chan, C.C., Hwang, S.J., Lee, F.Y., Wang, S.S., Chang, F.Y., Li, C.P., Chu, J.C., Lu, R.H., Lee, S.D. (1997). Prognostic value of plasma endotoxin levels in patients with cirrhosis. *Scandinavian Journal of Gastroenterology,* 32: 942–946.

Chang, K.M. (1998). The mechanisms of chronicity in hepatitis C virus infection. *Gastroenterology*, 115: 1015–1018.

Conlan, J.W. (1997). Critical roles of neutrophils in host defense against experimental systemic infections of mice by *Listeria monocytogens, Salmonella typhimurium* and *Yersinia enterocolitica . Infection and Immunity*, 65: 630–635.

Decreus, A., Abe, M., Lau, A.H., Hackstein, H., Raimondi, G., Thomson, A.W. (2005). Low TLR4 expression by liver dendritic cells correlates with reduced capacity to activate allogeneic T cells in response to endotoxin. *Journal of Immunology*, 174: 2037–2045.

Diehl, A.M. (1999). Cytokine and the molecular mechanisms of alcoholic liver disease. *Alcoholic Clinical Experimental Research*, 23: 1419–1424.

Engstad, C.S., Engstad, R.E., Olsen, J.O., Osterud, B. (2002). The effect of soluble beta-(1,3)-glucan and lipopolysaccharide on cytokine production and coagulation activation in whole blood. *International Immunopharmacology*, 11: 1585–1597.

Enomoto, N., Ikejima, K., Bradford, B.V., Rivera, C., Kono, H., Brenner, D.A., Thurman, R.G. (1998). Alcohol causes both tolerance and sensitization of rat Kupffer cells via mechanism dependent on endotoxin. *Gastroenterology*, 115: 443–451.

Fax, E.S., Thomas, P., Broitmans, A. (1999). Clearance of gut-derived endotoxins by the liver. Release and modifications of ^{3}H, ^{14}C-lipopolysaccharide by isolated Kupffer cells. *Gastroenterology*, 96: 456–461.

Fearns, C., Kravchenko, V.V., Ulevitch, R.J, Loskutoff, D.J. (1995). Murine CD14 gene expression in vivo: extramyeloid synthesis and regulation by lipopolysaccharides. *Journal of Experimental Medicine*, 181: 857–866.

Finkelman, M.A., Lempitski, S.J., Slater, J.E. (2006). B-glucans in standardized allergen extracts. *Journal of Endotoxin Research*, 12: 241–245.

Francis, J.N., Till, S.J., Durham, S.R. (2003). Induction of IL-10+ CD4+ CD25+ T cells by grass pollen immunotherapy. *Journal of Allergy and Clinical Immunology*, 111: 1255–1261.

Fraser, D.A., Bohlson, S.S., Jasinskiene, N., Rawal, N., Palmarini, G., Ruiz, S., Rochford, R., Tenner, A.J. (2006). C1q and MBL, components of the innate immune system, influence monocyte cytokine expression. *Journal of Leukocyte Biology*, 80: 107–116.

Galanos, C., Freudenberg, M.A., Reutter, W. (1979). Galactosamine-induced sensitization to the lethal effects of endotoxin. *Proceedings National Academy of Sciences USA*, 76: 5939–5943.

Gantner, B.N., Simmons, R.M., Canavera, S.J., Akira, S., Underhill, D.M. (2003). Collaborative induction of inflammatory responses by dectin-1 and Toll-like receptor 2. *Journal of Experimental Medicine*, 197: 1107–1117.

Goddard, S., Youster, J., Morgan, E., Adams, D.H. (2004). Interleukin-10 secretion differentiates dendritic cells from human liver and skin. *American Journal of Pathology*, 164: 511–519.

Godfrey, D.I., Kronenberg, M. (2004). Going both ways: immune regulation via CD1d-dependent NKT cells. *Journal of Clinical Investigation*, 114: 1379–1388.

Gregory, S.H., Sagnimeni, A.J., Wing, E.J. (1996). Bacteria in the blood stream are trapped in the liver and killed by immigrating neutrophils. *Journal of Immunology*, 157: 2514–2520.

Gregory, S.H., Cousens, L.P., Van Rooijen, N., Dopp, E.A., Carlos, T.M., Wing, E.J. (2002). Complementary adhesion molecules promote neutrophil-Kupffer cell interaction and the elimination of bacteria taken up by the liver. *Journal of Immunology*, 168: 308–315.

Harcourt, G.C., Lucas, M., Sheridan, I., Barnes, E., Phillips, R., Klenerman, P. (2004). Longitudinal mapping of protective CD4+ T cell responses against HCV: analysis of fluctuating dominant and subdominant HLA-DR11 restricted epitopes. *Journal of Viral Hepatitis*, 11: 324–331.

Hoffman, O.A., Olson, E.J., Limper, A.H. (1993). Fungal beta-glucans modulate macrophage release of tumor necrosis factor-alpha in response to bacterial lipopolysaccharide. *Immunology Letters*, 37(1): 19–25.

Iimuro, Y., Gallucci, R.M., Luster, M.I., Kono, H., Thurman, R.G. (1997). Antibodies to tumor necrosis factor-alpha attenuate hepatic necrosis and inflammation caused by chronic exposure to ethanol in the rat. *Hepatology*, 26: 1530–1537.

Jabbari, A., Harty, J.T. (2006). The generation and modulation of antigen-specific memory CD8 T cell responses. *Journal of Leukocyte Biology*, 80: 16–23.

Jellison, E.R., Kim, S.K., Welsh, R.M. (2005). Cutting edge: MHC class II-restricted killing in vivo during viral infection. *Journal of Immunology*, 174: 614–618.

Jirillo, E., Greco, B., Caradonna, L., Satalino, R., Pugliese, V., Cozzolongo, R., Cuppone, R., Manghisi, O.G. (1995). Evaluation of cellular immune responses and soluble mediators in patients with chronic hepatitis C virus (cHCV) infection. *Immunopharmacology and Immunotoxicology,* 17: 347–364.

Jirillo, E., Greco, B., Caradonna, L., Satalino, R., Amati, L., Cozzolongo, R., Cuppone, R., Manghisi, O.G. (1996). Immunological effects following administration of interferon-α in patients with chronic hepatitis C virus (cHCV) infection. *Immunopharmacology and Immunotoxicology,* 18: 355–374.

Jirillo, E., Amati, L., Caradonna, L., Greco, B., Cozzolongo, R., Cuppone, R., Piazzolla, G., Caccavo, D., Antonaci, S., Manghisi, O.G. (1998). Soluble (s) CD14 and plasmatic lipopolysaccharides (LPS) in patients with chronic hepatitis C before and after treatment with interferon (IFN)-α. *Immunopharmacology and Immunotoxicology,* 20: 1–14.

Jirillo, E., Pellegrino, N.M., Piazzolla, G., Caccavo, D., Antonaci, S. (2000). Hepatitis C virus infection: immune responsiveness and interferon-α treatment. *Current Pharmaceutical Design,* 6: 169–180.

Jirillo, E., Caccavo, D., Magrone, T., Piccigallo, E., Amati, L., Lembo, A., Kalis, C., Gumenscheimer, M. (2002). The role of the liver in the response to LPS: experimental and clinical findings. *Journal of Endotoxin Research,* 8: 319–327.

Kataoka, K., Muta, T., Yamazaki, S., Takeshige, K. (2002). Activation of macrophages by linear (1-right-arrow 3)-beta D-glucans. Implications for recognition of fungi by innate immunity. *Journal of Biological Chemistry,* 277: 36825–36831.

Knolle, P.A., Limmer, A. (2001). Neighborhood politics: the immunoregolatory function of organ-resident liver endothelial cells. *Trends in Immunology,* 22: 432–437.

Kondo, T., Suda, T. Fukuyama, H., Adachi, M., Nagata, S. (1997). Essential roles of the Fas ligand in the development of hepatitis. *Nature Medicine,* 3: 409–413.

Kronenberger, B., Herrmann, E., Hofmann, W.P., Wedemeyer, H., Sester, M., Mihm, V., Ghahai, T., Zeuzem, S., Sarrklin, C. (2006). Dynamics of CD81 expression on lymphocyte subsets during interferon-α-based antiviral treatment of patients with chronic hepatitis C. *Journal of Leukocyte Biology,* 80: 298–308.

Kuratsune, H., Koda, T., Kurahori, T. (1983). The relationship between endotoxin and the phagocytic activity of the reticuloendothelial system. *Hepatogastroenterology,* 30: 79–82.

Le Moine, O., Marchant, A., Degroote, D., Azar, C., Goldman, M., Deviere, J. (1995). Role of defective monocyte interleukin-10 release in tumor necrosis factor overproduction in alcoholic cirrhosis. *Hepatology,* 22: 1436–1439.

Liehr, H., Grun, M., Brunswig, D., Sautter, T. (1976). Endotoxinamie bei leberzirrhose. *Zeitschrift Gastroenterologie,* 14: 14–23.

Li, W., Krishnadas, D.K., Li, J., Tyrrell, D.L.J., Agrawal, B. (2006). Induction of primary human T cell responses against hepatitis C virus-derived antigens NS3 or core by autologous dendritic cells expressing hepatitis C virus antigens: potential for vaccine and immunotherapy. *Journal of Immunology,* 176: 6065–6075.

Li, X.K., Fujino, M., Sugioka, A., Morita, M., Okuyama, T., Guo, L., Funeshima, N., Rimura, H., Enosawa, S., Amemiya, H., Suzuki, S. (2001). Fulminant hepatitis by Fas-ligand expression in MLR-lpr/lpr mice grafted with Fas-positive livers and wild-type mice with Fas-mutant livers. *Transplantation,* 71: 503–507.

Liu, S., Khemlani, L.S., Shapiro, R.A., Johnson, M.L., Liu, K., Geller, D.A., Watkins, S.C., Goyert, S.M., Billiar, T.R. (1998). Expression of CD14 by hepatocytes: upregulation by cytokines during endotoxemia. *Infection and Immunity,* 66: 5089–5098.

MacPhee, P.J., Schmidt, E.E., Groom, A.C. (1992). Evidence for Kupffer cell migration along liver sinusoids from high-resolution in vivo microscopy. *American Journal of Physiology,* 263: G17–G23.

Maitra, S.K., Rachmilewitz, D., Eberle, D., Haplowitz, N. (1981). The hepatocellular uptake and biliary excretion of endotoxin in the rat. *Hepatology,* 8: 1550–1554.

Marr, K.A., Arunmozhi Balajee, S., Hawn, T.R., Ozinsky, A., Pham, U., Akira, S., Aderem, A., Liles, W.C. (2003). Differential role of MyD88 in macrophage-mediated responses to opportunistic fungal pathogens. *Infection and Immunity,* 71: 5280–5286.

Mignon, A., Rouquet, N., Fabre, M., Martin, S., Pages, J.C., Dhainaut, J.E., Kahn, A., Briand, P., Joulin, V. (1999). LPS challenge in D-galactosamine-sensitized mice accounts for caspase-dependent fulminant hepatitis, not for septic shock. *American Journal of Respiratory Critical Care Medicine,* 159: 1308–1315.

Miyake, K. (2006). Roles for accessory molecules in microbial recognition by Toll-like receptors. *Journal of Endotoxin Research,* 12: 195–204.

Nakagawa, Y., Ohno, N., Muras, T. (2003). Suppression by *Candida albicans* beta-glucan of cytokine release from activated human monocyte and from T cells in the presence of monocytes. *Journal of Infectious Disease,* 187: 710–713.

Nanji, A.A., Khettry, V., Sadrzaden, S.M. (1994). *Lactobacillus* feeding reduces endotoxemia and severity of experimental alcoholic liver disease. *Proceedings Society Experimental Biology Medicine,* 205: 243–247.

Napoli, J., Bishop, G.A., McGuinness, P.H., Painter, D.M., McCaughan, G.W. (1996). Progressive liver injury in chronic hepatitis C infection correlates with increased intrahepatic expression of Th-1-associated cytokines. *Hepatology,* 24: 759–765.

Neumann-Haefelin, C., Blum, H.E., Chisari, F.U., Thimme, R. (2005). T cell response in hepatitis C virus infection. *Journal of Clinical Virology,* 32: 75–85.

Ni, H.T., Deeths, M.J., Mescher, M.F. (1999). LFA-1-mediated costimulation of CD8+ T cell proliferation requires phosphatidylinositol 3-kinase activity. *Journal of Immunology,* 166: 6523–6529.

Nolan, J.P. (1975). The role of endotoxin in liver injury. *Gastroenterology,* 69: 1345–1356.

Oesterreicher, C., Pfefel, F., Petermann, D., Muller, C. (1995). Increased in vitro production and serum levels of the soluble lipopolysaccharide receptor sCD14 in liver disease. *Journal of Hepatology,* 23: 396–402.

Opal, S.M., Scannon, P.J., Vincent, J.L., White, M., Caroll, S.E., Palardy, J.E., Parejo, N.A., Pribble, J.P., Lemke, J.H. (1999). Relationship between plasma levels of lipopolysaccharide (LPS) and LPS-binding protein in patients with severe sepsis and septic shock. *Journal of Infectious Disease,* 180: 1584–1589.

Poltorak, A., He, X., Smirnova, I., Liu, M.Y., VanHuffel, C., Du, X., Birdwell, D., Alejos, E., Silva, M., Galanos, C., Freudenberg, M. (1998). Defective LPS signaling in C3H/HeJ and C57BL/10ScCr mice: mutations in Tlr4 gene. *Science,* 282: 2085–2088.

Prytz, H., Holtz-Christensen, J., Korner, B., Liehr, H. (1976). Portal venous and systemic endotoxaemia in patients with cirrhosis. *Scandinavian Journal of Gastroenterology,* 11: 857–863.

Rehermann, B., Nascimbeni, M. (2005). Immunology of hepatitis B virus and hepatitis C virus infection. *Nature Review in Immunology,* 5: 215–229.

Sandberg, J.K., Ljunggren, H.-G. (2005). Development and function of CD1d-restricted NKT cells: influence of sphingolipids, SAP and sex. *Trends in Immunology,* 26: 347–350.

Sansonno, D., DeVita, S., Iacobelli, A.R., Cornacchiulo, V., Boiocchi, M., Dammacco, F. (1998). Clonal analysis of intrahepatic B cells from HCV-infected patients with and without mixed cryoglobulinemia. *Journal of Immunology,* 160: 3594–3601.

Sato, T., Asanuma, Y., Tanaka, J., Koyama, K. (1997). Inflammatory cytokine production enhancement in the presence of lipopolysaccharide after hepatic resection in cirrhotic patients. *Therapeutic Apheresis,* 1: 75–78.

Sato, T., Iwabuchi, K., Nagaoka, I., Adachi, Y., Ohno, N., Tamura, H., Seyama, K., Fukuchi, Y., Nakayama, H., Yoshizaki, F., Takamori, K., Ogawa, H. (2006). Induction of human neutrophil chemotaxis by *Candida albicans*-derived β-1,6-long glycoside-chain-branched β-glucan. *Journal of Leukocyte Biology,* 80: 204–211.

Seljelid, R., Figenschau, Y., Bogwald, J., Rasmussen, L.T., Austgulen, R. (1989). Evidence that tumor necrosis induced by aminated beta 1-3 D polyglucose is mediated by a concerted action of local and systemic cytokines. *Scandinavian Journal of Immunology,* 30: 687–694.

Sher, A., Oswald, I.P., Hieny, S., Gazzinelli, R.T. (1993). *Toxoplasma gondii* induces a T-independent IFN-gamma response in natural killer cells that requires both adherent accessory cells and tumor necrosis factor-α. *Journal of Immunology,* 150: 3982–3989.

Sherwood, E.R., Varma, T.K., Fram, R.Y., Lin, C.Y., Koutrouvelis, A.P., Toliver-Kinsky, T.E. (2001). Glucan phosphate potentiates endotoxin-induced interferon-gamma expression in immunocompetent mice, but attenuates induction of endotoxin tolerance. *Clinical Science*, 101: 541–550.

Shoukry, N.H., Sidney, J., Sette, A., Walker, C.M. (2004). Conserved hierarchy of helper T cell responses in a chimpanzee during primary and secondary hepatitis C virus infection. *Journal of Immunology*, 172: 483–492.

Snyder, J.T., Alexander-Miller, M.A., Berzofskyl, J.A., Belyakov, I.M. (2003). Molecular mechanism and biological significance of CTL avidity. *Current HIV Research*, 1: 287–294.

Soltys, J., Quinn, M.T. (1999). Modulation of endotoxin-and enterotoxin-induced cytokine release by *in vivo* treatment with β-(1,6)-branched β-(1,3)-glucan. *Infection and Immunity*, 67: 244–252.

Sutmuller, R.P.M., Morgan, M.E., Netea, M.G., Graver. O, Adema, G.J. (2006). Toll-like receptors on regulatory T cells: expanding immune regulation. *Trends in Immunology*, 27: 387–393.

Tannahil, C.L., Fukuzuka, K., Marum, T., Abouhamze, Z., MacKay, S.L., Copeland, E.M. III, Moldawer, L.L. (1999). Discordant tumor necrosis factor-alpha superfamily gene expression in bacterial peritonitis and endotoxemic shock. *Surgery*, 126: 349–357.

Tian, Z., Shen, X., Feng, H., Gao, B. (2000). IL-1 beta attenuates IFN-alpha/beta-induced antiviral activity and Stat1 activation in the liver. *Journal of Immunology*, 165: 3959–3965.

Trobonjaca, Z., Leithauser, F., Moller, P., Schirmbeck, R., Reimann, J. (2001). Activating immunity in the liver. I. Liver dendritic cells (but not hepatocytes) are potent activators of IFN-γ release by liver NKT cells. *Journal of Immunology*, 167: 1413–1422.

Tsutsui, H. Matsui, K. Okamura, H., Nakanishi, K. (2000). Pathophysiological roles of interleukin-18 in inflammatory liver diseases. *Immunological Reviews*, 174: 192–209.

Uhrig, A., Banafsche, R., Kremer, M., Hegenbarth, S., Hamann, A., Neurath, M., Gerken, G., Limmer, A., Knolle, P.A. (2005). Development and functional consequences of LPS tolerance in sinusoidal endothelial cells of the liver. *Journal of Leukocyte Biology*, 77: 626–633.

Underhill, D.M., Olinsky, A. (2002). Toll-like receptors: key mediators of microbe detection. *Current Opinion in Immunology*, 14: 103–110.

Urbani, S., Amadei, B., Cariani, E., Fisicaro, P., Orlandini, A., Missale, G., Ferrari, C. (2005). The impairment of CD8 responses limits the selection of escape mutations in acute hepatitis C virus infection. *Journal of Immunology*, 175: 7519–7529.

Van Andel, R.A., Hook, R.R., Jr., Franklin, C.L., Besch-Williford, C.L., Van Roijen, N., Riley, L.K. (1997). Effects of neutrophil, natural killer cell, and macrophage depletion on murine *Clostridium piliforme* infection. *Infection and Immunity*, 65: 2725–2731.

Van Molle, W., Denecker, G., Rodriguez, I., Brouckaert, P., Vandenabeele, P., Libert, C. (1999). Activation of caspases in lethal experimental hepatitis and prevention by acute phase proteins. *Journal of Immunology*, 163: 5235–5241.

Vento, S. (2000). Fulminant hepatitis associated with hepatitis A virus superinfection in patients with chronic hepatitis C. *Journal of Viral Hepatitis*, 7: 7–8.

Verdrengh, M., Tarkowsky, A. (1997). Role of neutrophils in experimental septicemia and septic arthritis induced by *Staphylococcus aureus*. *Infection and Immunity*, 65: 2517–2521.

Vodovotz, Y., Liu, S., McCloskey, C., Shapiro, R., Green, A., Billiar, T.R. (2001). The hepatocyte as a microbial product- responsive cell. *Journal of Endotoxin Research*, 7: 365–373.

Weunsch, S.A., Pierce, R.H., Crispe, N.I. (2006). Local intrahepatic CD8+ T cell activation by a non-self-antigen results in full functional differentiation. *Journal of Immunology*, 177: 1689–1697.

Williams, D.L., Ha, T., Li, C., Kalbfleish, J.A., Laffan, J.J., Ferguson, D.A. (1999). Inhibiting early activation of tissue nuclear factor-Kappa B and interleukin-6. (1→3)-Beta-D-Glucan increases long-term survival in polymicrobial sepsis. *Surgery*, 126: 54–65.

Williams, D.L., Ha, T., Li, C., Laffan, J.J., Kalbfleish, J.A., Browder, W. (2002). Inhibition of LPS-induced NFKAPPAB activation by a glucan ligand involves down-regulation of IKKBETA kinase activity and altered phosphorylation and degradation of IKKABALPHA. *Shock*, 13: 446–452.

Williams, D.L.A., Mueller, A., Browder, W. (1998). Glucan-based macrophage stimulators: a review of their anti-infective potential. *Clinical Immunotherapy,* 5: 392–399.

Williams, R.C., Jr., Malone, C.C., Kao, K.J. (1994). IgM rheumatoid factors react with human class I HLA molecules. *Journal of Immunology,* 156: 1684–1694.

Wright, S.D., Ramos, R.A., Tobias, P.S., Ulevitch, R.J., Mathison, J.C. (1990). CD14, a receptor for complexes of lipopolysaccharide (LPS) and LPS binding protein. *Science,* 249: 1431–1433.

Yin, M., Bradford, B.V., Wheeler, M.D., Uesugi, T., Froh, M., Goyert, S.M., Thurman, R.G. (2001). Reduced early alcohol-induced liver injury in CD14-deficient mice. *Journal of Immunology,* 166: 4737–4742.

Interferon-Induced Effector Proteins and Hepatitis C Virus Replication

Michael Frese* and Eva Dazert

Abstract *Hepatitis C virus* (HCV) is a small, enveloped RNA virus that is often capable of establishing a persistent infection, which may lead to chronic liver disease, cirrhosis, hepatocellular carcinoma, and eventually death. For more than 20 years, hepatitis C patients have been treated with interferon-alpha (IFN-α). Current treatment usually consists of polyethylene glycol-conjugated IFN-α that is combined with ribavirin, but even the most advanced IFN-based therapies are still ineffective in eliminating the virus from a large proportion of individuals. Therefore, a better understanding of the IFN-induced innate immune response is urgently needed. By using selectable self-replicating RNAs (replicons) and, more recently, recombinant full-length genomes, many groups have tried to elucidate the mechanism(s) by which IFNs inhibit HCV replication. This chapter attempts to summarize the current state of knowledge in this interesting field of HCV research.

Introduction

Interferons and the Antiviral State

Interferons (IFNs) are a diverse class of cytokines with key functions in the innate immune response to viruses (reviewed in Pestka et al., 2004; Goodbourn et al., 2000; Samuel, 2001). Three types of IFNs can be distinguished that have partially overlapping biological properties. Type I IFNs are secreted by most virus-infected cells and by a highly specialized leukocyte population, termed natural IFN-producing cells or plasmacytoid dendritic cells (Colonna et al., 2002). The human genome contains many type I IFN genes encoding 12 IFN-α subtypes, IFN-β, IFN-ε, IFN-κ, and IFN-ω. The reason why the human genome encodes so many IFN-α subtypes is

* Corresponding Author: Dr. Michael Frese, School of Health Sciences, University of Canberra, ACT 2601, Australia, Phone: +61 (0)2 6201 2243, Fax: +61 (0)2 6201 5727, E-mail: michael.frese@canberra.edu.au

not known but has been speculated that different subtypes elicit a slightly different antiviral response. Furthermore, it is tempting to speculate that the most recent multiplication of IFN-α genes is a consequence of an ongoing arms race between viruses that encode soluble IFN receptors and the innate immune defense system. Type I IFN genes differ from all other IFN genes by the fact that they lack introns.

Recently, three distantly related cytokines have been identified that share sequence similarities with type I IFNs and the interleukin-10 (IL-10) family. Accordingly, these cytokines have been named IFN-λ1, IFN-λ2, and IFN-λ3 (Kotenko et al., 2003) or IL-29, IL-28A, and IL-28B, respectively (Sheppard et al., 2003). Although it seems that many biological properties of this most recently discovered group of IFN-like cytokines resemble those of type I IFNs, they are referred to as type III IFNs.

In contrast to types I and III IFNs, of which numerous genes have been identified, the human genome contains only one type II IFN gene. The gene product, IFN-γ, is only expressed in specialized immune cells such as activated T lymphocytes and natural killer (NK) cells.

All types of IFNs bind to highly specific cell surface receptors that trigger the phosphorylation and nuclear translocation of a family of latent transcription factors, known as signal transducers and activators of transcription (STATs). Type I IFNs bind to the IFN-α receptor (IFNAR), which leads to the formation of the IFN-stimulated gene factor-3 (ISGF-3), a heterotrimer consisting of STAT1, STAT2, and IFN-response factor-9 (IRF-9/p48). ISGF-3 activates gene transcription via the IFN-stimulated response element (ISRE). Type III IFNs bind to a different receptor complex consisting of the IFNLR1/IL-28Rα subunit and the IL-10β subunit (Donnelly et al., 2004) but nevertheless trigger a signaling cascade that is very similar to that of type I IFNs (Doyle et al., 2006). A slightly different signaling pathway has been described for IFN-γ. The type II IFN binds to the IFN-γ receptor (IFNGR) which leads to the phosphorytation of the gamma activation factor (GAF), a phosphorylated STAT1 homodimer, is translocated to the nucleus, where it enhances gene expression by binding to the gamma activation site (GAS). Beside these well-established signaling pathways, alternative pathways have been described, but their contribution to the antiviral activity of IFN remains to be further elucidated (Pestka et al., 2004).

Types I and III IFNs are believed to execute their antiviral activities through the induction of proteins that accumulate inside an infected host cell. These effector proteins may interfere with distinct steps in viral replication or trigger the degradation of viral RNAs. By contrast, IFN-γ predominantly induces the expression of proteins with systemic functions, such as those involved in antigen processing and presentation. In addition, IFN-γ induces the expression and release of chemokines that activate and orchestrate the adaptive immune response (e.g., IP-10). However, at least in some virus infections, IFN-γ may also contribute to the establishment of an antiviral state by the induction of proteins with direct antiviral activities (reviewed in Guidotti & Chisari, 2001). Of note, all IFNs that have been tested so far inhibit hepatitis C virus (HCV) RNA replication in cultured cells, although differences have been noted in respect to the IC_{50} and the kinetics of inhibition.

patients does not contain type I IFNs (Lanford et al., 2003). This enigma has recently been solved by Foy and co-workers, who reported that the HCV NS3/4A protease interferes with the ability of cells to sense double-stranded RNA (Foy et al., 2003). NS3/4A-mediated cleavage of the adaptor protein TRIF reduces its abundance and inhibits poly(I)-poly(C)-activated signaling through the Toll-like receptor 3 pathway before its bifurcation to IRF-3 and nuclear factor-κB (NFκB)-mediated gene activation pathways (Li et al., 2005). Furthermore, NS3/4A cleaves the adapter protein MAVS/IPS-1/VISA/Cardif, which interrupts the signaling between the double-stranded RNA binding protein RIG-I and kinases that phosphorylate the IFN regulatory factors IRF-3 and IRF-7 (Meylan et al., 2005). This act of sabotage efficiently prevents the nuclear import of the latent transcription factors IRF-3 and IRF-7, a crucial step in the activation of type I IFN gene transcription (reviewed by Hiscott et al., 2006). Taken together, these findings suggest that HCV-infected hepatocytes are prevented from producing the amount of type I IFN that is needed to assist virus clearance. Nevertheless, type I IFN-induced mRNAs/proteins are readily detectable in liver biopsies of hepatitis C patients, even in liver samples that do not contain detectable amounts of type I IFN mRNAs (Mihm et al., 2004). This begs the question as to where the IFN that is not locally produced comes from. According to Mihm and co-workers, natural IFN-producing cells or plasmacytoid dendritic cells may represent an important extrahepatic source of IFN in hepatitis C patients (Mihm et al., 2004). However, natural IFN-producing cells have the propensity to migrate to secondary lymphoid organs rather than to sites of inflammation (Penna et al., 2002). As a consequence, the expression of type I IFN-induced proteins may never reach levels required to eliminate the virus from already infected cells and/or to prevent the infection of new host cells.

Recombinant Interferon as an Antiviral Agent

All currently licensed HCV therapies rely on the antiviral activity of type I IFN (mostly polyethylene glycol-conjugated IFN-α2) that is given alone or in combination with ribavirin. The administration of recombinant IFN bypasses the block of IFN production in HCV-infected host cells and dramatically increases the expression of type I IFN-induced proteins throughout the body. This leads in most cases to a rapid decline of HCV RNA levels (first-phase response), which is believed to reflect an inhibition of virus replication. Later on, HCV RNA levels decline more gradually (second-phase response) as the liver is cleared of virus-infected cells (Neumann et al., 1998; Layden & Layden, 2002). Although IFN-α initially reduces the viral load in almost all patients, a sustained response (as defined by the loss of detectable HCV RNA during therapy and its continued absence for at least 6 months after the treatment has been ended) is not experienced by all patients. Especially those patients who suffer from an infection with genotype 1b viruses often fail to eradicate the virus (Manns et al., 2001; Fried et al., 2002). The correlation between therapy success and the infecting genotype suggests the involvement of viral factors, but the underlying molecular mechanisms are not yet understood.

With the development of HCV replicons (Lohmann et al., 1999), it became possible to analyze the role of individual cytokines in the innate immune response against HCV. Because most patients respond, at least initially, to a treatment with IFN-α, it was not unexpected that this IFN and other type I IFNs also block RNA replication of different HCV genotypes in human hepatoma cells (Blight et al., 2000; Frese et al., 2001; Guo et al., 2001; Cheney et al., 2002; Larkin et al., 2003; Okuse et al., 2005; Windisch et al., 2005; Miyamoto et al., 2006) and in cells of nonhepatic origin, e.g., HeLa cells (Guo et al., 2003) and 293 cells (Ali et al., 2004). Similar results were obtained by using type III IFNs (Robek et al., 2005; Marcello et al., 2006) and IFN-γ (Cheney et al., 2002; Frese et al., 2002) but not other antivirally active cytokines such as TNF-α (Frese et al., 2003). The idea that IFN-γ enforces the critical first line of defense in the HCV-infected liver was further elaborated by Li and co-workers, who demonstrated in a co-culture experiment that NK cells block HCV replication in Huh-7 cells through the secretion of IFN-γ (Li et al., 2004). Clinical data are limited and it is still controversial discussed whether hepatitis C patients benefit from IFN-γ administrations. Nevertheless, it is interesting to note that types I and II IFNs inhibit HCV RNA replication in Huh-7 cells in a highly synergistic manner (Larkin et al., 2003; Okuse et al., 2005). Given the power of combination therapies in the treatment of other persistent virus infections, it might be rewarding to elucidate the mechanism(s) responsible for the observed synergistic antiviral effects of different IFN types. For example, do IFN-α and IFN-γ enhance the expression level of one or more effector proteins in a synergistic manner as suggested by Tan et al. (2005), or do they induce the expression of different, IFN type-specific effector proteins that interfere with more than one step of the HCV life cycle as suggested by Windisch et al. (2005)? The answers to these questions may help physicians to predict the outcome of IFN therapies and lead to the improvement of IFN-based therapies.

IFN-Induced Proteins That May Inhibit HCV Replication

General Remarks

Several attempts have been made to analyze systematically the IFN-induced changes in the gene expression of HCV host cells. In one approach, liver biopsy samples were taken from expetimentally infected chimpanzees and the gene expression profile was monitored by using cDNA microarrays. The results revealed that the infection of the liver rapidly leads to the upregulation of numerous genes including those encoding well-known IFN-induced effector proteins such as the chimpanzee homologue of MxA (Bigger et al., 2001; Su et al., 2002). In both studies, the expression of MxA and that of other type I IFN-induced proteins correlated with the magnitude and duration of the infection. However, transient and sustained viral clearances were rather associated with the production of IFN-γ and the subsequent expression of type II IFN-induced genes, suggesting a biphasic course of the innate immune response and a crucial role for IFN-γ in virus clearance.

In another approach, cDNA microarrays were used to analyze IFN-induced changes in the gene expression profile of cultured human cells containing HCV replicons (Zhu et al., 2003; Hayashi et al., 2005). Even if these and similar studies did not lead to the identification of the effector proteins that inhibit HCV replication in IFN-stimulated cells, they will guide present and future investigations by suggesting potential candidate genes. The contribution of some of the most prominent IFN-induced effector proteins (Figure 2) to the IFN-induced inhibition of HCV replication is discussed in the following paragraphs.

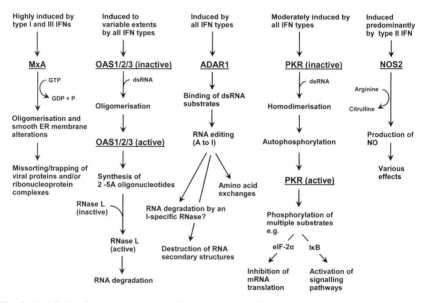

Fig. 2 Antiviral pathways that may contribute to the establishment of the so-called antiviral state in IFN-stimulated human cells. From left to right: The MxA GTPase inhibits viral replication by missorting and trapping of viral components into large membrane-associated complexes (the role of GTP hydrolysis in this process is not fully understood). Three different IFN-induced oligoadenylate synthetases (OAS1, OAS2, and OAS3) are encoded by the human genome. Binding to double-stranded RNA (dsRNA) leads to hetero- and/or homo-oligomerization and subsequently to the production of oligoadenylates with a 2',5'-phosphodiester bond linkage. These 2-5A oligonucleotides activate the latent endoribo-nuclease RNase L, which leads to the degradation of viral and cellular RNAs (in some cell types, the expression of RNase L is also regulated by IFNs). The p150 isoform of the adenosine deaminase ADAR1 binds to double-stranded RNA and catalyzes the conversion of adenosine to inosine (A to I). Such editing may occur selectivity at one or a few positions, or more frequently, at a large number of sites. Editing of viral RNAs may change the coding sequence, activate an I-specific RNase, and/or destroy RNA secondary structures by disrupting adenosine/uracil base pairings. The double-stranded RNA-activated protein kinase PKR may block viral protein translation by the phosphorylation and thereby inactivation of the eukaryotic initiation factor eIF2a. Furthermore, PKR may activate intracellular signaling pathways that contribute to the establishment of a robust antiviral response. The inducible nitric oxide synthetase NOS2 produces large amounts of nitric oxide (NO), which is implicated in a variety of immune functions such as the activation of macrophages

MxA

Mx proteins belong to the superfamily of dynamin-like large GTPases and their expression is tightly regulated by type I and type III IFNs (Holzinger et al, 2007; reviewed in Haller & Kochs, 2002; Haller et al, 2007). Of the two human Mx proteins, MxA and MxB, only MxA has demonstrable antiviral activity. MxA is a cytoplasmic protein with a size of $\sim$78 kDa that has been shown to inhibit the replication of a broad variety of RNA viruses. Cell culture experiments demonstrate that MxA inhibits orthomyxoviruses, bunyaviruses, rhabdoviruses, birnaviruses, reoviruses, and togaviruses (Mundt 2007 reviewed in Haller et al., 1999). In some cases, viral replication is almost completely blocked by MxA. For example, stably transfected Vero cells that constitutively express MxA produce up to 1,000,000-fold lower virus titers than control cells that did not express any Mx proteins (Frese et al., 1995). The antiviral effect of MxA has also been analyzed *in vivo* by using transgenic mice that constitutively express the human MxA protein but lack functional mouse Mx proteins and mice that constitutively express MxA but cannot mount a proper IFN-induced antiviral response due to a disruption in the gene for the β subunit of the IFN type I receptor. In both cases, MxA-expressing animals were found to be completely resistant to *Thogoto virus*, a tick-borne orthomyxovirus (Pavlovic et al., 1995; Hefti et al., 1999). Furthermore, MxA-expressing animals exhibited an enhanced resistance against … *Influenza A virus* (family *Orthomyxoviridae*), *Vesicular stomatitis virus* (VSV; family *Rhabdoviridae*), *LaCrosse virus* (family *Bunyaviridae*), and *Semliki Forest virus* (family *Togaviridae*) (Pavlovic et al., 1995; Hefti et al., 1999). Other reports suggest that MxA has an even wider antiviral activity, but the supporting data are less convincing.

The *modus operandi* of MxA is not completely understood, but accumulating data indicate that cytoplasmic Mx proteins missort and immobilize viral components. In cells that had been infected with *LaCrosse virus*, MxA binds and translocates the viral nucleocapsid protein into membrane-associated perinuclear complexes (Kochs et al., 2002; Reichelt et al., 2004). A similar phenomenon was observed in cells that had been infected with *Thogoto virus*. In this case, however, MxA inhibited the nuclear transport of incoming viral nucleocapsids (Kochs & Haller, 1999), thereby preventing primary transcription and leading to an early and very efficient block of virus replication.

In healthy individuals, MxA expression is below the detection limit, but expression levels increase dramatically during many viral infections and as a consequence of IFN-α treatment (Roers et al., 1994; Chieux et al., 1998). MxA mRNA quantification in peripheral blood mononuclear cells has even been used to monitor the bioavailability of administered type I IFNs in hepatitis C patients (Gilli et al., 2002; Jorns et al., 2006). Not surprisingly, elevated MxA expression levels have also been found in the liver of chronic hepatitis C patients, indicating an ongoing struggle between the innate immune system and HCV (MacQuillan et al., 2002, 2003; Patzwahl et al., 2001).

A genetic study from Japan addressing a single nucleotide polymorphism at position -88 in the promoter sequence of the MxA gene revealed that a thymidine (T) in that position favors a sustained response of hepatitis C patients to treatment with

IFN-α, whereas a guanosine (G) is more frequently found among nonresponders (Hijikata et al., 2000). Interestingly, a T at that position increases the homology of the first ISRE in the MxA promoter to the ISRE consensus sequence (Hijikata et al., 2000). Furthermore, experiments with reporter constructs suggest that the T allele has a higher transcriptional activity than the G allele when stimulated with IFN-α (Hijikata et al., 2001). A similar association between the G/T single nucleotide polymorphism at position -88 of the MxA gene and the response of hepatitis C patients to IFN-α therapy was found in a European study, in which the T genotype was also found to be associated with the ability to clear HCV naturally without the help of recombinant IFN (Knapp et al., 2003).

Since MxA has the ability to efficiently inhibit a variety of different RNA viruses, MxA was the first IFN-induced effector protein to be analyzed for its antiviral activity in the HCV replicon system (Frese et al., 2001). However, no evidence was found for an involvement of MxA in the IFN-induced inhibition of HCV RNA replication. The constitutive expression of MxA did not inhibit subgenomic HCV replicons, and the expression of a dominant-negative mutant of MxA did not restore HCV RNA replication during IFN-α treatment (Frese et al., 2001). These earlier observations are in line with the more recent finding that IFN-α inhibits HCV RNA replication in Huh-7 cells and HuH6 cells with a similar IC_{50} (3 to 5 IU/ml and 5 to 10 IU/ml, respectively), although the former produce nearly 75-fold more MxA mRNAs than the latter (Windisch et al., 2005). Taken together, the data indicate that IFN-α inhibits HCV RNA replication by MxA-independent pathways.

Most recently, it has been noted that brefeldin A, a Golgi apparatus disrupting agent, renders the replication of *Kunjin virus* susceptible to MxA (Hoemen et al., 2007). Since *Kunjin virus* and HCV are both flaviviruses that use host cell-derived membranes to establish replication factories, it is tempting to speculate whether a disruption of the membranous web in HCV-infected cells would expose HCV RNA-protein complexes to antivirally active proteins such as MxA.

OAS/RNase L

The OAS/RNase L pathway (also known as IFN-inducible 2–5A response) requires two types of enzymes, an oligoadenylate synthetase and a ribonuclease (reviewed in Samuel, 2001). The human genome contains four gene loci that encode IFN-induced oligoadenylate synthetases (OAS1, OAS2, and OAS3) and an OAS-like protein. This and alternative splicing leads to the expression of numerous isoforms with sizes ranging from 40 to 100 kDa (Rebouillat & Hovanessian, 1999). OAS protein expression is enhanced in response to most, if not all, IFNs, but the magnitude of induction can vary dramatically with different IFNs and the type of the producing cell. Newly produced OAS proteins are believed to be inactive, but binding to double-stranded RNA leads to their oligomerization and starts the production of oligoadenylates with a $2',5'$-phosphodiester bond linkage (2-5A oligonucleotides). These oligonucleotides bind to and activate the latent ribonuclease RNase L, a process that is associated with the formation of stable RNase L homodimers. Once activated, RNase L can degrade single-stranded RNAs of viral and cellular origin. Cleaving of target RNAs occurs preferentially on the $3'$ side of uracil-adenosine (UA) and UU dinucleotides (Floyd-Smith et al., 1981; Wreschner et al., 1981). The cleavage of mRNA and rRNA may trigger a general protein shut-off in virus-infected cells, thereby limiting virus replication and spread. Different OAS proteins are associated with different cellular compartments, vary with respect to the amount of double-stranded RNA needed for activation, and produce 2-5A oligonucleotides of different sizes (Samuel, 2001). It is therefore tempting to speculate that the diversity of OAS proteins and isoforms evolved to fight a rather wide spectrum of DNA and RNA viruses including poxviruses, reoviruses, and picornaviruses. Members of the family *Picornaviridae* seem to be especially sensitive to the OAS/RNase L pathway. For example, overexpression of the 40-kDa form of the human OAS1 protein confers resistance to *Mengovirus* but not VSV (Chebath et al., 1987), and the constitutive expression of the 69-kDa form of the OAS2 protein inhibited the replication of *Encephalomyocarditis virus* but not that of VSV, *Sendai virus*, and a reovirus (Ghosh et al., 2000). In this context it is interesting to note that one of the *Oas* gene loci has recently been identified to confer increased resistance to the *West Nile virus* (family *Flaviviridae*) in laboratory mice (Perelygin et al., 2002) and that the transcript of the *Oas1b* allele in susceptible mice contains a premature stop codon, which results in a truncated protein (Mashimo et al., 2002). Experiments with congenic mice and cells derived from those mice revealed that expression of the full-length OAS1 protein limited virus production *in vivo* and in cell culture. Surprisingly, however, RNase L activity was highest in susceptible cells, and downregulation of RNase L activity in resistant cells did not restore virus titers to levels observed in susceptible cells (Scherbik et al., 2006).

As with many other IFN-induced proteins, OAS protein expression is slightly upregulated in hepatitis C patients (MacQuillan et al., 2003) and further enhanced in response to the administration of recombinant IFN-α (Murashima et al., 2000). Rather indirect evidence that the OAS/RNase L pathway may indeed target HCV replication/translation was recently provided by Taguchi and co-workers, who reported that the N-terminal portion of NS5A (amino acids 1 to 148), which lacks the so-called PKR-binding domain, binds to OAS proteins and there by counteract the antiviral activity of IFN-α (Taguchi et al., 2004). Furthermore, it has been demonstrated that purified recombinant RNAse L and that from HeLa cell extracts efficiently cleaves HCV RNA *in vitro* (Han et al., 2004). However, further investigations are needed to determine whether RNase L also cleaves HCV RNAs in infected HCV host cells and to what extent an OAS-induced block of HCV protein translation contributes to the IFN-induced inhibition of HCV RNA replication.

ADAR1

ADAR1 forms together with ADAR2 and the less extensively studied ADAR3 protein a small family of constitutively expressed adenosine deaminases that act on

RNA (reviewed in Valente & Nishikura, 2005; Toth et al., 2006). ADAR1 and ADAR2 bind highly structured RNAs and catalyze the hydrolytic C6 deamination of adenosine, a reaction that converts adenosine to inosine (A to I editing). ADAR-mediated editing may occur selectively at one or a few positions or, more frequently, at a larger number of sites (hyperediting or hypermutation). A to I exchanges may have severe consequences: (1) editing of coding sequences may lead to amino acid exchanges because I is recognized as G by the translational machinery (of note, A to I editing does not create stop codons); (2) editing of noncoding regions may affect RNA splicing, stability, or translational efficiency (e.g., by disrupting AU base pairs); (3) editing may regulate gene silencing (e.g., by disrupting AU base pairs); and (4) hyperedited RNA may be recognized and cleaved by an I-specific RNase (Scadden & Smith, 1997, 2001).

A prominent example of a cellular RNA that is edited by ADAR proteins is the mRNA of the alpha-amino-3-hydroxy-5-methyl-4-isoxazole propionate (AMPA) receptor subunit GluR-2. ADAR2 edits a codon in exon 11, which results in an amino acid exchange that changes the Ca^{2+} permeability of the receptor. This highly specific editing event has far-reaching consequences. ADAR2 knockout mice are prone to seizures and die young. The impaired phenotype appears to result entirely from a single underedited position in the GluR-2 mRNA, as it reverted to normal when both alleles for the underedited transcript were substituted with alleles encoding the edited version exonically (Higuchi et al., 2000). Likewise, genetic targeting of the *Adar1* locus revealed an essential requirement for this ADAR protein in the embryogenesis of mice (Wang et al., 2000, 2004; Hartner et al., 2004), and it has been suggested that its expression protects against stress-induced apoptosis (Wang et al., 2004).

A closer look at the *Adar1* gene locus revealed that protein expression is controlled by three promoters and alternative splicing (reviewed in Toth et al., 2006). Two constitutively active promoters drive the expression of a ~110-kDa protein (p110), whereas an IFN-regulated promoter with an ISRE controls the expression of a larger isoform (p150) that is expressed in response to inflammation or IFN treatment (Patterson & Samuel, 1995; George & Samuel, 1999; Yang et al., 2003a, 2003b). Both isoforms contain multiple nuclear localization signals, but only the IFN-induced p150 isoform has a nuclear export signal. Accordingly, p110 is a nuclear protein and p150 has been detected in both nuclear and cytoplasmic compartments.

Hypermutation of viral RNAs has been observed for several RNA viruses, including *Measles virus, Parainfluenza virus 3* (both family *Paramyxoviridae*), and VSV (O'Hara et al., 1984; Cattaneo et al., 1988; Murphy et al., 1991). It has been speculated that subacute sclerosing panencephalitis (SSPE), a fatal necropathic response in patients with a persistent measles virus infection of the brain, is associated with extensive editing of the matrix protein mRNA. This prevents virion assembly and release because these steps in the viral life cycle require a functional matrix protein. Other transcripts, however, are less frequently edited, which is thought to result in a persistent virus replication (Cattaneo et al., 1989; Baczko et al., 1993). Thus, an incomplete ADAR-mediated innate immune response might contribute to the pathology of SSPE.

Interestingly, certain viruses abuse ADAR proteins to control important check-points in replication and particle formation. A well-known example is the *Hepatitis D virus* (HDV), a subviral human pathogen that depends on *Hepatitis B virus* as a helper virus (reviewed in Casey, 2006). HDV has a small, circular RNA genome that encodes only a single protein, the hepatitis delta antigen (HDAg). Without editing, a 195-amino acid version of HDAg is made that is essential for virus replication (Kuo et al., 1989). Later on, in the viral life cycle a highly specific A to I editing event changes a UAG amber stop codon to an UIG tryptophan codon, and a 214-amino acid HDAg-L is produced that mediates genome packaging (Chang et al., 1991).

HCV RNAs may also be subject to ADAR-mediated modifications, but in this case, editing seems to be less specific and to inhibit virus replication. It has recently been reported that the silencing of ADAR1 expression in HCV replicon cells increases the amount of HCV RNA about 40-fold (Taylor et al., 2005). Moreover, Taylor and co-workers noted that IFN-α increases the frequency of A to G mutations in subgenomic replicon RNAs and that the transfection of ADAR-specific siRNAs rescues HCV RNA replication in the presence of moderate IFN-α concentrations. Based on these findings, Taylor et al. concluded that IFN-α inhibits HCV replication through ADAR-mediated hyperediting of viral RNA. In our laboratory, we have used specific antibodies to determine the intracellular localization of p150 and — despite its nuclear export signal —, we observed that p150 accumulates predom-inantly in the nucleus of IFN-treated Huh-7 cells. In the presence of subgenomic or full-length HCV RNAs, however, we observed that p150 localizes to distinct cytoplasmic structures (E. Dazert, R. Bartenschlager, and M. Frese, unpublished results). We also analyzed the antiviral effect of constitutively expressed p150 on HCV replication and found that the overexpression of p150 in Huh-7 cells did not block HCV RNA replication. Taken together, our findings support the idea of Taylor et al. that p150 interacts with HCV RNAs, but we argue that p150 does that only in the context off other IFN-induced proteins. Further studies are under way to fully elucidate the role of p150 in the IFN-induced inhibition of HCV replication.

PKR

Another prominent protein of the innate immune defense that has long been sus-pected of interfering with HCV replication is the double-stranded RNA-activated protein kinase PKR. This serine/threonine kinase is constitutively expressed and has multiple functions in the control of host cell transcription and translation (reviewed in Garcia et al., 2006). Upon stimulation with IFNs, most cells respond by increas-ing the expression of PKR. IFN-induced PKR accumulates in the cytoplasm and was found in association with ribosomes (Thomis et al., 1992). PKR may exert its antivi-ral activity through different pathways (reviewed in Toth et al., 2006). First, PKR is able to control the cellular translation machinery through phosphorylation of the α subunit of the eukaryotic translation initiation factor eIF-2α, which would affect the production of both host and virus proteins. Second, PKR-mediated phosphoryla-tion is implicated in several signaling pathways that contribute to the establishment of a robust antiviral response. For example, PKR has been shown to activate the latent transcription factor NFκB, which may lead to the enhanced expression of pro-inflammatory genes (Gil et al., 2004). In addition, PKR may activate other kinases such as the p38 mitogen-activated protein (MAP) kinase, which further intensifies and diversifies the innate immune response (Goh et al., 2000).

The concept that PKR-mediated phosphorylation events play an important role in the innate immune response against viral infections is largely based on the fact that many RNA and DNA viruses try to inhibit PKR by (1) overexpressing small RNAs that bind to but do not activate PKR, (2) producing eIF-2α decoys, and (3) enhancing PKR degradation (reviewed in Langland et al., 2006). If PKR is a key player in IFN-induced antiviral defense, genetically targeted knockout mice that lack functional PKR proteins should be extremely sensitive to viral infections. Two lines of PKR$^{-/-}$ mice have been generated in which the coding sequences of either the N-terminal or the C-terminal part of the protein have been disrupted (Yang et al., 1995; Abraham et al., 1999, respectively). PKR$^{-/-}$ mice are indeed more susceptible to certain virus infections than wild-type animals (Balachandran et al., 2000; Stojdl et al., 2000; Carr et al., 2006; Samuel et al., 2006), but at least in some cases, this seems to depend on the mouse strain used, and other experimental conditions (Murphy et al., 2003). Additional experiments have been conducted by using MEFs from PKR$^{-/-}$ mice, but a direct antiviral activity of PKR (e.g., the inhibition of virus multiplication by blocking protein translation) is still controversial. Interestingly, priming of PKR$^{-/-}$ mice with poly(I)-poly(C) or IFNs before the virus challenge points to a rather indirect mode of PKR action, such as the enhancement of double-stranded RNA-induced signaling events (Yang et al., 1995). However, it should be noted that most of these experiments have been performed with viruses that encode PKR inhibitors. It would be interesting to re-evaluate the phenotype of PKR$^{-/-}$ mice with genetically modified viruses that cannot express functional PKR inhibitors. Another problem in the characterization of PKR$^{-/-}$ mice is the presence of related kinases that also phosphorylate eIF-2α (Toth et al., 2006). Even if these kinases differ from PKR in their response to double-stranded RNA and/or other physiological stress signals, they may partially substitute for the lack of PKR in PKR$^{-/-}$ mice, thereby making it difficult to quantify the contribution of PKR to the innate immune response (as exemplified in Smith et al., 2005).

Two HCV proteins have been described as interacting with the kinase. By analyzing HCV sequences from Japanese hepatitis C patients, mutations within a discrete region of NS5A, the so-called IFN sensitivity determining region (ISDR), were proposed to confer resistance to IFN-α (Enomoto et al., 1995, 1996). Since the original reports by Enomoto and co-workers, numerous studies have been conducted in Japan as well as in other countries to determine the predictive value of NS5A sequences in the outcome of IFN-based therapies, but the existence of an ISDR is still controversial (reviewed in Tan & Katze, 2001; reinvestigated by Pascu et al., 2004; Brillet et al., 2007). Whether or not an ISDR really exists, the description of such a sequence put NS5A in the focus of HCV research. The subsequent finding that mutations in the ISDR affect the ability of NS5A to bind to and inhibit PKR (Gale et al., 1998) led to the hypothesis that PKR blocks HCV replication and that NS5A is able to counteract the antiviral activity of PKR. However, experiments with HCV replicons provided no further evidence for an involvement of NS5A in IFN resistance. On the contrary, point mutations within the ISDR or a deletion of 47 amino acids encompassing the entire ISDR enhanced viral replication without affecting the IFN sensitivity of HCV replicons (Blight et al., 2000; Guo et al., 2001). These findings were extended by A. Kaul and R. Bartenschlager, who analyzed the function of NS5A by using two subgenomic genotype 1b replicons that differ only in the NS5A coding sequence. In one replicon, the ISDR was identical to that of IFN-susceptible strains, whereas the ISDR sequence of the other replicon contained mutations that have been suspected to confer PKR binding and IFN resistance (Gale et al., 1998). Despite these differences, both replicons were found to be equally sensitive to IFN-α (unpublished results). This result argues against the hypothesis that NS5A counteracts an IFN-induced and PKR-mediated block of viral protein translation. However, the result does not contradict the idea that NS5A inhibits other activities of PKR (e.g., a PKR-mediated priming of intracellular signaling pathways). Of note, several reports suggest that NS5A may sabotage the innate immune response through PKR-independent activities (discussed in MacDonald & Harris, 2004). For example, it has been reported that NS5A increases the production of interleukin (IL)-8, thereby attenuating the antiviral properties of IFNs (Polyak et al., 2001a, 2001b). It would be interesting to study the immunomodulatory activities of NS5A in an immunocompetent small animal model, which might finally put an end to the discussion about the role of PKR in HCV pathology.

A second HCV protein has been reported to interact with PKR. It was found that E2 binds to PKR through its PKR-eIF2α homology domain (PePHD) (Taylor et al., 1999, 2001; Pavio et al., 2002). However, the significance of this observation has been questioned because an increasing number of clinical studies demonstrate that the PePHD is a highly conserved region with no conspicuous mutations accumulating during IFN-α therapy (reviewed in Tan & Katze, 2001). Furthermore, E2 expression does not increase the resistance of HCV genotype 1b replicons toward IFNs. A genomic replicon that encodes an E2 protein with the PePHD sequence of a resistant HCV isolate had a similar degree of susceptibility as a subgenomic replicon lacking E2 (Frese et al., 2002; A. Kaul and R. Bartenschlager, unpublished results).

The adenovirus-associated RNA I (VA$_I$), a small, highly structured RNA that binds to PKR and but does not trigger its dimerization and activation, has recently been found to stimulate HCV RNA replication in the replicon system (Taylor et al., 2005). It was also reported that recombinant VA$_I$ RNA efficiently rescues HCV RNA replication in the presence of as much as 500 IU/ml of IFN-α (Taylor et al., 2005). Since VA$_I$ RNAs may also bind to other proteins of the innate immune response such as ADAR1, more research is needed to define the role of PKR in limiting HCV protein translation.

A more direct approach to the question of PKR interference with HCV RNA replication/translation has recently been undertaken by using RNA silencing. A. Kaul and R. Bartenschlager transfected cells containing subgenomic HCV replicons with siRNAs that target PKR mRNAs for degradation and subsequently treated the cells with different concentrations of IFN-α. In no case did they observe that a downregulation of PKR expression levels results in a restoration of HCV replication in the presence of IFN (unpublished results).

With the establishment of a new generation of HCV replicons that contain the consensus sequence from a Japanese genotype 2a isolate and replicate efficiently without the need for adaptive mutations, it became possible to study HCV RNA replication in a variety of new host cells including those of nonhepatic and non-human origin (Kato et al., 2005; Uprichard et al., 2006). Most recently, genotype 2a replicons were employed by Chang and co-workers, who set out to analyze the antiviral effect of type I IFNs on HCV replication in MEFs from $PKR^{-/-}$ mice and congenic wild-type mice. Interestingly, IFN-α as well as IFN-β inhibited HCV RNA replication in $PKR^{-/-}$ MEFs as efficiently as in $PKR^{+/+}$ MEFs (Chang et al., 2006), suggesting that PKR-mediated translational control plays only a minor role in the IFN-induced inhibition of HCV RNA replication.

NOS2 and Other Effector Proteins

The inducible nitric oxide (NO) synthetase, originally named iNOS but also abbreviated as NOS2, belongs to a small family of NO-producing enzymes. In unstimulated cells, NOS2 is virtually absent, but expression levels increase rapidly in response to pro-inflammatory cytokines, especially IFN-γ. The expression of NOS2 results in a long-lasting production of NO (Karupiah et al., 2000). The NO free radical has been recognized for its strong antimicrobial activity against various protozoa, bacteria, and viruses. For example, the replication of a coxsackievirus is suppressed by NO through inactivation of the viral cysteine protease by S-nitrosylation (Saura et al., 1999). Furthermore, NO production is essential for the T cell-mediated noncytopathic inhibition of *Hepatitis B virus* replication in virus-transgenic mice (Guidotti et al., 2000). Other viruses that have been reported as sensitive to NO include *Severe acute respiratory coronavirus* (Akerstrom et al., 2005), *Respiratory syncytial virus* (Stark et al., 2005), *Mouse hepatitis virus* (Pope et al., 1998), and *Herpes simplex virus type 1* (Adler et al., 1997). However, the use of NO by the infected host as an antimicrobial substance is a double-edged sword. NO-induced oxidative stress may cause severe cellular and organ dysfunction. Influenza A virus-infected wild-type mice, for example, suffer from an excessive production of NO in the lungs, which often leads to respiratory failure and death, whereas knockout mice that cannot express functional NOS2 survive the infection with little evidence of pneumonitis (Akaike et al., 1996; Karupiah et al., 1998).

In the liver of most HCV-infected individuals, NOS2 is easily detectable (Mihm et al., 1997; Majano et al., 1998; Schweyer et al., 2000). The enhanced expression of NOS2 in the liver of hepatitis C patients has largely been attributed to IFN-γ that is released by resident and infiltrating immune cells. In addition, it has been speculated that the HCV replication itself may stimulate the expression of NOS2 in infected hepatocytes (Machida et al., 2004). Based on genetic studies, it has been suggested that the production of NO is involved in HCV clearance, as certain NOS2 haplotypes were more frequently found among HCV-infected individuals who spontaneously cleared the infection and in IFN-treated hepatitis C patients who could mount a sustained antiviral response (Yee et al., 2004). This hypothesis, however, lacks supporting evidence from cell culture experiments. The treatment of Huh-7 cells with the NO donor (Z)-1-[2-(aminoethyl)-N-(2-ammonioethyl)-amino]diazen-1-ium-1,2-diolate (DETA NONOate) or the arginase inhibitor NG-hydroxy-L-arginine (NOHA) did not result in an inhibition of HCV RNA replication (Frese et al., 2002). Furthermore, the NOS inhibitor L-N6-(1-iminoethyl)-lysine (L-NIL) did not even partially restore HCV replication in the presence of IFN-γ (Frese et al., 2002). One should, however, not overinterpret these results. The *in vivo* production of NO may have more complex consequences than those that could be investigated in cell culture. NO may act as a messenger rather than as an effector molecule, or NO may induce DNA damage and apoptosis (Jaiswal et al., 2000). Thus, further studies are needed to fully elucidate the role of NO in hepatitis C pathology.

So far, only a few further IFN-induced effector proteins have been investigated with respect to their potential to inhibit HCV RNA replication, most notably indoleamine 2,3-dioxygenase (IDO). The IDO-mediated depletion of tryptophan is well known as a defense mechanism against certain intracellular parasites (Carlin et al., 1989). Rather recently, this pathway has also been recognized as an IFN-γ-induced antiviral defense mechanism against herpesviruses (Bodaghi et al., 1999) and poxviruses (Terajima & Leporati, 2005). If IDO inhibits HCV replication as well, inhibition of the effector protein or addition of tryptophan to the cell culture medium should restore viral protein synthesis. However, the IDO inhibitor α-methyl-DL-tryptophan did not restore the replication of subgenomic HCV replicons in the presence of IFN-γ (Frese et al., 2002). Likewise, increased concentrations of L-tryptophan could not rescue the viral protein synthesis (Frese et al., 2002), suggesting that the depletion of tryptophan is not—or is not the only—mechanism by which IFN-γ inhibits the replication of HCV RNAs.

Beside the induction of direct antiviral activities in infected host cells, IFNs may enhance and direct the activities of NK cells and T cells. Such indirect effects are usually ascribed to IFN-γ, but Shin and co-workers noted that type I IFNs also stimulate the generation of immunoproteasomes in the liver of hepatitis C patients (Shin et al., 2006). It would be interesting to determine the extent to which these indirect effects contribute to the antiviral activity of type I IFNs.

Acknowledgments We are indebted to Kerry Mills, Sandra Thomas, Ali Zaid, Friedemann Weber, Brett Lidbury and Ralf Bartenschlager for helpful suggestions and careful reading of the manuscript; and Artur Kaul and R. Bartenschlager for the communication of unpublished results.

References

Abraham, N., Stojdl, D.F., Duncan, P.I., Methot, N., Ishii, T., Dube, M., Vanderhyden, B.C., Atkins, H.L., Gray, D.A., McBurney, M.W., Koromilas, A.E., Brown, E.G., Sonenberg, N., Bell, J.C. (1999). Characterization of transgenic mice with targeted disruption of the catalytic domain of the double-stranded RNA-dependent protein kinase, PKR. *Journal of Biological Chemistry*, 274: 5953–5962.

Adler, H., Beland, J.L., Del-Pan, N.C., Kobzik, L., Brewer, J.P., Martin, T.R., Rimm, I.J. (1997). Suppression of herpes simplex virus type 1 (HSV-1)-induced pneumonia in mice by inhibition of inducible nitric oxide synthase (iNOS, NOS2). *Journal of Experimental Medicine*, 185: 1533–1540.

Akaike, T., Noguchi, Y., Ijiri, S., Setoguchi, K., Suga, M., Zheng, Y.M., Dietzschold, B., Maeda, H. (1996). Pathogenesis of influenza virus-induced pneumonia: involvement of both nitric oxide and oxygen radicals. *Proceedings of the National Academy of Sciences of the United States of America*, 93: 2448–2453.

Akerstrom, S., Mousavi-Jazi, M., Klingstrom, J., Leijon, M., Lundkvist, A., Mirazimi, A. (2005). Nitric oxide inhibits the replication cycle of severe acute respiratory syndrome coronavirus. *Journal of Virology*, 79: 1966–1969.

Ali, S., Pellerin, C., Lamarre, D., Kukolj, G. (2004). Hepatitis C virus subgenomic replicons in the human embryonic kidney 293 cell line. *Journal of Virology*, 78: 491–501.

Balachandran, S., Roberts, P.C., Brown, L.E., Truong, H., Pattnaik, A.K., Archer, D.R., Barber, G.N. (2000). Essential role for the dsRNA-dependent protein kinase PKR in innate immunity to viral infection. *Immunity*, 13: 129–141.

Bartenschlager, R., Frese, M., Pietschmann, T. (2004). Novel insights into hepatitis C virus replication and persistence. *Advances in Virus Research*, 63: 71–180.

Baczko, K., Lampe, J., Liebert, U.G., Brinckmann, U., ter Meulen, V., Pardowitz, I., Budka, H., Cosby, S.L., Isserte, S., Rima, B.K. (1993). Clonal expansion of hypermutated measles virus in an SSPE brain. *Virology*, 197: 188–195.

Bigger, C.B., Brasky, K.M., Lanford, R.E. (2001). DNA microarray analysis of chimpanzee liver during acute resolving hepatitis C virus infection. *Journal of Virology*, 75: 7059–7066.

Blight, K.J., Rice, C.M. (1997). Secondary structure determination of the conserved 98-base sequence at the 3′ terminus of hepatitis C virus genome RNA. *Journal of Virology*, 71: 7345–7352.

Blight, K.J., Kolykhalov, A.A., Rice, C.M. (2000). Efficient initiation of HCV RNA replication in cell culture. *Science*, 290: 1972–1974.

Bodaghi, B., Goureau, O., Zipeto, D., Laurent, L., Virelizier, J.-L., Michelson, S. (1999). Role of IFN-induced indoleamine 2,3 dioxygenase and inducible nitric oxide synthase in the replication of human cytomegalovirus in retinal pigment epithelial cells. *Journal of Immunology*, 162: 957–964.

Brillet, R., Penin, F., Hezode, C., Chouteau, P., Dhumeaux, D., Pawlotsky, J.M. (2007). The nonstructural 5A protein of hepatitis C virus genotype 1b does not contain an interferon sensitivity-determining region. *Journal of Infectious Diseases*, 195: 432–441.

Carlin, J.M., Ozaki, Y., Byrne, G.I., Brown, R.R., Borden, E.C. (1989). Interferons and indoleamine 2,3-dioxygenase: role in antimicrobial and antitumor effects. *Experientia*, 45: 535–541.

Carr, D.J., Wuest, T., Tomanek, L., Silverman, R.H., Williams, B.R. (2006). The lack of RNA-dependent protein kinase enhances susceptibility of mice to genital herpes simplex virus type 2 infection. *Immunology*, 118: 520–526.

Casey, J.L. (2006). RNA editing in hepatitis delta virus. *Current Topics in Microbiology and Immunology*, 307: 67–89.

Cattaneo, R., Schmid, A., Eschle, D., Baczko, K., ter Meulen, V., Billeter, M.A. (1988). Biased hypermutation and other genetic changes in defective measles viruses in human brain infections. *Cell*, 55: 255–265.

Cattaneo, R., Schmid, A., Spielhofer, P., Kaelin, K., Baczko, K., ter Meulen, V., Pardowitz, J., Flanagan, S., Rima, B.K., Udem, S.A., Billeter, B.A. (1989). Mutated and hypermutated genes of persistent measles viruses which caused lethal human brain diseases. *Virology*, 173: 415–425.

Chang, F.L., Chen, P.J., Tu, S.J., Wang, C.J., Chen, D.S. (1991). The large form of hepatitis delta antigen is crucial for assembly of hepatitis delta virus. *Proceedings of the National Academy of Sciences of the United States of America*, 88: 8490–8494.

Chang, K.S., Cai, Z., Zhang, C., Sen, G.C., Williams, B.R., Luo, G. (2006). Replication of hepatitis C virus (HCV) RNA in mouse embryonic fibroblasts: protein kinase R (PKR)-dependent and

PKR-independent mechanisms for controlling HCV RNA replication and mediating interferon activities. *Journal of Virology,* 80: 7364–7374.

Chebath, J., Benech, P., Revel, M., Vigneron, M. (1987). Constitutive expression of $(2'-5')$ oligo A synthetase confers resistance to picornavirus infection. *Nature,* 330: 587–588.

Cheney, I.W., Lai, V.C., Zhong, W., Brodhag, T., Dempsey, S., Lim, C., Hong, Z, Lau, J.Y., Tam, R.C. (2002). Comparative analysis of anti-hepatitis C virus activity and gene expression mediated by alpha, beta, and gamma interferons. *Journal of Virology,* 76: 11148–11154.

Chieux, V., Hober, D., Harvey, J., Lion, G., Lucidarme, D., Forzy, G., Duhamel, M., Cousin, J., Ducoulombier, H., Wattre, P. (1998). The MxA protein levels in whole blood lysates of patients with various viral infections. *Journal of Virological Methods,* 70: 183–191.

Colonna, M., Krug, A., Cella, M. (2002). Interferon-producing cells: on the front line in immune responses against pathogens. *Current Opinion in Immunology,* 14: 373–379.

Donnelly, R.P., Sheikh, F., Kotenko, S.V., Dickensheets, H. (2004). The expanded family of class II cytokines that share the IL-10 receptor-2 (IL-10R2) chain. *Journal of Leukocyte Biology,* 76: 314–321.

Doyle, S.E., Schreckhise, H., Khuu-Duong, K., Henderson, K., Rosler, R., Storey, H., Yao, L., Liu, H., Barahmandpour, F., Sivakumar, P., Chan, C., Birks, C., Foster, D., Clegg, C.H., Wietzke-Braun, P., Mihm, S., Klucher, K.M. (2006). Interleukin-29 uses a type 1 interferon-like program to promote antiviral responses in human hepatocytes. *Hepatology,* 44: 896–906.

Enomoto, N., Sakuma, I., Asahina, Y., Kurosaki, M., Murakami, T., Yamamoto, C., Izumi, N., Marumo, F., Sato, C. (1995). Comparison of full-length sequences of interferon-sensitive and resistant hepatitis C virus 1b. Sensitivity to interferon is conferred by amino acid substitutions in the NS5A region. *Journal of Clinical Investigation,* 96: 224–230.

Enomoto, N., Sakuma, I., Asahina, Y., Kurosaki, M., Murakami, T., Yamamoto, C., Ogura, Y., Izumi, N., Marumo, F., Sato, C. (1996). Mutations in the nonstructural protein 5A gene and response to interferon in patients with chronic hepatitis C virus 1b infection. *New England Journal of Medicine,* 334: 77–81.

Floyd-Smith, G., Slattery, E., Lengyel, P. (1981). Interferon action: RNA cleavage pattern of a $(2'-5')$oligoadenylate-dependent endonuclease. *Science,* 212: 1030–1032.

Foy, E., Li, K., Wang, C., Sumpter, R., Jr., Ikeda, M., Lemon, S.M., Gale, M., Jr. (2003). Regulation of interferon regulatory factor-3 by the hepatitis C virus serine protease. *Science,* 300: 1145–1148.

Frese, M., Kochs, G., Meier-Dieter, U., Siebler, J., Haller, O. (1995). Human MxA protein inhibits tick-borne Thogoto virus but not Dhori virus. *Journal of Virology,* 69: 3904–3909.

Frese, M., Pietschmann, T., Moradpour, D., Haller, O., Bartenschlager, R. (2001). Interferon-alpha inhibits hepatitis C virus subgenomic RNA replication by an MxA-independent pathway. *Journal of General of Virology,* 82: 723–733.

Frese, M., Schwärzle, V., Barth, K., Krieger, N., Lohmann, V., Mihm, S., Haller, O., Bartenschlager, R. (2002). Interferon-gamma inhibits replication of subgenomic and genomic hepatitis C virus RNAs. *Hepatology,* 35: 694–703.

Frese, M., Barth, K., Kaul, A., Lohmann, V., Schwärzle, V., Bartenschlager, R. (2003). Hepatitis C virus RNA replication is resistant to tumour necrosis factor-alpha. *Journal of General of Virology,* 84: 1253–1259.

Friebe, P., Boudet, J., Simorre, J.P., Bartenschlager, R. (2005). Kissing-loop interaction in the 3′ end of the hepatitis C virus genome essential for RNA replication. *Journal of Virology,* 79: 380–392.

Fried, M.W., Shiffman, M.L., Reddy, K.R., Smith, C., Marinos, G., Goncales, F.L., Jr., Haussinger, D., Diago, M., Carosi, G., Dhumeaux, D., Craxi, A., Lin, A., Hoffman, J., Yu, J. (2002). Peginterferon alfa-2a plus ribavirin for chronic hepatitis C virus infection. *New England Journal of Medicine,* 347: 975–982.

Gale, M., Jr., Blakely, C.M., Kwieciszewski, B., Tan, S.L., Dossett, M., Tang, N.M., Korth, M.J., Polyak, S.J., Gretch, D.R., Katze, M.G. (1998). Control of PKR protein kinase by hepatitis C virus nonstructural 5A protein: molecular mechanisms of kinase regulation. *Molecular and Cellular Biology,* 18: 5208–5218.

Garcia, M.A., Gil, J., Ventoso, I., Guerra, S., Domingo, E., Rivas, C., Esteban, M. (2006). Impact of protein kinase PKR in cell biology: from antiviral to antiproliferative action. *Microbiology and Molecular Biology Reviews*, 70: 1032–1060.

George, C.X., Samuel, C.E. (1999). Human RNA-specific adenosine deaminase ADAR1 transcripts possess alternative exon 1 structures that initiate from different promoters, one constitutively active and the other interferon inducible. *Proceedings of the National Academy of Sciences of the United States of America*, 96: 4621–4626.

Ghosh, A., Sarkar, S.N., Sen, G.C. (2000). Cell growth regulatory and antiviral effects of the P69 isozyme of 2-5 (A) synthetase. *Virology*, 266: 319–328.

Gil, J., Garcia, M.A., Gomez-Puertas, P., Guerra, S., Rullas, J., Nakano, H., Alcami, J., Esteban, M. (2004). TRAF family proteins link PKR with NF-kappa B activation. *Molecular and Cellular Biology*, 24: 4502–4012.

Gilli, F., Sala, A., Bancone, C., Salacone, P., Gallo, M., Gaia, E., Bertolotto, A. (2002). Evaluation of IFN-alpha bioavailability by MxA mRNA in HCV patients. *Journal of Immunological Methods*, 262: 187–190.

Goh, K.C., deVeer, M.J., Williams, B.R. (2000). The protein kinase PKR is required for p38 MAPK activation and the innate immune response to bacterial endotoxin. *EMBO Journal*, 19: 4292–4297.

Goodbourn, S., Didcock, L., Randall, R.E. (2000). Interferons: cell signalling, immune modulation, antiviral response and virus countermeasures. *Journal of General Virology*, 81: 2341–2364.

Guidotti, L.G., Chisari, F.V. (2001). Noncytolytic control of viral infections by the innate and adaptive immune response. *Annual Review of Immunology*, 19: 65–91.

Guidotti, L.G., McClary, H., Loudis, J.M., Chisari, F.V. (2000). Nitric oxide inhibits hepatitis B virus replication in the livers of transgenic mice. *Journal of Experimental Medicine*, 191: 1247–1252.

Guo, J.T., Bichko, V.V., Seeger, C. (2001). Effect of alpha interferon on the hepatitis C virus replicon. *Journal of Virology*, 75: 8516–8523.

Guo, J.T., Zhu, Q., Seeger, C. (2003). Cytopathic and noncytopathic interferon responses in cells expressing hepatitis C virus subgenomic replicons. *Journal of Virology*, **77**: 10769–10779.

Haller, O., Frese, M., Kochs, G. (1999). Mx proteins: mediators of innate resistance to RNA viruses. *Revue Scientific et Technique*, 17: 220–230.

Haller, O., Staeheli, P., Kochs, G. (2007). Interferon-induced Mx proteins in antiviral host defense. *Biochimie* (in press).

Haller, O., Kochs, G. (2002). Interferon-induced Mx proteins: dynamin-like GTPases with antiviral activity. *Traffic*, 3: 710–717.

Han, J.Q., Wroblewski, G., Xu, Z., Silverman, R.H., Barton, D.J. (2004). Sensitivity of hepatitis C virus RNA to the antiviral enzyme ribonuclease L is determined by a subset of efficient cleavage sites. *Journal of Interferon and Cytokine Research*, 24: 664–676.

Hartner, J.C., Schmittwolf, C., Kispert, A., Muller, A.M., Higuchi, M., Seeburg, P.H. (2004). Liver disintegration in the mouse embryo caused by deficiency in the RNA-editing enzyme ADAR1. *Journal of Biological Chemistry*, 279: 4894–4902.

Hayashi, J., Stoyanova, R., Seeger, C. (2005). The transcriptome of HCV replicon expressing cell lines in the presence of alpha interferon. *Virology*, 335: 264–275.

Hefti, H.P., Frese, M., Landis, H., Di Paolo, C., Aguzzi, A., Haller, O., Pavlovic, J. (1999). Human MxA protein protects mice lacking a functional alpha/beta interferon system against La Crosse virus and other lethal viral infections. *Journal of Virology*, 73: 6984–6991.

Higuchi, M., Maas, S., Single, F.N., Hartner, J., Rozov, A., Burnashev, N., Feldmeyer, D., Sprengel, R., Seeburg, P.H. (2000). Point mutation in an AMPA receptor gene rescues lethality in mice deficient in the RNA-editing enzyme ADAR2. *Nature*, 406: 78–81.

Hijikata, M., Ohta, Y., Mishiro, S. (2000). Identification of a single nucleotide polymorphism in the *MxA* gene promoter (G/T at nt -88) correlated with the response of hepatitis C patients to interferon. *Intervirology*, 43: 124–127.

Hijikata, M., Mishiro, S., Miyamoto, C., Furuichi, Y., Hashimoto, M., Ohta, Y. (2001). Genetic polymorphism of the *MxA* gene promoter and interferon responsiveness of hepatitis C patients: revisited by analyzing two SNP sites (-123 and -88) *in vivo* and *in vitro*. *Intervirology*, 44: 379–382.

Hiscott, J., Lin, R., Nakhaei, P., Paz, S. (2006). MasterCARD: a priceless link to innate immunity. *Trends in Molecular Medicine,* 12: 53–56.

Hoenen, A., Liu, W., Kochs, G., Khromykh, A.A., Mackenzie, J.M. (2007). West Nile virus-induced cytoplasmic membrane structures provide partial protection aginst the interferon-induced antiviral MxA protein. *Journal of General Virology,* (in press).

Holzinger, D., Jorns, C., Sterz, S. Boisson-Dupuis, S. Thimme, R., Weidmann, M., Casanova, J.L., Haller, O., Kochs, G. (2007). Induction of MxA gene expression by influenza A virus requires type I or type II interferon signaling. Journal of Virology, 81: 7776–7785.

Honda, M., Beard, M.R., Ping, L.H., Lemon, S.M. (1999). A phylogenetically conserved stem-loop structure at the 5′ border of the internal ribosome entry site of hepatitis C virus is required for cap-independent viral translation. *Journal of Virology,* 73: 1165–1174.

Jaiswal, M., LaRusso, N.F., Burgart, L.J., Gores, G.J. (2000). Inflammatory cytokines induce DNA damage and inhibit DNA repair in cholangiocarcinoma cells by a nitric oxide-dependent mechanism. *Cancer Research,* 60: 184–190.

Jorns, C., Holzinger, D., Thimme, R., Spangenberg, H.C., Weidmann, M., Rasenack, J., Blum, H.E., Haller, O., Kochs, G. (2006). Rapid and simple detection of IFN-neutralizing antibodies in chronic hepatitis C non-responsive to IFN-alpha. *Journal of Medical Virology,* 78: 74–82.

Karupiah, G., Chen, J.H., Mahalingam, S., Nathan, C.F., MacMicking, J.D. (1998). Rapid interferon gamma-dependent clearance of influenza A virus and protection from consolidating pneumonitis in nitric oxide synthase 2-deficient mice. *Journal of Experimental Medicine,* 188: 1541–1546.

Karupiah, G., Hunt, N.H., King, N.J.C., Chaudhri, G. (2000). NADPH oxidase, Nramp1 and nitric oxide in the host microbiol response. *Reviews in Immunogenetics,* 2: 387–415.

Kato, T., Date, T., Miyamoto, M., Zhao, Z., Mizokami, M., Wakita, T. (2005). Nonhepatic cell lines HeLa and 293 support efficient replication of the hepatitis C virus genotype 2a subgenomic replicon. *Journal of Virology,* 79: 592–659.

Keskinen, P., Nyqvist, M., Sareneva, T., Pirhonen, J., Melen, K., Julkunen, I. (1999). Impaired antiviral response in human hepatoma cells. *Virology,* 263: 364–375.

Kim, Y.K., Lee, S.H., Kim, C.S., Seol, S.K., Jang, S.K. (2003). Long-range RNA-RNA interaction between the 5′ nontranslated region and the core-coding sequences of hepatitis C virus modulates the IRES-dependent translation. *RNA,* 9: 599–606.

Knapp, S., Yee, L.J., Frodsham, A.J., Hennig, B.J., Hellier, S., Zhang, L., Wright, M., Chiaramonte, M., Graves, M., Thomas, H.C., Hill, A.V., Thursz, M.R. (2003). Polymorphisms in interferon-induced genes and the outcome of hepatitis C virus infection: roles of MxA, OAS-1 and PKR. *Genes and Immunity,* 4: 411–419.

Kochs, G., Haller, O. (1999). Interferon-induced human MxA GTPase blocks nuclear import of Thogoto virus nucleocapsids. *Proceedings of the National Academy of Sciences of the United States of America,* 96: 2082–2086.

Kochs, G., Janzen, C., Hohenberg, H., Haller, O. (2002). Antivirally active MxA protein sequesters La Crosse virus nucleocapsid protein into perinuclear complexes. *Proceedings of the National Academy of Sciences of the United States of America,* 99: 3153–3158.

Kotenko, S.V., Gallagher, G., Baurin, V.V., Lewis-Antes, A., Shen, M., Shah, N.K., Langer, J.A., Sheikh, F., Dickensheets, H., Donnelly, R.P. (2003). IFN-lambdas mediate antiviral protection through a distinct class II cytokine receptor complex. *Nature Immunology,* 4: 69–77.

Kuo, M.Y., Chao, M., Taylor, J. (1989). Initiation of replication of the human hepatitis delta virus genome from cloned DNA: role of delta antigen. *Journal of Virology,* 63: 1945–1950.

Lanford, R.E., Guerra, B., Lee, H., Averett, D.R., Pfeiffer, B., Chavez, D., Notvall, L., Bigger, C. (2003). Antiviral effect and virus-host interactions in response to alpha interferon, gamma interferon, poly(I)-poly(C), tumor necrosis factor alpha, and ribavirin in hepatitis C virus subgenomic replicons. *Journal of Virology,* 77: 1092–1104.

Langland, J.O., Cameron, J.M., Heck, M.C., Jancovich, J.K., Jacobs, B.L. (2006). Inhibition of PKR by RNA and DNA viruses. *Virus Research,* 119: 100–110.

Larkin, J., Jin, L., Farmen, M., Venable, D., Huang, Y., Tan, S.L., Glass, J.I. (2003). Synergistic antiviral activity of human interferon combinations in the hepatitis C virus replicon system. *Journal of Interferon and Cytokine Research,* 23: 247–257.

Layden, J.E., Layden, T.J. (2002). Viral kinetics of hepatitis C: new insights and remaining limitations. *Hepatology,* 35: 967–970.

Li, K., Foy, E., Ferreon, J.C., Nakamura, M., Ferreon, A.C., Ikeda, M., Ray, S.C., Gale, M., Jr., Lemon, S.M. (2005). Immune evasion by hepatitis C virus NS3/4A protease-mediated cleavage of the Toll-like receptor 3 adaptor protein TRIF. *Proceedings of the National Academy of Sciences of the United States of America,* 102: 2992–2997.

Li, Y., Zhang, T., Ho, C., Orange, J.S., Douglas, S.D., Ho, W.Z. (2004). Natural killer cells inhibit hepatitis C virus expression. *Journal of Leukocyte Biology,* 76: 1171–1179.

Lohmann, V., Körner, F., Koch, J.O., Herian, U., Theilmann, L., Bartenschlager, R. (1999). Replication of subgenomic hepatitis C virus RNAs in a hepatoma cell line. *Science,* 285: 110–113.

MacDonald, A., Harris, M. (2004). Hepatitis C virus NS5A: tales of a promiscuous protein. *Journal of General Virology,* 85: 2485–2502.

Machida, K., Cheng, K.T., Sung, V.M., Lee, K.J., Levine, A.M., Lai, M.M. (2004). Hepatitis C virus infection activates the immunologic (type II) isoform of nitric oxide synthase and thereby enhances DNA damage and mutations of cellular genes. *Journal of Virology,* 78: 8835–8843.

MacQuillan, G.C., de Boer, W.B., Platten, M.A., McCaul, K.A., Reed, W.D., Jeffrey, G.P., Allan, J.E. (2002). Intrahepatic MxA and PKR protein expression in chronic hepatitis C virus infection. *Journal of Medical Virology,* 68: 197–205.

MacQuillan, G.C., Mamotte, C., Reed, W.D., Jeffrey, G.P., Allan, J.E. (2003). Upregulation of endogenous intrahepatic interferon stimulated genes during chronic hepatitis C virus infection. *Journal of Medical Virology,* 70: 219–227.

Majano, P.L., Garcia-Monzon, C., Lopez-Cabrera, M., Lara-Pezzi, E., Fernandez-Ruiz, E., Garcia-Iglesias, C., Borque, M.J., Moreno-Otero, R. (1998). Inducible nitric oxide synthase expression in chronic viral hepatitis. Evidence for a virus-induced gene upregulation. *Journal of Clinical Investigation,* 101: 1343–1352.

Manns, M.P., McHutchison, J.G., Gordon, S.C., Rustgi, V.K., Shiffman, M., Reindollar, R., Goodman, Z.D., Koury, K., Ling, M., Albrecht, J.K. (2001). Peginterferon alfa-2b plus ribavirin compared with interferon alfa-2b plus ribavirin for initial treatment of chronic hepatitis C: a randomised trial. *Lancet,* 358: 958–965.

Marcello, T., Grakoui, A., Barba-Spaeth, G., Machlin, E.S., Kotenko, S.V., Macdonald, M.R., Rice, C.M. (2006). Interferons alpha and lambda inhibit hepatitis C virus replication with distinct signal transduction and gene regulation kinetics. *Gastroenterology,* 131: 1887–1898.

Mashimo, T., Lucas, M., Simon-Chazottes, D., Frenkiel, M.P., Montagutelli, X., Ceccaldi, P.E., Deubel, V., Guenet, J.L., Despres, P. (2002). A nonsense mutation in the gene encoding 2′-5′-oligoadenylate synthetase/L1 isoform is associated with West Nile virus susceptibility in laboratory mice. *Proceedings of the National Academy of Sciences of the United States of America,* 99: 11311–11316.

McHutchison, J.G., Fried, M.W. (2003). Current therapy for hepatitis C: pegylated interferon and ribavirin. *Clinics in Liver Disease,* 7: 149–161.

Meylan, E., Curran, J., Hofmann, K., Moradpour, D., Binder, M., Bartenschlager, R., Tschopp, J. (2005). Cardif is an adaptor protein in the RIG-I antiviral pathway and is targeted by hepatitis C virus. *Nature,* 437: 1167–1172.

Mihm, S., Fayyazi, A., Ramadori, G. (1997). Hepatic expression of inducible nitric oxide synthase transcripts in chronic hepatitis C virus infection: relation to hepatic viral load and liver injury. *Hepatology,* 26: 451–458.

Mihm, S., Frese, M., Meier, V., Wietzke-Braun, P., Scharf, J.-G., Bartenschlager, R., Ramadori, G. (2004). Interferon type I gene expression in chronic hepatitis C. *Laboratory Investigation,* 84: 1148–1159.

Miyamoto, M., Kato, T., Date, T., Mizokami, M., Wakita, T. (2006). Comparison between subgenomic replicons of hepatitis C virus genotypes 2a (JFH-1) and 1b (Con1 NK5.1). *Intervirology,* 49: 37–43.

Mundt, E. (2007). Human MxA confers resistance to double-standed RNA viruses of two virus families. *Journal of General Virology,* 88: 1319–1323.

Murashima, S., Kumashiro, R., Ide, T., Miyajima, I., Hino, T., Koga, Y., Ishii, K., Ueno, T., Sakisaka, S., Sata, M. (2000). Effect of interferon treatment on serum 2′,5′-oligoadenylate synthetase levels in hepatitis C-infected patients. *Journal of Medical Virology,* 62: 185–190.

Murphy, D.G., Dimock, K., Kang, C.Y. (1991). Numerous transitions in human parainfluenza virus 3 RNA recovered from persistently infected cells. *Virology,* 181: 760–763.

Murphy, J.A., Duerst, R.J., Smith, T.J., Morrison, L.A. (2003). Herpes simplex virus type 2 virion host shutoff protein regulates alpha/beta interferon but not adaptive immune responses during primary infection in vivo. *Journal of Virology,* 77: 9337–9345.

Neumann, A.U., Lam, N.P., Dahari, H., Gretch, D.R., Wiley, T.E., Layden, T.J., Perelson, A.S. (1998). Hepatitis C viral dynamics in vivo and the antiviral efficacy of interferon-alpha therapy. *Science,* 282: 103–107.

O'Hara, P.J., Nichol, S.T., Horodyski, F.M., Holland, J.J. (1984). Vesicular stomatitis virus defective interfering particles can contain extensive genomic sequence rearrangements and base substitutions. *Cell,* 36: 915–924.

Okuse, C., Rinaudo, J.A., Farrar, K., Wells, F., Korba, B.E. (2005). Enhancement of antiviral activity against hepatitis C virus in vitro by interferon combination therapy. *Antiviral Research,* 65: 23–34.

Pascu, M., Martus, P., Hohne, M., Wiedenmann, B., Hopf, U., Schreier, E., Berg, T. (2004). Sustained virological response in hepatitis C virus type 1b infected patients is predicted by the number of mutations within the NS5A-ISDR: a meta-analysis focused on geographical differences. *Gut,* 53: 1345–1351.

Patterson, J.B., Samuel, C.E. (1995). Expression and regulation by interferon of a double-stranded-RNA-specific adenosine deaminase from human cells: evidence for two forms of the deaminase. *Molecular and Cellular Biology,* 15: 5376–5388.

Patzwahl, R., Meier, V., Ramadori, G., Mihm, S. (2001). Enhanced expression of interferon-regulated genes in the liver of patients with chronic hepatitis C virus infection: detection by suppression-subtractive hybridization. *Journal of Virology,* 75: 1332–1338.

Pavio, N., Taylor, D.R., Lai, M.M. (2002). Detection of a novel unglycosylated form of hepatitis C virus E2 envelope protein that is located in the cytosol and interacts with PKR. *Journal of Virology,* 76: 1265–1272.

Pavlovic, J., Arzet, H.A., Hefti, H.P., Frese, M., Rost, D., Ernst, B., Kolb, E., Staeheli, P., Haller, O. (1995). Enhanced virus resistance of transgenic mice expressing the human MxA protein. *Journal of Virology,* 69: 4506–4510.

Penna, G., Vulcano, M., Sozzani, S., Adorini, L. (2002). Differential migration behavior and chemokine production by myeloid and plasmacytoid dendritic cells. *Human Immunology,* 63: 1164–1171.

Perelygin, A.A., Scherbik, S.V., Zhulin, I.B., Stockman, B.M., Li, Y., Brinton, M.A. (2002). Positional cloning of the murine flavivirus resistance gene. *Proceedings of the National Academy of Sciences of the United States of America,* 99: 9322–9327.

Pestka, S., Krause, C.D., Walter, M.R. (2004). Interferons, interferon-like cytokines, and their receptors. *Immunological Reviews,* 202: 8–32.

Polyak, S.J., Khabar, K.S., Paschal, D.M., Ezelle, H.J., Duverlie, G., Barber, G.N., Levy, D.E., Mukaida, N., Gretch, D.R. (2001a). Hepatitis C virus nonstructural 5A protein induces interleukin-8, leading to partial inhibition of the interferon-induced antiviral response. *Journal of Virology,* 75: 6095–6106.

Polyak, S.J., Khabar, K.S., Rezeiq, M., Gretch, D.R. (2001b). Elevated levels of interleukin-8 in serum are associated with hepatitis C virus infection and resistance to interferon therapy. *Journal of Virology,* 75: 6209–6211.

Pope, M., Marsden, P.A., Cole, E., Sloan, S., Fung, L.S., Ning, Q., Ding, J.W., Leibowitz, J.L., Phillips, M.J., Levy, G.A. (1998). Resistance to murine hepatitis virus strain 3 is dependent on production of nitric oxide. *Journal of Virology,* 72: 7084–7090.

Rebouillat, D., Hovanessian, A.G. (1999). The human 2′,5′-oligoadenylate synthetase family: interferon-induced proteins with unique enzymatic properties. *Journal of Interferon and Cytokine Research,* 19: 295–308.

Reichelt, M., Stertz, S., Krijnse-Locker, J., Haller, O., Kochs, G. (2004). Missorting of LaCrosse virus nucleocapsid protein by the interferon-induced MxA GTPase involves smooth ER membranes. *Traffic,* 5: 772–784.

Robek, M.D., Boyd, B.S., Chisari, F.V. (2005). Lambda interferon inhibits hepatitis B and C virus replication. *Journal of Virology,* 79: 3851–3854.

Roers, A., Hochkeppel, H.K., Horisberger, M.A., Hovanessian, A., Haller, O. (1994). MxA gene expression after live virus vaccination: a sensitive marker for endogenous type I interferon. *Journal of Infectious Diseases,* 169: 807–813.

Samuel, C.E. (2001). Antiviral actions of interferons. *Clinical Microbiological Reviews,* 14: 778–809.

Samuel, M.A., Whitby, K., Keller, B.C., Marri, A., Barchet, W., Williams, B.R., Silverman, R.H., Gale, M., Jr., Diamond, M.S. (2006). PKR and RNase L contribute to protection against lethal West Nile Virus infection by controlling early viral spread in the periphery and replication in neurons. *Journal of Virology,* 80: 7009–7019.

Saura, M., Zaragoza, C., McMillan, A., Quick, R.A., Hohenadl, C., Lowenstein, J.M., Lowenstein, C.J. (1999). An antiviral mechanism of nitric oxide: inhibition of a viral protease. *Immunity,* 10: 21–28.

Scadden, A.D.J., Smith, C.W.J. (1997). A ribonuclease specific for inosine-containing RNA: a potential role in antiviral defense? *EMBO Journal,* 16: 2140–2149.

Scadden, A.D.J., Smith, C.W.J. (2001). Specific cleavage of hyper-edited dsRNAs. *EMBO Journal,* 20: 4243–4252.

Scherbik, S.V., Paranjape, J.M., Stockman, B.M., Silverman, R.H., Brinton, M.A. (2006). RNase L plays a role in the antiviral response to West Nile virus. *Journal of Virology,* 80: 2987–2999.

Schweyer, S., Mihm, S., Radzun, H.J., Hartmann, H., Fayyazi, A. (2000). Liver infiltrating T lymphocytes express interferon-gamma and inducible nitric oxide synthase in chronic hepatitis C virus infection. *Gut,* 46: 255–259.

Sheppard, P., Kindsvogel, W., Xu, W., Henderson, K., Schlutsmeyer, S., Whitmore, T.E., Kuestner, R., Garrigues, U., Birks, C., Roraback, J., Ostrander, C., Dong, D., Shin, J., Presnell, S., Fox, B., Haldeman, B., Cooper, E., Taft, D., Gilbert, T., Grant, F.J., Tackett, M., Krivan, W., McKnight, G., Clegg, C., Foster, D., Klucher, K.M. (2003). IL-28, IL-29, and their class II cytokine receptor IL-28R. *Nature Immunology,* 4: 63–68.

Shin, E.C., Seifert, U., Kato, T., Rice, C.M., Feinstone, S.M., Kloetzel, P.M., Rehermann, B. (2006). Virus-induced type I IFN stimulates generation of immunoproteasomes at the site of infection. *Journal of Clinical Investigation,* 116: 3006–3014.

Smith, J.A., Schmechel, S.C., Williams, B.R., Silverman, R.H., Schiff, L.A. (2005). Involvement of the interferon-regulated antiviral proteins PKR and RNase L in reovirus-induced shutoff of cellular translation. *Journal of Virology,* 79: 2240–2250.

Stark, J.M., Khan, A.M., Chiappetta, C.L., Xue, H., Alcorn, J.L., Colasurdo, G.N. (2005). Immune and functional role of nitric oxide in a mouse model of respiratory syncytial virus infection. *Journal of Infectious Diseases,* 191: 387–395.

Stojdl, D.F., Abraham, N., Knowles, S., Marius, R., Brasey, A., Lichty, B.D., Brown, E.G., Sonenberg, N., Bell, J.C. (2000). The murine double-stranded RNA-dependent protein kinase PKR is required for resistance to vesicular stomatitis virus. *Journal of Virology,* 74: 9580–9585.

Su, A.I., Pezacki, J.P., Wodicka, L., Brideau, A.D., Supekova, L., Thimme, R., Wieland, S., Bukh, J., Purcell, R.H., Schultz, P.G., Chisari, F.V. (2002). Genomic analysis of the host response to hepatitis C virus infection. *Proceedings of the National Academy of Sciences of the United States of America,* 99: 15669–15674.

Taguchi, T., Nagano-Fujii, M., Akutsu, M., Kadoya, H., Ohgimoto, S., Ishido, S., Hotta, H. (2004). Hepatitis C virus NS5A protein interacts with 2',5'-oligoadenylate synthetase and inhibits antiviral activity of IFN in an IFN sensitivity-determining region-independent manner. *Journal of General Virology,* 85: 959–969.

Tan, H., Derrick, J., Hong, J., Sanda, C., Grosse, W.M., Edenberg, H.J., Taylor, M., Seiwert, S., Blatt, L.M. (2005). Global transcriptional profiling demonstrates the combination of type I and type II interferon enhances antiviral and immune responses at clinically relevant doses. *Journal of Interferon and Cytokine Research,* 25: 632–649.

Tan, S.-L., Katze, M.G. (2001). How hepatitis C virus counteracts the interferon response: the jury is still out on NS5A. *Virology,* 284: 1–12.

Taylor, D.R., Shi, S.T., Romano, P.R., Barber, G.N., Lai, M.C. (1999). Inhibition of the interferon-inducible protein kinase PKR by HCV E2 protein. *Science*, 285: 107–110.

Taylor, D.R., Tian, B., Romano, P.R., Hinnebusch, A.G., Lai, M.M., Mathews, M.B. (2001). Hepatitis C virus envelope protein E2 does not inhibit PKR by simple competition with autophosphorylation sites in the RNA-binding domain. *Journal of Virology*, 75: 1265–1273.

Taylor, D.R., Puig, M., Darnell, M.E., Mihalik, K., Feinstone, S.M. (2005). New antiviral pathway that mediates hepatitis C virus replicon interferon sensitivity through ADAR1. *Journal of Virology*, 79: 6291–6298.

Terajima, M., Leporati, A.M. (2005). Role of indoleamine 2,3-dioxygenase in antiviral activity of interferon-gamma against vaccinia virus. *Viral Immunology*, 18: 722–729.

Thomis, D.C., Doohan, J.P., Samuel, C.E. (1992). Mechanism of interferon action: cDNA structure, expression, and regulation of the interferon-induced, RNA-dependent P1/eIF-2 alpha protein kinase from human cells. *Virology*, 188: 33–46.

Toth, A.M., Zhang, P., Das, S., George, C.X., Samuel, C.E. (2006). Interferon action and the double-stranded RNA-dependent enzymes ADAR1 adenosine deaminase and PKR protein kinase. *Progress in Nucleic Acid Research and Molecular Biology*, 81: 369–434.

Uprichard, S.L., Chung, J., Chisari, F.V., Wakita, T. (2006). Replication of a hepatitis C virus replicon clone in mouse cells. *Virology Journal*, 3: 89.

Valente, L., Nishikura, K. (2005). ADAR gene family and A-to-I RNA editing: diverse roles in posttranscriptional gene regulation. *Progress in Nucleic Acid Research and Molecular Biology*, 79: 299–338.

Van Regenmortel, M.H.V., Fauquet, C.M., Bishop, D.H.L., Carstens, E.B., Estes, M.K., Lemon, S.M., Maniloff, J., Mayo, M.A., McGeoch, D.J., Pringle, C.R., Wickner, R.B. (2000). Virus Taxonomy: The VIIth Report of the International Committee on Taxonomy of Viruses. Academic Press, San Diego.

Wang, Q., Khillan, J., Gadue, P., Nishikura, K. (2000). Requirement of the RNA editing deaminase ADAR1 gene for embryonic erythropoiesis. *Science*, 290: 1765–1768.

Wang, Q., Miyakoda, M., Yang, W., Khillan, J., Stachura, D.L., Weiss, M.J., Nishikura, K. (2004). Stress-induced apoptosis associated with null mutation of ADAR1 RNA editing deaminase gene. *Journal of Biological Chemistry*, 279: 4952–4961.

Westaway, E.G., Mackenzie, J.M., Khromykh, A.A. (2002). Replication and gene function in Kunjin virus. *Current Topics in Microbiology and Immunology*, 267: 323–351.

Windisch, M.P., Frese, M., Kaul, A., Trippler, M., Lohmann, V., Bartenschlager, R. (2005). Dissecting the interferon-induced inhibition of hepatitis C virus replication by using a novel host cell line. *Journal of Virology*, 79: 13778–13793.

Wreschner, D.H., McCauley, J.W., Skehel, J.J., Kerr, I.M. (1981). Interferon action -sequence specificity of the ppp(A2'p)nA-dependent ribonuclease. *Nature*, 289: 414–417.

Yang, J.H., Luo, X., Nie, Y., Su, Y., Zhao, Q., Kabir, K., Zhang, D., Rabinovici, R. (2003a). Widespread inosine-containing mRNA in lymphocytes regulated by ADAR1 in response to inflammation. *Immunology*, 109: 15–23.

Yang, J.H., Nie, Y., Zhao, Q., Su, Y., Pypaert, M., Su, H., Rabinovici, R. (2003b). Intracellular localization of differentially regulated RNA-specific adenosine deaminase isoforms in inflammation. *Journal of Biological Chemistry*, 278: 45833–45842.

Yang, Y.L., Reis, L.F., Pavlovic, J., Aguzzi, A., Schäfer, R., Kumar, A., Williams, B.R., Aguet, M., Weissmann, C. (1995). Deficient signaling in mice devoid of double-stranded RNA-dependent protein kinase. *EMBO Journal*, 14: 6095–6106.

Yee, L.J., Knapp, S., Burgner, D., Hennig, B.J., Frodsham, A.J., Wright, M., Thomas, H.C., Hill, A.V., Thursz, M.R. (2004). Inducible nitric oxide synthase gene (NOS2A) haplotypes and the outcome of hepatitis C virus infection. *Genes and Immunity*, 5: 183–187.

You, S., Stump, D.D., Branch, A.D., Rice, C.M. (2004). A cis-acting replication element in the sequence encoding the NS5B RNA-dependent RNA polymerase is required for hepatitis C virus RNA replication. *Journal of Virology*, 78: 1352–1366.

Zhu, H., Zhao, H., Collins, C.D., Eckenrode, S.E., Run, Q., McIndoe, R.A., Crawford, J.M., Nelson, D.R., She, J.X., Liu, C. (2003). Gene expression associated with interferon alfa antiviral activity in an HCV replicon cell line. *Hepatology*, 37: 1180–1188.

Treatment of Chronic Hepatitis C with Different Genotypes

James Fung, Ching-Lung Lai, and Man-Fung Yuen*

Background

An estimated 175 million people worldwide are infected with the hepatitis C virus (HCV), making it the most common cause of liver cirrhosis and hepatocellular carcinoma in both Europe and the United States (World Health Organization, 2000). Each year, up to 4 million people will be newly infected, of which the majority (75–85%) will go on to develop chronic infection. The characterization of HCV has been dependent on genetic sequencing in the absence of an appropriate small animal model or cell culture system. In particular, genotyping of HCV has become routine in day-to-day clinical management of chronic hepatitis C patients as a result of differences in treatment response to interferon (IFN) therapy between genotypes. Population studies of HCV genotypes may also give insight into the origin, evolution, and migration of HCV. However, the role of genotypes in disease progression, outcome, and association with other disease states remains to be fully determined.

Classification of Genotypes

The HCV RNA genome is made up of approximately 9,400 nucleotides, which encode a single polypeptide protein of just over 3,000 amino acids. The genome is organized into three structural proteins at the N-terminal and four functional proteins at the C-terminal, which encodes several different enzymes, including a zinc-dependent metalloprotease, a helicase, and an RNA-dependent RNA polymerase. The structural proteins are the core (C) and envelope 1 (E1) and 2 (E2) proteins, whereas the nonstructural 2 (NS2), 3 (NS3), 4 (NS4), and 5 (NS5) make up the functional proteins, as shown in Figure 1. The RNA structure of HCV resembles other viruses in the *Flaviviridae* family, which includes a number of arthropod-borne viruses.

* Reprints address and correspondence: Dr Man-Fung Yuen, Department of Medicine, The University of Hong Kong, Queen Mary Hospital, Pokfulam Road, Hong Kong SAR, Tel: 852 28553984, Fax: 852 28725828, E-mail: mfyuen@hkucc.hku.hk

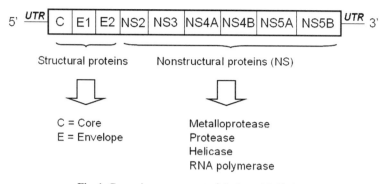

Fig. 1 Genomic arrangement of the hepatitis C virus

After the discovery of HCV in 1989, the first complete HCV genome was sequenced in 1991. Comparison of genome sequences from isolates originating from different parts of the world revealed different genotypes of HCV, with as much as 35% sequence diversity over the whole genome (Choo et al., 1991). This genetic variability is concentrated in the E1 and E2 regions, not as concentrated in conserved regions such as the core regions and NS3, and even less so in highly conserved regions in the 5' untranslated (UTR) region. In addition, the length of the open reading frame between each genotype is different, being approximately 9,400 nucleotides in genotype 1, 9,099 nucleotides in genotype 2, and 9,063 nucleotides in genotype 3 (Bukh et al., 1995).

The initial classification system of HCV genotypes is confusing due to the lack of standardization, with different investigators developing their own classification systems and methods of genotyping. It was not until 1994 when a consensus nomenclature system was developed to provide a more standardized approach to the classification of genotypes based on phylogenetic analysis (Simmonds et al., 1994). Using this system, the genotype numbers are assigned in their order of discovery, with subtypes depicted by a lowercase letter, showing closely related strains within some types.

Currently, all HCV isolates can be separated into six phylogenetically distinct groups, known as clades. HCV genotypes 1, 2, 4, and 5 are referred to as clades 1, 2, 4, and 5, respectively. Clade 3 consists of genotypes 3 and 10, and genotypes 6, 7, 8, 9, and 11 make up clade 6. The clade classification system is based on the combination of sequence homology in at least two regions of the viral genome and phylogenetic tree analysis. Using this consensus approach, genotypes 7 to 11 have been classified to fit the existing six clades (Robertson et al., 1998). The recent consensus for a unified system of nomenclature of HCV genotypes is currently based on the clade classification, recognizing that current HCV variants should be classified into six genotypes (Simmonds et al., 2005).

Between each genotype isolates, the entire viral genome is approximately 60–70% homologous. Viral isolates from the same subtypes will have no more than 5–15% variation in their nucleotide sequence, and this is increased to 10–30% between isolates from two different subtypes. Furthermore, HCV in an infected individual does not exist as a homogenous population, but as a mixture of genetically

Table 1 Genomic Heterogeneity in Hepatitis C

Classification	Nucleotide Homology of Entire HCV Genome (Approximation)
Genotype	50–70%
Subtype	70–85%
Quasispecies	90–100%

different but related quasispecies. The existence of quasispecies is likely maintained by the high production rate of viral particles of up to 10^{12} virions per day along with the error-prone RNA polymerase of HCV giving rise to a high mutation rate at 1.5–2.0×10^{-3} nucleotide substitutions per site per year (Bukh et al., 1995). As part of evolution, the selection of variant strains leads to emergence and divergence of HCV genotypes. Patients with a large population of different quasispecies are less responsive to treatment, as they will have an increased chance of variants escaping from the effects of IFN treatment. The genomic heterogeneity of HCV genotypes, subtypes, and quasispecies is shown in Table 1.

Geographical Distribution

With the six currently known genotypes, there are geographical differences in the distribution of these different genotypes. Although HCV genotypes 1 to 3 are found globally, there are variations in their regional prevalence worldwide. The distribution of the six HCV genotypes is shown in Table 2. Genotype 1a is most commonly found in the United States and Europe and is frequently associated with intravenous drug abuse (Zein et al., 1996a). Genotype 1b is distributed worldwide and is the most common type, accounting for over 70% of overall HCV infection. Genotypes 1a and 1b are the most common genotypes in the United States (Alter et al., 1999). Genotypes 2a and 2b are common in North America, Europe, and Japan and account for 10–30% of worldwide HCV. Genotype 2c is mainly found in Northern Italy. Genotype 3 is prominent in Southeast Asia and Indonesia, whereas subtype 3a is found predominantly in intravenous drug users from Western Europe and the United States (Pawlotsky et al., 1995). Genotype 4 is common in North Africa and the Middle East (Chamberlain et al., 1997), whereas genotypes 5 and 6

Table 2 Geographical Distribution of Hepatitis C Genotypes

Genotype	Regions
1a	United States, Europe
1b	Worldwide
2a	Japan, Europe, North America
2b	Japan, Europe, North America
2c	Northern Italy
3a	United States, Europe
4	North Africa, Middle East
5	South Africa
6	Southeast Asia

are found predominantly in South Africa (Smuts & Kannemeyer, 1995) and Hong Kong (Zhang et al., 1995; Prescott et al., 1996; Wong et al., 1998), respectively. Genotypes 7 to 11, described previously by some investigators, have been limited to patients in Vietnam, Indonesia, Thailand, and Burma (Tokita et al., 1994, 1996).

Clinical Implications

Currently, there appears to be no conclusive differences in clinical presentation and outcome among the six HCV genotypes, and therefore HCV genotype is not a useful prognostic marker of disease severity. There are few studies associating genotype 1b with more severe liver disease and hepatocellular carcinoma - or more aggressive disease following liver transplantation compared with other genotypes (Tanaka et al., 1998; Bruno et al., 1997; Ikeda et al., 2002; Zein et al., 1996b). However, other studies have found no association between HCV genotypes and the severity of liver disease (Reid et al., 1999; Naoumov et al., 1997).

There is a more definite association among liver steatosis, chronic HCV infection, and HCV genotypes. Steatosis is found in approximately 50% of people chronically infected with HCV and occurs over twice as much as would be expected in the general population (Lonardo et al., 2006). Although the accumulation of fat within hepatocytes, or steatosis, can occur across all HCV genotypes, there is a higher prevalence with increasing severity in those patients infected with HCV genotype 3 (Adinolfi et al., 2001; Rubbia-Brandt et al., 2000) . In this setting, the amount of steatosis correlates with the viral load, and steatosis typically regresses with eradication of HCV and reappears with viral relapses (Kumar et al., 2002). This suggests that, in these patients, steatosis can be considered as a genotype-related HCV-induced lesion. The exact pathogenic mechanism for steatosis mediated by HCV remains unknown, although the core and NS5A protein have been implicated in both animal and cell culture models. In the transgenic mouse model, expression of HCV core protein has been shown to induce steatosis (Moriya et al., 1997). In addition, the HCV NS5A protein has been shown to have an association with lipid droplets and apolipoprotein A1 and, in combination with core protein expression, may have a role in inducing steatosis through deranged lipid metabolism (Shi et al., 2002).

However, steatosis can also occur in chronic HCV patients independent of viral factors. Host-dependent factors such as obesity, insulin resistance, diabetes mellitus, alcohol consumption, and certain medications are likely factors and are similar to those found in patients with nonalcoholic fatty liver disease. Factors such as increased body mass index are more associated with patients infected with HCV genotype 1 (Adinolfi et al., 2001).

Response to Antiviral Therapy

The main objective of treating chronic HCV is to achieve a sustained virological response (SVR), which is characterized by nondetectable serum HCV RNA at the end of therapy, maintained for six months after the completion of antiviral therapy.

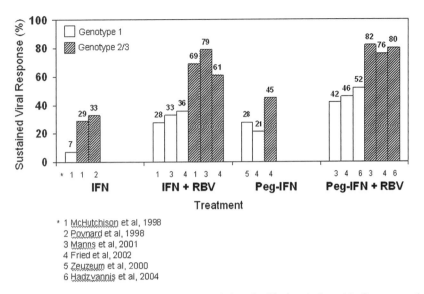

Fig. 3 Sustained viral response rates in patients infected with chronic hepatitis C genotype 1, 2, or 3

with a lower dose of ribavirin (800 mg/day) (National Institute of Health, 2002). The SVR rates in patients infected with HCV genotype 1, 2, or 3 with different antiviral therapies are shown in Figure 3.

Shorter Duration of Therapy with HCV Genotypes 2 and 3

With the high SVR rates achieved with HCV genotypes 2 and 3, the possibility of an even shorter duration of therapy has been recently studied. In a pilot study of 95 patients infected with HCV genotype 2 or 3, those who achieved early virological response, as defined by negative HCV RNA at weeks 4 and 8, were treated for 16 weeks with peg-IFN-α2b and ribavirin. SVR was achieved in 90%, with the presence of liver fibrosis being an independent predictor of SVR rate (Dalgard et al., 2004). In another study of HCV genotypes 2 and 3 patients, those with rapid response, as defined by HCV RNA below 600 IU/ml at week 4, achieved SVR rates of 82% and 80% after treatment with peg-IFN-α2a plus ribavirin for 16 and 24 weeks, respectively. However, a significantly lower SVR rate was seen in genotype 3 patients with a high viral load (>800,000 IU/ml) compared with patients with lower viral loads (59% vs. 85%, respectively, $p = 0.003$). Therefore, patients infected with HCV genotype 3 who have high viral loads may need a longer duration of therapy (von Wagner et al., 2005). In patients with undetectable HCV RNA at 4 weeks, an even shorter duration of therapy with peg-IFN-α2b at 1.0 mcg/kg and weight-based ribavirin for 12 weeks achieved an SVR rate of 85%, similar to current standard treatment of 24 weeks (Mangia et al., 2005).

This recent evidence suggests that in patients infected with HCV genotype 2 or 3, the length of treatment can be reduced depending on the early viral response after 4 weeks of treatment. The advantages of short-duration therapy include improved compliance of patients to treatment schedule, less dose modification, less premature withdrawal to therapy, and reduced cost of therapy. The disadvantages include the potential for higher relapse rate. In addition, the SVR rate may be reduced in patients infected with HCV genotype 3 with underlying fibrosis.

Induction Therapy

The decline in HCV RNA with IFN therapy follows a biphasic pattern, with the initial decline being highly predictive of SVR. Attempts to maximize this early viral response have been made using induction therapy with standard IFN given daily at a higher dosage. Although using an induction dose was associated with improved initial response, this was not maintained at the end of treatment. Furthermore, induction treatment with standard IFN was accompanied by a higher rate of drop-out and adverse effects (Bjoro et al., 2002; Layden et al., 2002; van Vlierberghe et al., 2003; Perez et al., 2003).

The use of induction therapy has also been investigated using peg-IFN. In genotype 1 infected patients, high dose induction therapy with peg-IFN-α2b, at dosages of 3 µg/kg for 1 week, 1.5 µg/kg for 3 weeks, and 1.0 µg/kg for 44 weeks, showed a more pronounced decline of HCV RNA during induction treatment compared to peg-IFN-α2b at a dose of 0.5 µg/kg for 48 weeks. This initial benefit became progressively less after reducing the peg-IFN dose in the induction group (Buti et al., 2002). The effectiveness of induction therapy was also demonstrated in patients infected with HCV genotype 1 using peg-IFN-α2b at 80–100 µg/week (depending on body weight) plus ribavirin at 1000–1200 mg/day for 8 weeks followed by peg-IFN-α2b at 50 µg/week plus ribavirin at 1000–1200 mg/day for 40 weeks (Bruno et al., 2004). However, the response was compared with patients treated with standard IFN and not peg-IFN.

Currently, the exact role of induction therapy with peg-IFN remains to be determined. In patients infected with genotype 1 HCV and a high baseline viral load and in patients who have failed previous combination therapy, induction therapy has a potential role, perhaps in combination with more prolonged therapy.

Genotype 4

HCV genotype 4 is common in Africa and the Middle East. In Egypt and Saudi Arabia, HCV genotype 4 accounts for up to 91% and 75% of HCV infections, respectively. Because of the low prevalence of HCV genotype 4 in North America and Europe, this genotype is under-represented in large, major multicenter clinical trials, accounting mostly for only 1–3% of the total study population.

Patients infected with HCV genotype 4 have a poor response to standard IFN therapy. A study of 100 Egyptian patients treated with IFN-α2a at a dose of 3 MU thrice weekly for six months resulted in an SVR rate of only 4%. However, 45% of the study patients had cirrhosis, which may lower the SVR rate (el-Zayadi et al., 1996). Two studies of 20 and 17 patients with genotype 4 reported SVR rates of 5% and 11%, respectively (Zylberberg et al., 2000; Remy et al., 1998).

Combination therapy with standard IFN and ribavirin results in a higher SVR in patients with genotype 4. In three randomized control trials comparing IFN-α2b plus ribavirin versus IFN monotherapy for 24 weeks, the SVR rates were 20% vs. 8%, 42% vs. 8%, and 14% vs. 0%, respectively (el-Zayadi et al., 1999; Koshy et al., 2000,, 2002). The latter study was performed on cirrhotic patients, explaining the low SVR achieved, whereas 30% and 0% had underlying cirrhosis in the first and second studies, respectively.

An open label prospective study of 66 patients with genotype 4 treated with peg-IFN-a2b at 1.5 µg/kg/week plus ribavirin for 48 weeks resulted in an SVR rate of 68% (Hasan et al., 2004). In a randomized double-blind study using peg-IFN-α2b (1.5 µg/kg/week) plus ribavirin in 287 patients for 24, 36, or 48 weeks, the SVR rates were 29%, 66%, and 69%, respectively. There was no significant difference in patients treated for 36 weeks or 48 weeks (Kamal et al., 2005).

The treatment responses with IFN plus ribavirin, peg-IFN, and combination peg-IFN plus ribavirin are summarized in Figure 4. A meta-analysis of 6 randomized controlled trials of 424 patients showed an SVR rate of 55% in patients treated with peg-IFN and ribavirin compared to 30% for those treated with standard IFN plus ribavirin (Khuroo et al., 2004). The results to date suggest that treatment of HCV genotype 4 using standard IFN with or without ribavirin produced similar SVR rates as seen in genotype 1. With the current optimal treatment using combination peg-IFN plus ribavirin, the observed SVR rate in HCV genotype 4 is higher than genotype 1, but lower than genotypes 2 and 3. Treatment durations of 48 weeks and 36 weeks were superior to 24 weeks, although further studies are required to determine the optimal treatment duration.

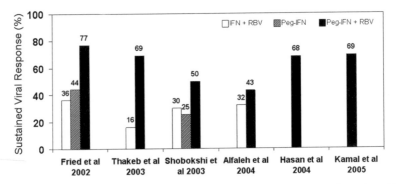

Fig. 4 Sustained viral response in chronic hepatitis C genotype 4 (IFN = interferon, Peg-IFN = pegylated interferon, RBV = ribavirin)

Genotype 5

Although HCV genotype 5 has a worldwide distribution, it is predominantly found in South Africa, accounting for up to 30% of HCV infections. A higher-than-expected prevalence rate of up to 14% has also been found in areas of southern Europe, including France and Spain (Henquell et al., 2004). In a small study of 16 patients from southern Belgium, 6 received combination IFN and ribavirin, with an overall SVR rate of 83%. Four patients received standard IFN-α2b at 3 MU thrice weekly and ribavirin, of which three patients achieved SVR. Two patients treated with peg-IFN and ribavirin achieved SVR. One patient treated with IFN-a2b monotherapy did not respond to treatment (Delwaide et al., 2005). A case-control study of 12 patients from southern France treated with either IFN-α2b at 3 MU thrice weekly plus ribavirin or peg-IFN-α2b (1.5 μg/kg/week) plus ribavirin for 48 weeks showed greater SVR rates for patients infected with genotype 5 compared to genotype 1 (64% vs. 23%, respectively, $p < 0.05$) (Legrand-Abravanel et al., 2004).

Given the restricted geographical distribution of genotype 5, limited existing data suggest that genotype 5 patients have a favorable treatment profile, one that is comparable to genotypes 2 and 3. Further large trials are required to confirm the findings of these small studies.

Genotype 6

As HCV genotype 6 is mainly found only in Hong Kong, southern China, Taiwan, and Southeast Asia, there is also limited data availability with respect to its response to treatment. In a study of 61 Southeast Asian patients using standard IFN-α2b at a dose of 5 MU/day for 8 weeks followed by 3 MU thrice weekly for 44 weeks combined with ribavirin, the SVR for genotypes 6-9 patients was 83% compared with 62% in genotype 1 (Dev et al., 2002). A study of 40 Hong Kong patients treated with IFN-α2b at 5 MU thrice weekly and ribavirin for 12 months showed better SVR in genotype 6 compared with genotype 1 (63% vs. 29%, respectively, $p = 0.04$) (Hui et al., 2003).

Taken together, these results suggest that response to antiviral therapy in patients infected with HCV genotype 6 is similar to that of genotypes 2 and 3 and superior to genotype 1. Further studies will be needed to determine the efficacy of pegylated interferon plus ribavirin and to determine the optimal duration of therapy.

Differences in Response in Different Genotypes

Despite the abundance of data regarding genotypic differences in antiviral therapy response, the exact mechanism whereby some genotypes are more difficult to treat than others remains unknown and is likely to involve both host and viral factors. This is partly contributed by the lack of understanding in the precise antiviral actions of both IFN and ribavirin *in vivo*. Furthermore, in the absence of a reliable serological or cell culture model, future studies into the antiviral mechanisms of these agents are likely to be dependent on replicon models.

The difference in genotypic response to IFN implicates HCV as having a direct role in IFN treatment. Furthermore, within each genotype, different strains of HCV may display different sensitivity to IFN therapy (Pawlotsky, 2003). Potential mechanisms include specific regions within the HCV genome and viral effects on the host, both of which may alter sensitivity to IFN therapy.

HCV NS5A Protein

Recent studies have focused on the NS5A protein as having a potential role in affecting response to IFN treatment in patients infected with HCV genotype 1. An early study comparing full-length sequences of HCV genotype 1b sensitive and resistant to IFN treatment identified amino acid differences at the carboxy-terminal half of the NS5A region spanning codons 2209 to 2248. Missense mutations were identified in this region in IFN-sensitive HCV, whereas in IFN-resistant HCV, the amino acid sequence was identical to the prototype Japanese HCV genotype 1b. The term "interferon sensitivity determining region" (ISDR) was designated to refer to this region (Enomoto et al., 1995). A subsequent study by the same group also showed that the number of amino acid mutations within the ISDR correlated with the success of IFN therapy. All patients with wild-type ISDR sequences did not respond to IFN, whereas all patients with four or more amino acid substitutions responded to IFN. Those patients with one to three amino acid substitutions had an intermediate response to IFN (Enomoto et al., 1996). Numerous studies have been published since then, investigating the relationship between ISDR mutations and the response to IFN therapy, with contradictory results. Initial findings have confirmed the correlations between ISDR mutations and IFN response by other Japanese studies, although non-Japanese studies including studies in the United States and Europe have failed to confirm this (McKechnie et al., 2000; Khorsi et al., 1997; Zeuzem et al., 1997; Chung et al., 1999). The discrepancy in results raises the possibilities of geographical factors and viral factors intrinsic to the Japanese viral isolates, which may account for the difference in IFN sensitivities. Two recent meta-analyses have supported the existence of an ISDR in HCV genotype 1b, with mutant-type ISDR strains having a more favorable response toward IFN therapy. In addition, Schinkel and colleagues (2004) in their meta-analysis have shown that the discrepant findings between Japanese and non-Japanese studies can be explained by differences in dosing regimens.

Despite the identification of an ISDR, the biological effect of different mutations remains to be determined. It is possible that ISDR mutations may affect viral replication or binding to other antiviral proteins mediated by IFN. In the latter, NS5A has been shown to bind and suppress IFN-inducible protein kinase (PKR), thereby evading the antiviral effects of IFN (Gale et al., 1998).

HCV E2 Protein

Another viral factor that may contribute to IFN resistance is the HCV E2 protein, an outer protein of the virus envelope involved in HCV binding to target cells. The E2 protein has been shown to inhibit the function of IFN-inducible PKR (Taylor, 1999). The amino acid 276-287 sequence of the E2 protein is similar to the autophosphorylation site of PKR and to the phosphorylation site of the translation initiation factor eIF2a, which is a target of PKR. This 12-amino acid domain is known as the PKR-eIF2a phosphorylation homology domain (PePHD), with increased homology among HCV genotypes 1a/1b and PKR/eIF2a sequences compared with HCV genotypes 2a/2b and 3a. Furthermore, this sequence is highly conserved within each genotype and may therefore contribute to the difference in genotypic response to IFN therapy via interaction between E2 and PKR. However, other studies have not shown any significant correlation between the E2 sequence and genotypes (Saito et al., 2003; Watanabe et al., 2003; Quer et al., 2004) or treatment outcomes (Gaudy et al., 2005).

Viral Kinetics

In patients treated with IFN, viral kinetic studies have shown slower turnover of hepatocytes infected with HCV genotype 1 compared with those infected with non-genotype 1 HCV (Neumann et al., 2000; Zeuzem et al., 2001). This suggests that a longer duration of treatment is required and is consistent with the fact that 48 weeks of treatment is superior to 24 weeks of treatment in patients with HCV genotype 1. Also, with slower turnover, constant antiviral pressure may be required and may explain the poor response to standard IFN, with its large peak-trough fluctuations in plasma concentration. With pegylated IFN, antiviral pressure is maintained more consistently, resulting in an improved response to treatment in genotype 1 patients.

Steatosis

The associations among steatosis, chronic hepatitis C, and HCV genotypes have been discussed earlier. Underlying steatosis is associated with a lower response to antiviral therapy, as shown in an evaluation of 1,428 patients in whom the SVR rate was significantly reduced to 35% in patients with underlying steatosis compared with 57% in those without steatosis (Poynard et al., 2003).

The mechanisms for decreased response to antiviral therapy in patients with steatosis remain unclear although there are several proposed explanations, including more severe fibrosis and alteration of IFN binding to hepatocyte induced by fat deposition. Reduced sensitivity to IFN therapy in patients with steatosis appears to be specific for metabolic steatosis associated with risk factors such as diabetes, obesity, and hyperlipidemia and therefore occurs predominantly in patients infected with non-genotype 3. Patients who have viral steatosis associated with genotype 3 appear to have less of a response to IFN therapy, although in the presence of other host-dependent factors such as obesity, a reduction in SVR rates may be observed.

Summary

The determination of the HCV genotype has become an essential part of management of chronic infection with HCV. The duration of treatment is dependent on the genotype, as a variation in response to antiviral therapy is observed among different genotypes. With the current knowledge, patients infected with genotypes 1 and 4 have an inferior response to IFN-based therapy compared to patients with genotypes 2, 3, 5, and 6. Therefore, patients infected with genotype 1 are treated for 48 weeks, whereas those infected with genotype 2 or 3 are treated for 24 weeks. The optimal duration of treatment for patients infected with genotypes 5 and 6 has yet to be defined, and further large trials involving these latter genotypes are awaited.

The differences in response to IFN-based therapy observed among genotypes are likely due to both viral and host factors. Although a number of these have been described, the exact mechanism remains unknown. The fact that the majority of patients infected with acute hepatitis C genotype 1 are cured with IFN therapy suggests that viral-host interactions are important rather than an intrinsic genotype-specific resistance to IFN.

References

Adinolfi, L.E., Gambardella, M., Andreana, A., Tripodi, M.F., Utili, R., Ruggiero, G. (2001). Steatosis accelerates the progression of liver damage of chronic hepatitis C patients and correlates with specific HCV genotype and visceral obesity. *Hepatology*, 33(6): 1358–1364.

Alter, M.J., Kruszon-Moran, D., Nainan, O.V., McQuillan, G.M., Gao, F., Moyer, L.A., Kaslow, R.A., Margolis, H.S. (1999). The prevalence of hepatitis C virus infection in the United States, 1988 through 1994. *New England Journal of Medicine*, 341(8): 556–562.

Bjoro, K., Bell, H., Hellum, K. B., Skaug, K., Raknerud, N., Sandvei, P., Doskeland, B., Maeland, A., Lund-Tonnesen, S., Myrvang, B. (2002). Effect of combined interferon-alpha induction therapy and ribavirin on chronic hepatitis C virus infection: a randomized multicentre study. *Scandinavian Journal of Gastroenterology*, 37(2): 226–232.

Bruno, S., Silini, E., Crosignani, A., Borzio, F., Leandro, G., Bono, F., Asti, M., Rossi, S., Larghi, A., Cerino, A., Podda, M., Mondelli, M.U. (1997). Hepatitis C virus genotypes and risk of hepatocellular carcinoma in cirrhosis: a prospective study. *Hepatology*, 25(3): 754–758.

Bruno, S., Camma, C., Di Marco, V., Rumi, M., Vinci, M., Camozzi, M., Rebucci, C., Di Bona, D., Colombo, M., Craxi, A., Mondelli, M.U., Pinzello, G. (2004). Peginterferon alfa-2b plus ribavirin for naive patients with genotype 1 chronic hepatitis C: a randomized controlled trial. *Journal of Hepatology*, 41(3): 474–481.

Bukh, J., Miller, R.H., Purcell, R.H. (1995). Genetic heterogeneity of hepatitis C virus: quasispecies and genotypes. *Seminars in Liver Disease*, 15: 41–63.

Buti, M., Sanchez-Avila, F., Lurie, Y., Stalgis, C., Valdes, A., Martell, M., Esteban, R. (2002). Viral kinetics in genotype 1 chronic hepatitis C patients during therapy with 2 different doses of peginterferon alfa-2b plus ribavirin. *Hepatology*, 35(4): 930–936.

Chamberlain, R.W., Adams, N., Saeed, A.A., Simmonds, P., Elliott, R.M. (1997). Complete nucleotide sequence of a type 4 hepatitis C virus variant, the predominant genotype in the Middle East. *Journal of General Virology*, 78: 1341–1347.

Choo, Q.L., Richman, K.H., Han, J. H., Berger, K., Lee, C., Dong, C., Gallegos, C., Coit, D., Medina-Selby, R., Barr, P.J., Weiner, P.J., Bradley, D.W., Kuo, G., Houghton, M. (1991). Genetic organization and diversity of the hepatitis C virus. *Proceedings of the National Academy of Sciences USA*, 88: 2451–2455.

Chung, R.T., Monto, A., Dienstag, J.L., Kaplan, L.M. (1999). Mutations in the NS5A region do not predict interferon-responsiveness in American patients infected with genotype 1b hepatitis C virus. *Journal of Medical Virology*, 58(4): 353–358.

Dalgard, O., Bjoro, K., Block Hellum, K., Myrvang, B., Ritland, S., Skaug, K., Raknerud, N., Bell, H. (2004). Treatment with pegylated interferon and ribavirin in HCV infection with genotype 2 or 3 for 14 weeks: a pilot study. *Hepatology*, 40(6): 1260–1265.

Delwaide, J., Gerard, C., Reenaers, C., Vaira, D., Bastens, B., Bataille, C., Servais, B., Maes, B., Belaiche, J., Hepatotropes GL; Groupe Liegeois d'Etudes des Virus Hepatotropes (GLEVE) (2005). Hepatitis C virus genotype 5 in southern Belgium: epidemiological characteristics and response to therapy. *Digestive Diseases and Sciences*, 50(12): 2348–2351.

Dev, A.T., McCaw, R., Sundararajan, V., Bowden, S., Sievert, W. (2002). Southeast Asian patients with chronic hepatitis C: the impact of novel genotypes and race on treatment outcome. *Hepatology*, 36(5): 1259–1265.

el-Zayadi, A., Simmonds, P., Dabbous, H., Prescott, L., Selim, O., Ahdy, A. (1996). Response to interferon-alpha of Egyptian patients infected with hepatitis C virus genotype 4. *Journal of Viral Hepatology*, 3: 261–264.

el-Zayadi, A., Selim, O., Haddad, S., Simmonds, P., Hamdy, H., Badran, H.M., Shawky, S. (1999). Combination treatment of interferon alpha-2b and ribavirin in comparison to interferon monotherapy in treatment of chronic hepatitis C genotype 4 patients. *Italian Journal of Gastroenterology and Hepatology*, 31(6): 472–475.

Enomoto, N., Sakuma, I., Asahina, Y., Kurosaki, M., Murakami, T., Yamamoto, C., Izumi, N., Marumo, F., Sato, C. (1995). Comparison of full-length sequences of interferon-sensitive and resistant hepatitis C virus 1b. Sensitivity to interferon is conferred by amino acid substitutions in the NS5A region. *Journal of Clinical Investigation*, 96(1): 224–230.

Enomoto, N., Sakuma, I., Asahina, Y., Kurosaki, M., Murakami, T., Yamamoto, C., Ogura, Y., Izumi, N., Marumo, F., Sato, C. (1996). Mutations in the nonstructural protein 5A gene and response to interferon in patients with chronic hepatitis C virus 1b infection. *New England Journal of Medicine*, 334(2): 77–81.

Fried, M.W., Shiffman, M.L., Reddy, K.R., Smith, C., Marinos, G., Goncales, F.L., Jr., Haussinger, D., Diago, M., Carosi, G., Dhumeaux, D., Craxi, A., Lin, A., Hoffman, J., Yu, J. (2002). Peginterferon alpha-2a plus ribavirin for chronic hepatitis C virus infection. *New England Journal of Medicine*, 347(13): 975–982.

Gale, M., Jr., Blakely, C.M., Kwieciszewski, B., Tan, S.L., Dossett, M., Tang, N.M., Korth, M.J., Polyak, S.J., Gretch, D.R., Katze, M.G. (1998). Control of PKR protein kinase by hepatitis C virus nonstructural 5A protein: molecular mechanisms of kinase regulation. *Molecular and Cellular Biology*, 18(9): 5208–5218.

Gaudy, C., Lambele, M., Moreau, A., Veillon, P., Lunel, F., Goudeau, A. (2005). Mutations within the hepatitis C virus genotype 1b E2-PePHD domain do not correlate with treatment outcome. *Journal of Clinical Microbiology*, 43(2): 750–754.

Hadziyannis, S.J., Sette, H., Jr., Morgan, T.R., Balan, V., Diago, M., Marcellin, P., Ramadori, G., Bodenheimer, H.Jr., Bernstein, D., Rizzetto, M., Zeuzem, S., Pockros, P.J., Lin, A., Ackrill, A.M.; PEGASYS International Study Group (2004). Peginterferon-alpha2a and ribavirin combination therapy in chronic hepatitis C: a randomized study of treatment duration and ribavirin dose. *Annals of Internal Medicine*, 140(5): 346–355.

Smith, D., Stuyver, L., Weiner, A. (1998). Classification, nomenclature, and database development for hepatitis C virus (HCV) and related viruses: proposals for standardization. International Committee on Virus Taxonomy. *Archives in Virology*, 143: 2493–2503.

Rubbia-Brandt, L., Quadri, R., Abid, K., Giostra, E., Male, P.J., Mentha, G., Spahr, L., Zarski, J.P., Borisch, B., Hadengue, A., Negro, F. (2000). Hepatocyte steatosis is a cytopathic effect of hepatitis C virus genotype 3. *Journal of Hepatology*, 33(1): 106–115.

Saito, T., Ito, T., Ishiko, H., Yonaha, M., Morikawa, K., Miyokawa, A., Mitamura, K. (2003). Sequence analysis of PePHD within HCV E2 region and correlation with resistance of interferon therapy in Japanese patients infected with HCV genotypes 2a and 2b. *American Journal of Gastroenterology*, 98: 1377–1383.

Schinkel, J., Spoon, W.J., Kroes, A.C. (2004). Meta-analysis of mutations in the NS5A gene and hepatitis C virus resistance to interferon therapy: uniting discordant conclusions. *Antiviral Therapy*, 9(2): 275–286.

Shi, S.T., Polyak, S.J., Tu, H., Taylor, D.R., Gretch, D.R., Lai, M.M. (2002). Hepatitis C virus NS5A colocalizes with the core protein on lipid droplets and interacts with apolipoproteins. *Virology*, 292(2): 198–210.

Shobokshi, O.A., Serebour, F.E., Skakni, L., Al Khalifa, M. (2003). Combination therapy of peginterferon alfa-2a and ribavirin significantly enhance sustained virological and biochemical response rate in chronic hepatitis C genotype 4 patients in Saudi Arabia. *Hepatology*, 38: 996A.

Simmonds, P., Alberti, A., Alter, H.J., Bonino, F., Bradley, D.W., Brechot, C., Brouwer, J.T., Chan, S.W., Chayama, K., Chen, D.S. (1994). A proposed system for the nomenclature of hepatitis C viral genotypes. *Hepatology*, 19(5): 1321–1324.

Simmonds, P., Bukh, J., Combet, C., Deleage, G., Enomoto, N., Feinstone, S., Halfon, P., Inchauspe, G., Kuiken, C., Maertens, G., Mizokami, M., Murphy, D.G., Okamoto, H., Pawlotsky, J.M., Penin, F., Sablon, E., Shin-I, T., Stuyver, L.J., Thiel, H.J., Viazov, S., Weiner, A.J., Widell, A. (2005). Consensus proposals for a unified system of nomenclature of hepatitis C virus genotypes. *Hepatology*, 42(4): 962–973.

Smuts, H.E, Kannemeyer, J. (1995). Genotyping of hepatitis C virus in South Africa. *Journal of Clinical Microbiology*, 33: 1679–1681.

Squadrito, G., Leone, F., Sartori, M., Nalpas, B., Berthelot, P., Raimondo, G., Pol, S., Brechot, C. (1997). Mutations in the nonstructural 5A region of hepatitis C virus and response of chronic hepatitis C to interferon alpha. *Gastroenterology*, 113(2): 567–572.

Tanaka, H., Tsukuma, H., Yamano, H., Okubo, Y., Inoue, A., Kasahara, A., Hayashi, N. (1998). Hepatitis C virus 1b(II) infection and development of chronic hepatitis, liver cirrhosis and hepatocellular carcinoma: a case-control study in Japan. *Journal of Epidemiology*, 8(4): 244–249.

Taylor, D.R., Shi, S.T., Romano, P.R., Barber, G.N., Lai, M.M.C. (1999). Inhibition of the interferon-inducible protein kinase PKR by HCV E2 protein. *Science*, 285: 107–109.

Thakeb, F.A.I., Omar, M.M., El Awady, M.M., Isshak, S.Y. (2003). Randomized controlled trial of peginterferon alfa-2a plus ribavirin for chronic hepatitis C virus genotype 4 among Egyptian patients. Abstract 37, Proceedings of the 54th Annual Meeting of the AASLD, Boston, MA, October 24–28.

Tokita, H., Okamoto, H., Iizuka, H., Kishimoto, J., Tsuda, F., Lesmana, L.A., Miyakawa, Y., Mayumi, M. (1996). Hepatitis C virus variants from Jakarta, Indonesia, classifiable into novel genotypes in the second (2e and 2f), tenth (10a) and eleventh (11a) genetic groups. *Journal of General Virology*, 77: 293–301.

Tokita, H., Okamoto, H., Tsuda, F., Song, P., Nakata, S., Chosa, T., Iizuka, H., Mishiro, S., Miyakawa, Y., Mayumi, M. (1994). Hepatitis C virus variants from Vietnam are classifiable into the seventh, eighth, and ninth major genetic groups. *Proceedings of the National Academy of Sciences USA*, 91: 11022–11026.

van Vlierberghe, H., Leroux-Roels, G., Adler, M., Bourgeois, N., Nevens, F., Horsmans, Y., Brouwer, J., Colle, I., Delwaide, J., Brenard, R., Bastens, B., Henrion, J., de Vries, R.A., de Galocsy, C., Michielsen, P., Robaeys, G., Bruckers, L. (2003). Daily induction combination treatment with alpha 2b interferon and ribavirin or standard combination treatment in naive chronic hepatitis C patients. A multicentre randomized controlled trial. *Journal of Viral Hepatology*, 10(6): 460–466.

von Wagner, M., Huber, M., Berg, T., Hinrichsen, H., Rasenack, J., Heintges, T., Bergk, A., Bernsmeier, C., Haussinger, D., Herrmann, E., Zeuzem, S. (2005). Peginterferon-alpha-2a (40KD) and ribavirin for 16 or 24 weeks in patients with genotype 2 or 3 chronic hepatitis C. *Gastroenterology*, 129(2): 522–527.

Watanabe, H., Nagayama, K., Enomoto, N., Itakura, J., Tanabe, Y., Sato, C., Izumi, N., Watanabe, M. (2003). Amino acid substitutions in PKR-eIF2 phosphorylation homology domain (PePHD) of hepatitis C virus E2 protein in genotype 2a/2b and 1b in Japan and interferon efficacy. *Hepatology Research*, 26: 268–274.

Wong, D.A., Tong, L.K., Lim, W. (1998). High prevalence of hepatitis C virus genotype 6 among certain risk groups in Hong Kong. *European Journal of Epidemiology*, 14: 421–426.

World Health Organization. Hepatitis C. Factsheet No 164 (2000). http://www.who.int/ mediacentre/factsheets.

Zein, N.N., Rakela, J., Krawitt, E.L., Reddy, K.R., Tominaga, T., Persing, D.H. (1996a). Hepatitis C virus genotypes in the United States: epidemiology, pathogenicity, and response to interferon therapy. Collaborative Study Group. *Annals of Internal Medicine*, 125(8): 634–639.

Zein, N.N., Poterucha, J.J., Gross J,B, Jr., Wiesner, R.H., Therneau, T.M., Gossard, A.A., Wendt, N.K., Mitchell, P.S., Germer, J.J., Persing, D.H. (1996b). Increased risk of hepatocellular carcinoma in patients infected with hepatitis C genotype 1b. *American Journal of Gastroenterology*, 91: 2560–2562.

Zhang, Y.Y., Lok, A.S., Chan, D.T., Widell, A. (1995). Greater diversity of hepatitis C virus genotypes found in Hong Kong than in mainland China. *Journal of Clinical Microbiology*, 33: 2931–2934.

Zeuzem, S., Lee, J.H., Roth, W.K. (1997). Mutations in the nonstructural 5A gene of European hepatitis C virus isolates and response to interferon alpha. *Hepatology*, 25(3): 740–744.

Zeuzem, S., Feinman, S.V., Rasenack, J., Heathcote, E.J., Lai, M.Y., Gane, E., O'Grady, J., Reichen, J., Diago, M., Lin, A., Hoffman, J., Brunda, M.J. (2000). Peginterferon alfa-2a in patients with chronic hepatitis C. *New England Journal of Medicine*, 343(23):1666–1672.

Zeuzem, S., Herrmann, E., Lee, J.H., Fricke, J., Neumann, A.U., Modi, M., Colucci, G., Roth, W.K. (2001). Viral kinetics in patients with chronic hepatitis C treated with standard or peginterferon alpha2a. *Gastroenterology*, 120(6): 1438–1447.

Zeuzem, S., Hultcrantz, R., Bourliere, M., Goeser, T., Marcellin, P., Sanchez-Tapias, J., Sarrazin, C., Harvey, J., Brass, C., Albrecht, J. (2004). Peginterferon alpha-2b plus ribavirin for treatment of chronic hepatitis C in previously untreated patients infected with HCV genotypes 2 or 3. Journal of Hepatology, 40(6): 993–939. Erratum in *Journal of Hepatology*, 42(3): 434.

Zylberberg, H., Chaix, M.L., Brechot, C. (2000). Infection with hepatitis C virus genotype 4 is associated with a poor response to interferon-a (Letter). *Annals of Internal Medicine*, 132: 845–846.

Interferon Treatment of Hepatitis C Virus Infection: From Basic Biology to Clinical Application

Norihiro Furusyo*, Masayuki Murata, and Jun Hayashi

Summary

Tremendous progress has been made in the field of antiviral treatment for hepatitis C virus (HCV) infection since the identification of the virus in 1989. Although early treatment regimens with interferon (IFN) alone achieved only limited success, the addition of the broad-spectrum antiviral agent ribavirin has greatly improved response. The primary goal of treatment for HCV infection—viral eradication—is best achieved when the viral level is reduced substantially during the early phase of treatment. Viral eradication is expressive of sustained virological response, the benefits of which are multifactorial and include improved hepatic histology: a decreased occurrence of hepatocellular carcinoma or liver failure and a lower probability of liver-related mortality. Treatment of HCV infection with the current "gold standard" of care—pegylated IFN in combination with ribavirin—is associated with an approximately 50% overall rate of viral eradication, a great improvement over previous IFN treatment regimens. However, more effective and better-tolerated treatments are needed for patients with unfavorable treatment profiles, such as genotype 1, a high viral level at baseline, hepatic steatosis, and poor adherence to treatment due to severe side effects.

Introduction

The hepatitis C virus (HCV), an approximately 9,600-nt single-stranded RNA virus of the *Flaviviridae* family, was found to be the causative agent of post-transfusion non-A, non-B hepatitis in 1989 (Choo et al., 1989). It has recently been classified as the sole member of the genus *Hepacivirus* (Robertson et al., 1998). An estimated 3% of the world's population, 170 million people, is infected with HCV. Chronic HCV infection is well known to be a major cause of chronic liver diseases worldwide and represents a major public health problem (Hayashi et al., 1991a). The virus is

* Address to the corresponding: Norihiro Furusyo, Associate Professor, Department of General Medicine, Kyushu University Hospital, Higashi-Ku, Fukuoka, 812-8582, Japan, Phone: +81-92-642-5909, Fax: +81-92-642-5916, E-mail: furusyo@genmedpr.med.kyushu-u.ac.jp

hepatotrophic, but not directly cytopathic, and elicits slowly progressive liver injury that results in end-stage liver disease unless effectively eradicated (Liang et al., 2000). Internists and primary care physicians need to be made aware that HCV infection is closely associated with hepatocellular carcinoma (HCC) and death due to chronic liver disease.

The eradication of HCV by antiviral treatment that leads to a sustained virological response (SVR) results in improved liver histology and a higher survival rate (Marcellin et al., 1997; Niederau et al., 1998). Patients who achieved a sustained response have maintained it for 7 to 10 years in almost all cases, and HCV RNA levels are undetectable in the liver of such successful patients, suggesting that an SVR is tantamount to a cure (Lau et al., 1998). Therefore, the primary goal of antiviral treatment of patients with chronic HCV infection is an SVR, defined as undetectable serum HCV RNA by a sensitive molecular assay 24 weeks after the end of the treatment (Hagiwara et al., 1992). Interferon (IFN), the only antiviral agent capable of eradicating HCV, has been widely used for the treatment of patients with chronic HCV infection. A treatment regimen currently in wide use combines IFN with ribavirin and has dramatically improved treatment outcomes.

This review focuses on historical and recent developments in the field and on the use of antiviral drugs in the treatment of patients with HCV infection.

Indications for IFN Treatment

Chronic HCV infection causes mild chronic inflammation of the liver. Ongoing cycles of inflammation, necrosis, and apoptosis eventually lead to fibrosis and, ultimately, cirrhosis, a severe bridging fibrosis with nodular regression (Yano et al., 1996; Poynard et al., 1997; Hayashi et al., 1997a, 2000; Ghany et al., 2003). Although the progression of liver fibrosis may not be linear and the determinants of the progression rate are not known definitively, treatment indications can be based on histological assessment of hepatic lesions. Potential contributing factors to progressive fibrosis include excessive alcohol intake, concomitant diseases associated with liver injury, for example, hepatitis B, steatohepatitis, hemochromatosis, co-infection with HIV, male gender, older age, obesity, immunosuppression, and certain major histocompatibility complex haplotypes (Hayashi et al., 1994a, 1998a; Yano et al., 1996; Furusyo et al., 1997, 2005; Poynard et al., 1997; Alric et al., 1997; Hourigan et al., 1999; Thomas et al., 2000; Monga et al., 2001; Ghany et al., 2003; Crosse et al., 2004; Kubo et al., 2005).

Table 1 shows the clinical features of candidates for successful IFN treatment. Patients who experience a biochemical and virological response to antiviral treatment have considerable improvement in the necroinflammatory components of their liver histopathology, which leads to a decrease in HCC incidence (Nishiguchi et al., 1995; Kasahara et al., 1998; Yoshida et al., 1999; Okanoue et al., 1999; Kashiwagi et al., 2003; Murata et al., 2006). The Consensus Panel of the National Institutes of Health (NIH) in the United States recommended that all patients with chronic HCV infection be considered as potential candidates for antiviral treatment (Sheeff

Table 1 Candidates for antiviral treatment of chronic HCV infection

Widely accepted candidates
1. 18 or more years of age
2. Elevated aminotransferase activity (abnormal ALT level)
3. Presence of moderate to severe fibrosis by biopsy (METAVIR stage 2 or more; Ishak satge 3 or more)
4. Absence of jaundice, acites, encephalopathy
5. Absence of uncontrolled seizure or psychiatric disorder
6. Good compliance and willingness to be treated
7. Infection with genotype 2 or 3 regardless of ALT abonormality

Individually considered candidates
1. Infection with genotype 1 and persistently normal ALT level
2. Presence of no or mild fibrosis by biopsy (METAVIR stage less than 2; Ishak stage less than 3)
3. Recent record of alcohol abuse (Abstinence will be necessary)
4. Injection drug user (Goodl compliance and a substance abuse program will be necessary)
5. Acute hepatitis C (Observation after 2 to 4 months of the onset for waiting the spontaneous clearance)
6. Less than 18 years of age
7. Coinfection with HIV
8. Chronic renal disease
9. Liver transplantation recipient

and Hoofnagle 2002). Treatment should be recommended especially for patients who are at risk of progression to cirrhosis, such as those who are characterized by the presence of HCV viremia, generally persistent elevation of the serum alanine aminotransferase (ALT) level, portal or bridging fibrosis, and moderate inflammation and necrosis of the liver. The following particular patients should be taken into account for antiviral treatment: those having mild liver disease, those with recurrence after transplantation, those who received a liver transplant, those with acute HCV infection, and those with co-infection with HIV. These recommendations are consistent with guidelines set by the American Association for the Study of the Liver, Strader et al. (2004), and the European Association for the Study of the Liver (Alberti & Benvegnu, 2003).

Appropriate Evaluations Before IFN Treatment

In addition to standard tests, special attention should be paid to extrahepatic mani-festations, psychiatric disorders, HIV co-infection, excessive alcohol consumption, and excess body weight. As recommended previously (Sheeff and Hoofnagle 2002; Strader et al., 2004; Alberti & Benvegnu, 2003), HCV genotypes and serum HCV RNA levels must be determined before treatment (Table 2). The HCV genotype influences both the treatment indications and the therapeutic strategy, because treatment is more effective and shorter in patients infected with HCV genotypes 2 and 3, for which the efficacy was approximately 80% in clinical trials (Hayashi et al., 1994a, 1998a; Furusyo et al., 2002, 2006). Although the measurement of the HCV RNA level by a qualitative polymerase reaction (PCR) test at baseline is not

Table 2 Appropriate testing before IFN treatment

1.   HCV genotyping
2.   Serum HCV RNA level
3.   Hepatic histology by biopsy (not mandatory)
4.   Testing for HIV infection

commonly done to determine the length of treatment, it may be useful, at least for patients infected with HCV genotype 1 for whom a low HCV RNA level can provide the expectation of an early response, which is a good indicator of the probability of an SVR (Hayashi et al., 1998b; Yamaji et al., 1998; Furusyo et al., 2002).

Liver biopsy, despite the possibility of sampling error, remains the gold standard for evaluating fibrosis and hepatitis activity (Dienstag, 2002). For patients infected with genotypes 2 and 3 who have the probability of a favorable IFN treatment, response is extremely high, and IFN treatment may outweigh considerations of disease severity and the potential for progression in the future. Therefore, some authorities have suggested that it is not necessary to obtain a liver biopsy before treating patients with genotypes 2 and 3. Although a liver biopsy is not absolutely necessary in all cases, it is a useful tool because it is a key parameter for assessing the current status of the liver and because it provides prognostic information concerning disease progression (Dienstag, 2002; National Institutes of Health Consensus Development Conference statement, 2002).

The Background of IFN Treatment

In 1986, before the discovery of HCV, IFN was reported to have a biochemical response as an inflammatory agent in non-A, non-B hepatitis. Hoofnagle et al. reported the normalization of ALT following the administration of IFN-alpha for patients with non-A, non-B hepatitis (Hoofnagle et al., 1986). In the 1990s, IFN-alpha became the most widely accepted form of treatment for chronic HCV infection. The use of IFN was shown to result in a decrease in the serum ALT level and to cause HCV RNA to decline to the undetectable level (Shindo et al., 1991; Hayashi et al., 1994a). However, in many cases the ALT and HCV RNA levels promptly returned to pretreatment levels after cessation of IFN treatment (Hayashi et al., 1994a). An optimal response, an SVR, can be defined as a persistently normal serum ALT level and the absence of HCV RNA from the serum at the end of treatment and for at least 6 months thereafter. Since the first observations of these IFN-produced biochemical and virological effects, studies have reported several host and viral characteristics associated with an SVR (Brouwer et al., 1998; Castro et al., 2002; Berg et al., 2003). The most important predictors of an SVR following IFN treatment are hepatic fibrosis, the HCV genotype, and the pretreatment serum HCV RNA level (Hayashi et al., 1994a, 1998a; Furusyo et al., 2002, 2006). Table 3 shows factors correlated with an SVR to a combination treatment of pegylated IFN (peg-IFN) and ribavirin for chronic hepatitis C. Most of the patients with a good response had only a mild or moderate degree of fibrosis on liver biopsy, had HCV genotype 2 or 3, and had a low baseline HCV RNA level.

Table 3 Predictors of successful response to IFN treatment for patients with chronic HCV infection

Non-genotype 1
Low HCV RNA level
Absence of severe fiblosis and cirrhosis
Age 40 years or younger
Male
Lighter body weight
Non-black ethnocity
Absence of liver steatosis
Good adherence
Avoiding of discontined treatment

IFN Monotherapy

IFNs are multifunctional immunomodulatory cytokines whose effects include antiviral activity, inhibition of angiogenesis, regulation of cell differentiation, growth regulatory properties, and enhancement of major histocompatibility complex antigen expression. They have an anti-inflammatory through what has come to be called a cytokine cascade (Kirchner, 1984; Tilg, 1997; Kawakami et al., 2000; Murata et al., 2002; Furusyo et al., 2005). Several types of IFN, recombinant IFN-alpha-2a, recombinant IFN-alpha-2b, natural IFN-alpha, natural IFN-beta, recombinant IFN-beta, and consensus IFN (IFN-alfacon-1), are available for the treatment for patients with HCV infection. Consensus IFN was designed by selecting the most frequently occurring amino acid at each site of the amino acid sequences of 13 known IFN-alpha subtypes. Broadly speaking, IFN-alpha and IFN-beta have been the most widely used IFNs for the treatment of HCV infection. Like IFN-alpha and IFN-beta, IFN-gamma is classified as a type 1 IFN and has shown activity against HCV in cell culture systems (Frese et al., 2002) but does not effectively reduce the HCV RNA level in humans (Soza et al., 2005).

There are differences in the specific activities and potencies of IFNs. The dosage and duration of IFN treatment may vary, but only a few of the IFNs and their approved regimens have been compared head to head. However, monotherapy outcomes, in terms of response rates, generally appear to be similar for the different regimens commonly used to treat patients. An SVR occurs in about 15% to 20% of patients treated with IFN monotherapy for 6 months (about 5% for genotype 1 patients and about 50% for non-genotype 1 patients) (Marcellin et al., 1994; Poynard et al., 1996; Hayashi et al., 1994a, 1998a; Furusyo et al., 1997). Several analyses have found that prolonged courses of IFN, 12 to 18 months, appear to be needed to maximize the chances of having an SVR to treatment, with 25% to 30% of patients responding to prolonged treatment (Poynard et al., 1995), but higher dosages of IFN, greater than 3 million units three times per week, do not seem to substantially improve the rates of sustained response, and a higher dosage has been associated with increased adverse effects (Bennett et al., 1997). IFN monotherapy, especially for patients infected with HCV genotype 1, has had limited success.

Combined Treatment with IFN and Ribavirin

Treatment regimens using a combination of an IFN and ribavirin have significantly improved the treatment outcome (Figure 1) (Lindsay et al., 2001; Luxon et al., 2002; Scott & Perry, 2002; Hugle & Cerny, 2003; Sanchez-Tapias et al., 2006). Combining weekly subcutaneous peg-IFN-alpha treatment with daily oral ribavirin is more effective than monotherapy with standard IFN or peg-IFN-alpha or a combination treatment with a standard IFN and ribavirin (Poynard et al., 1998; McHutchison et al., 1998; Heathcote et al., 2000; Manns et al., 2001; Reddy et al., 2001; Fried et al., 2002; Hadziyannis et al., 2004; Mangia et al., 2005; Furusyo et al., 2006). Pegylation is defined as modification of a drug by the addition of an artificial polymer, polyethylene glycol, for the purposes of delaying drug elimination, lowering its antigenicity, and modifying the drug's effect. Although a standard IFN requires a dosing interval of 1 or 2 days to maintain an effective blood concentration because of its approximately 8-hour elimination half-life, peg-IFN has the great advantage of making it possible to maintain a stable blood concentration with a single weekly administration. At present, two peg-IFNs are available: a weight-based, 1.5-mcg/kg dose of peg-IFN-alpha-2b, and a fixed, 180-mcg dose of peg-IFN-alpha-2a.

Ribavirin, initially synthesized in 1970, is an orally administered nucleoside analogue (a guanosine analogue) with a broad spectrum of antiviral properties that possess activity against several RNA and DNA viruses. When monotherapy with ribavirin is used for chronic HCV infection, the serum ALT level of most patients declines without any significant change in the serum HCV RNA level, even with prolonged treatment (Di Bisceglie et al., 1995; Hoofnagle et al., 1996; Furusyo et al., 2006), suggesting that the biologically beneficial effect is not associated with antiviral activity (Furusyo et al., 2005). The probable beneficial roles of ribavirin in the treatment of chronic HCV infection are an immunologic modulation (a shift from a Th2 to a Th1 response and suppression of interleukin-10 synthesis),

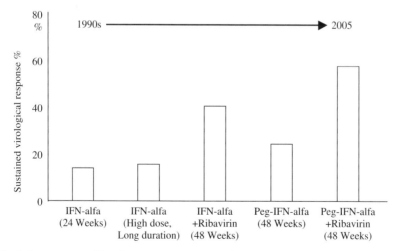

Fig. 1 Improvement of interferon treatment outcome for patients with chronic HCV infection

inhibition of host inosine monophosphate dehydrogenase activity, depletion of intracellular guanosine triphosphate pools, induction of mutational catastrophe, and a moderate, transient, early direct antiviral effect (Crotty et al., 2000; Cramp et al., 2000; Lau et al., 2002; Dixit et al., 2004). Surprisingly, the addition of ribavirin to IFN treatment leads to a marked improvement in the rate of sustained response (McHutchison et al., 1998; Lau et al., 2002). Patients treated with peg-IFN plus a ribavirin dose of 10.6 mg/kg/day or more had a greater chance of developing an SVR than those treated with peg-IFN plus a lower daily dose of ribavirin (Manns et al., 2001).

According to the most recent large clinical trials, a uniform 48 weeks of a combination treatment with peg-IFN and ribavirin yields the highest rate of sustained response (Strader et al., 2004; Dienstag & McHutchison, 2006). The response to a 48-week peg-IFN plus ribavirin treatment can be divided into three general patterns: a sustained virological response (SVR), relapse, and non-response (Figure 2). The overall sustained response rates were 54–56%. The response rate for genotype 1 patients exceeded 40% for the first time, and some rates were recorded as high as 42–46%. The rates of 76–82% for genotypes 2 and 3 are also impressive. Patients with genotypes 2 and 3 can be treated with a shorter duration (24 weeks) of treatment and with a lower dose of ribavirin with no sacrifice to the response rate (Hadziyannis et al., 2004).

Serum HCV RNA level at baseline is another determinant of the antiviral treatment outcome of patients infected with genotype 1 but is not useful for the other genotypes (Furusyo et al., 2006). A sustained response is consistently higher in patients with a low HCV RNA level, usually defined as 800,000 or fewer IU/ml (Manns et al., 2001; Fried et al., 2002).

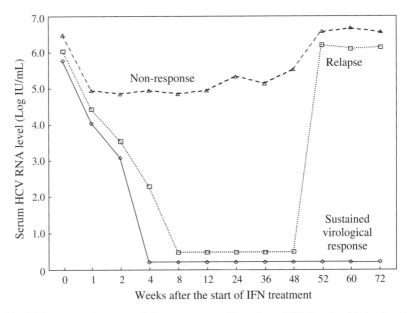

Fig. 2 Three general patterns of the response to a 48-week peg-IFN plus ribavirin treatment

Host factors affect the chance of a sustained response, albeit less so than the genotype. These factors include age, race, gender, obesity, and the degree of hepatic fibrosis and steatosis. African Americans have been shown to have responses rates only one-half to one-third those of Caucasians (Muir et al., 2004). The reasons for the racial differences in response rates to peg-IFN plus ribavirin treatment are not well known.

The sustained response rates for genotypes 2 and 3 are not significantly higher than those achieved for these favorable genotypes with non-peg-IFN combined with ribavirin. However, the combination of peg-IFN and ribavirin is currently considered to be the standard of care for the treatment of previously untreated patients with chronic HCV infection, even for those with genotypes 2 and 3, because the reduction of the injection frequency favors this combination treatment.

Monitoring Serum HCV RNA Level During IFN Treatment

Response to the IFN treatment regimens, including IFN monotherapy, peg-IFN monotherapy, and the combined treatments with ribavirin, is characterized by a two-phase pattern of decreases in the serum HCV RNA level, with an initial rapid decline seen from 24–72 hours after the start of treatment, followed by a gradual decline for several weeks (Neumann et al., 1998). This pattern is believed to reflect an initial inhibition of HCV replication and/or release followed by a different antiviral mechanism (i.e., the loss of infected hepatocytes) (Zeuzem et al., 1998; Buti et al., 2002; Layden-Almer et al., 2003).

In the early 1990s, comparison of assays was problematic because they did not use the same units to represent the amount of HCV RNA (Hayashi et al., 1998b; Furusyo et al., 2002). Recently, this problem has been overcome by the World Health Organization's establishment of an international standard unit for the universal standardization of HCV RNA quantification units (Neumann et al., 1998). Several assays for the quantification of HCV RNA, both PCR and branched DNA techniques, have been developed and have become available for clinical use, especially for the monitoring of the antiviral response to antiviral treatment. These assays are especially useful because the early monitoring of favorable viral kinetics has a direct bearing on the possibility of a sustained response by IFN treatment (Yamaji et al., 1998; Davis, 2002). The monitoring of the serum HCV RNA levels at baseline and at 12 weeks after the start of treatment is most effective when the same quantitative assays are used for both tests. A sustained response can be confirmed by a 2 or more $\log_{10}$ reduction in the HCV RNA level during the first 12 weeks of treatment, which is called an "early virological response" (EVR). The probability of an ultimate sustained response is approximately 70% for patients with an EVR, while the probability is less than 3% for those without an EVR (Davis et al., 2002, 2003). Moreover, even in these reports of patients with a 2 or more $\log_{10}$ reduction of the HCV RNA level at 12 weeks after the start of treatment, 84% of those with undetectable HCV RNA by PCR achieved a sustained response, but a sustained response was achieved by only 21% of those with detectable HCV RNA. These findings suggest that an SVR is more likely after a rapid and profound reduction of the serum HCV RNA level by IFN treatment.

Management of Side Effects and Educational Guidance

To protect against and control side effects, it is important to carefully monitor the clinical course and laboratory findings during IFN plus ribavirin treatment (Table 4). Flu-like symptoms are often found during treatment but are usually not severe and can be managed with analgesics such as acetaminophen or non-steroidal anti-inflammatory drugs. Marrow suppression, especially leukopenia and thrombocy-topenia, which can be induced by IFN, is very important for judging whether or not to continue treatment. Ribavirin contributes additional side effects, the most important of which is hemolytic anemia. When leukopenia, thrombocytopenia, or anemia is found, dosage adjustments may be required to avoid having to discontinue treatment (Nomura et al., 2004a).

Alcohol intake in HCV-infected patients appears to increase the higher fibrosis progression rate (Kubo et al., 2005). Avoiding alcohol during IFN treatment is essential, because alcohol affects the treatment response (National Institutes of Health Consensus Development Conference statement, 2002; Corrao & Arico, 1998). Patients should limit their alcohol intake to fewer than four drinks per week.

IFN has been shown to have an abortifacient effect in animal studies and thus should not be used during pregnancy. Ribavirin is embrytoxic and teratogenic. Strict birth control is necessary by both men and women during treatment and for more than 6 months after treatment.

Table 4 Side effects of antiviral treatment

Interferon-induced
1. Marrow suppression (leukopenia, thrombocytopenia, anemia)
2. Flu-like symptoms (fever, myalgia, headache)
3. General fatigue and irritability
4. Depression and insomnia
5. Weight loss and anorexia
6. Autoimmone diseases (Hypothyroidism and hyperthyroidism, Diabetes mellitus)
7. Rash and pruritis
8. Retinopathy (especially in daibetes patients and hypertention patients)
9. Nausea, vomiting, and diarrhea
10. Hair loss
11. Cough and Plumonary interstitial fibrosis

Ribavirin-induced
1. Marrow suppression (anemia, leukopenia, thrombocytopenia)
2. General fatigue and irritability
3. Weight loss and anorexia
4. Rash and pruritis
5. Nausea, vomiting, and diarrhea
6. Cough

Approach to Other Patient Populations

Acute Hepatitis C

Acute infection with HCV is marked by a high rate of viral persistence, with chronic infection seen in 50–80% of these patients. An important clinical observation about IFN treatment, without ribavirin, of HCV infection is the very high rate of response of patients with acute hepatitis C (Jaeckel et al., 2001). IFN treatment for acute infection with HCV reduces the chronicity rate to 10% or lower (Nomura,et al., 2004b). Strikingly, almost all patients with acute hepatitis C, regardless of genotype or initial viral level, rapidly become serum HCV RNA negative on treatment. A recent study of antiviral treatment of patients with acute hepatitis C reported that a high-dose, short-term treatment was effective (Nomura et al., 2004b). The results of a randomized control study of high-dose, short-term, early treatment (beginning 8 weeks after the onset of acute hepatitis) versus late treatment (beginning after 1 year of observation) showed that 87% of the patients with early intervention achieved a sustained response by the intramuscular, daily administration of 6 million units (MU) of IFN-alpha for four weeks, but only 40% of patients with late intervention had a sustained response (Nomura et al., 2004b). Moreover, all of the patients who experienced a relapse after the initial 4 weeks of treatment received an additional 20 weeks of treatment (6 MU, 3 times weekly) and became SVR, meaning that the early intervention resulted in total 100% SVR, while total 53% of patients achieved SVR after the additional 20 weeks of treatment (Nomura et al., 2004b). Similar outcomes were found for patients on maintained hemodialysis being treated with IFN monotherapy for acute hepatitis C (Furusyo et al., 2004). These findings suggest that nonresponse to IFN treatment might be acquired during the establishment of chronic infection.

Liver Transplantation

Although HCV-related end-stage liver disease represents the most frequent indication for liver transplantation, transplantation is not a clinical cure for hepatitis C. Viral recurrence, as documented by detectable viremia, is universal, and damage to the new liver occurs routinely. Recurrent HCV remains a persistent problem and a leading cause of graft loss. Attempts to prevent reinfection with immune globulin or other agents have not been successful (Charlton, 2003). Histological recurrence with allograft hepatitis owing to HCV occurs in up to 90% of patients by the fifth year after transplantation (Berenguer, 2003). Moreover, up to 42% of patients with HCV-reinfected cirrhosis after transplantation develop decompensation, manifested as ascites, encephalopathy, or hepatic hydrothorax, and less than 50% of these patients survive for one year after they develop decompensation (Berenguer et al., 2000). Clearly, the progression of hepatitis C is accelerated after transplantation as compared with non-transplantation patients. Thus, finding a way to use liver transplantation as a treatment for hepatitis C will be difficult, but if such a treatment is found, it will be clinically important.

Treating patients waiting for transplant when they are on the waiting list and pre-transplant viral eradication represent the ideal. Unfortunately, the results of a peg-IFN plus ribavirin treatment in an NIH trial showed that patients with compensated cirrhosis had a sustained response rate of only 11% (Shiffman et al., 2004). However, the trial results also showed that an initial treatment of low-dose IFN, including peg-IFN plus ribavirin treatment, followed by a slow escalation in dose may be associated with improved tolerability and efficacy in patients with compensated cirrhosis (Shiffman et al., 2004). Additionally, such cirrhosis patients who achieved a sustained response before transplantation or who are transplanted while on treatment but who are undetectable for HCV RNA have good outcomes, with a less than 10% probability of HCV recurrence (Everson, 2005). Thus, IFN treatment could potentially cure some of these patients, but the high discontinuation rate, 27%, should be taken into account when treating patients on a waiting list. After liver transplantation, the tolerability of IFN plus ribavirin treatment is suboptimal, with very high discontinuation of treatment, 37%, because of severe leukopenia and anemia arising from drug-induced bone marrow suppression and renal insufficiency (Gane, 2002; Rodriguez-Luna et al., 2004). A sustained response is achieved by less than 30% of patients after liver transplantation (Rodriguez-Luna et al., 2004). For patients undergoing liver transplantation for chronic HCV infection, the development of new classes of potent, well-tolerated antiviral agents merits a high priority.

Co-infection with HCV and HIV

Because both HCV and HIV are blood-borne viruses and share routes of transmission, HCV and HIV co-infection is particularly common in injection drug users. HIV infection has a detrimental effect on the natural history of HCV infection. The acceleration of liver disease, progression of fibrosis, frequency of cirrhosis, liver failure, and HCC have become substantial sources of mortality and morbidity in patients with HCV and HIV co-infection (Monga et al., 2001; Mohsen et al., 2003).

Ideally, HIV infection should be well controlled with highly active antiretroviral therapy (HAART) before the treatment of HCV infection is initiated. In addition, the management of the chronic hepatitis C of HCV and HIV co-infected patients can be confounded by the difficulty in distinguishing among hepatitis caused by the HCV infection itself, HAART hepatotoxicity, and opportunistic infection involving the liver (Laskus et al., 1998; Kottilil et al., 2004). Patients with HCV and HIV co-infection have lower response rates to peg-IFN plus ribavirin treatment for HCV than do patients with HCV infection alone (Perez-Olmeda et al., 2003). However, HCV and HIV co-infected patients with genotype 2 or 3 achieve a sufficient sustained response, 62–73%, while patients with genotype 1 have a sustained response rate of only 14–29% (Torriani et al., 2004; Chung et al., 2004). Thus, peg-IFN plus ribavirin treatment is optimal for HCV and HIV co-infected patients. However, ribavirin should be avoided if didanosine is critical to the HIV treatment regimen because of the possibility of a drug interaction. Ribavirin can increase the activity and potentiate the toxicity of didanosine (Lafeuillade et al., 2001).

The human T-lymphotropic virus (HTLV-I) is a human retrovirus, as is HIV, and prevalence studies have shown that it infects 10 to 20 million people worldwide

(Kashiwagi et al., 1990). HTLV-1 infection appears to modify the natural progression of HCV infection by leading to a more severe and rapid progression of liver diseases (Kishihara et al., 2001).This occurs because HTLV-1 causes impairment of host immunity and induces functional impairment of cellular immune response (Hayashi et al., 1997b). Moreover, the rate of sustained response to IFN treatment of patients with HTLV-1 and HCV co-infection is significantly lower than for patients with HCV alone (Kishihara et al., 2001). Taken together, further modification of the currently popular treatments is needed for HCV patients co-infected with such human retoviruses.

End-Stage Renal Disease

HCV infection is highly prevalent in patients with end-stage renal disease (ESRD). The prevalence shows a consideration variation, 3–80%, between countries and centers (Hayashi et al., 1991b, 1994a; Furusyo et al., 2001; Fabrizi et al., 2002). A detrimental effect of HCV on patients and graft survival after kidney transplantation has been reported (Bruchfeld et al., 2004). The main complications for such patients are an increased risk of severe infection, liver disease, de novo glomeruonephritis with or without crioglobulinemia, and diabetes (Pereira et al., 1998; Furusyo et al., 2000a; Cruzado et al., 2001; Bloom et al., 2002). As a result of these potential risks of HCV infection of patients with ESRD, the current recommendation is to give antiviral treatment before transplantation with the objective of eradicating the virus.

IFN-alpha monotherapy is generally well tolerated and is more effective for hemodialysis patients with chronic HCV infection than for those with normal renal function (Fabrizi et al., 2003; Russo et al., 2003). This can be partly explained by the phenomenon in which serum HCV RNA levels are lower in hemodialysis patients than in patients with normal renal function (Furusyo et al., 2000b). Maintained hemodialysis affects a lower HCV RNA level in sera. However, a fairly high rate of discontinuation of IFN treatment, over 30%, due to serious adverse events has been reported for these patients (Fabrizi et al., 2003; Russo et al., 2003). The role of treatment for this population and the safety and utility of small doses ribavirin in combination with peg-IFN are currently under investigation. Monitoring plasma ribavirin concentration during ribavirin treatment, ribavirin-induced anemia can be handled by injecting an erythropoietin and supplying adequate iron (Bruchfeld et al., 2003). In several cases, the use of peg-IFN plus ribavirin in ESRD patients was found to be safe, even though side effects were fairly frequent (Sporea et al., 2004; Annicchiarico & Siciliano, 2004; Bruchfeld et al., 2006).

Problems to Be Solved

Despite the great improvement in IFN treatment response, the rate of SVR is far from ideal, and many problems remain: about a 50% rate of nonresponse (especially for genotype 1 patients); adverse effects causing patients to have difficulty tolerating the treatment regimens; and contraindicated patients in some special settings. Moreover, the probability of success depends on viral and host factors that are often beyond the control of patients and physicians. Recently, it has been reported that extension of treatment with peg-IFN plus ribavirin from 48 to 72 weeks significantly increases the rate of SVR in genotype 1 infected patients with detectable HCV RNA in sera at the week of treatment (Sanchez-Tapias et al., 2006). However, more effective and easily tolerated treatments are desirable.

The epidemiology of chronic HCV infection has seen great change over the last decade, and great progress has been made in the development of new diagnostic methods and treatment strategies, thanks to the combined efforts of academic- and industry-sponsored research. However, a number of issues that have not been completely solved remain, such as the management of the substantial number of nonresponders to IFN treatment, a full understanding of the pathogenesis of liver disease and the mechanisms of chronicity, and severe end-stage liver disease.

Conclusions

Treatments using peg-IFN plus ribavirin have yielded improved rates of SVR, but studies continue to show that the SVR rate is low for patients with HCV genotype 1 infection, for patients with high HCV RNA levels, and for patients with more advanced stages of fibrosis. Future advances in the management of HCV infection will require tremendous efforts to develop more effective treatments for the currently nonresponding populations.

Acknowledgments We wish to thank Norihiko Kubo, MD, Kazuhiro Toyoda, MD, Hiroaki Takeoka, MD, Masataka Etoh, BS, and the members of our laboratory for their discussion and critical evaluation of this manuscript.

References

Alberti, A., Benvegnu, L. (2003). Management of hepatitis C. *Journal of Hepatology*, 36: S104–S118.

Alric, L., Fort, M., Izopet, J., Vinel, J.P., Charlet, J.P., Selves, J., Puel, J., Pascal, J.P., Duffaut, M., Abbal, M. (1997). Genes of the major histocompatibility complex class II influence the outcome of hepatitis C virus infection. *Gastroenterology*, 113: 1675–1681.

Annicchiarico, B.E., Siciliano, M. (2004). Pegylated interferon-alpha 2b monotherapy for haemodialysis patients with chronic hepatitis C. *Alimentary Pharmacology Therapy*, 20: 123–124.

Bennett, W.G., Inoue, Y., Beck, J.R., Wong, J.B., Pauker, S.G., Davis, G.L. (1997). Estimates of the cost-effectiveness of a single course of interferon-alpha 2b in patients with histologically mild chronic hepatitis C. *Annals of Internal Medicine*, 127: 855–865.

Berenguer, M., Prieto, M., Rayon, J.M., Mora, J., Pastor, M., Ortiz, V., Carrasco, D., San Juan, F., Burgueno, M.D., Mir, J., Berenguer, J. (2000). Natural history of clinically compensated hepatitis C virus-related graft cirrhosis after liver transplantation. *Hepatology*, 32: 852–858.

Berenguer, M. (2003). Host and donor risk factors before and after liver transplantation that impact HCV recurrence. *Liver Transplantation*, 9: S44–S47.

Berg, T., Sarrazin, C., Herrmann, E., Hinrichsen, H., Gerlach, T., Zachoval, R., Wiedenmann, B., Hopf, U., Zeuzem, S. (2003). Prediction of treatment outcome in patients with chronic hepatitis C: significance of baseline parameters and viral dynamics during therapy. *Hepatology*, 37: 600–609.

Bloom, R.D., Rao, V., Weng, F., Grossman, R.A., Cohen, D., Mange, K.C. (2002). Association of hepatitis C with posttransplant diabetes in renal transplant patients on tacrolimus. *Journal of the American Society of Nephrology*, 13: 1374–1380.

Brouwer, J.T., Nevens, F., Kleter, B., Elewaut, A., Adler, M., Brenard, R., Chamuleau, R.A., Michielsen, P.P., Pirotte, J., Hautekeete, M.L., Weber, J., Bourgeois, N., Hansen, B.E., Bronkhorst, C.M., ten Kate, F.J., Heijtink, R.A., Fevery, J., Schalm, S.W. (1998). Efficacy of interferon dose and prediction of response in chronic hepatitis C: Benelux study in 336 patients. *Journal of Hepatology*, 28: 951–959.

Bruchfeld, A., Lindahl, K., Stahle, L., Soderberg, M., Schvarcz, R. (2003). Interferon and ribavirin treatment in patients with hepatitis C-associated renal disease and renal insufficiency. *Nephrology Dialysis Transplantation*, 18: 1573–1580.

Bruchfeld, A., Wilczek, H., Elinder, C.G. (2004). Hepatitis C infection, time in renal-replacement therapy, and outcome after kidney transplantation. *Transplantation*, 78: 745–750.

Bruchfeld, A., Lindahl, K., Reichard, O., Carlsson, T., Schvarcz, R. (2006). Pegylated interferon and ribavirin treatment for hepatitis C in haemodialysis patients. *Journal of Viral Hepatitis*, 13: 316–321.

Buti, M., Sanchez-Avila, F., Lurie, Y., Stalgis, C., Valdes, A., Martell, M., Esteban, R. (2002). Viral kinetics in genotype 1 chronic hepatitis C patients during therapy with 2 different doses of peginterferon alpha-2b plus ribavirin. *Hepatology*, 35: 930–936.

Castro, F.J., Esteban, J.I., Juarez, A., Sauleda, S., Viladomiu, L., Martell, M., Moreno, F., Allende, H., Esteban, R., Guardia, J. (2002). Early detection of nonresponse to interferon plus ribavirin combination treatment of chronic hepatitis C. *Journal of Viral Hepatitis*, 9: 202–207.

Charlton, M. (2003). Natural history of hepatitis C and outcomes following liver transplantation, *Clinical Liver Disease*, 7: 585–602.

Choo, Q.L., Kuo, G., Weiner, A.J., Overby, L.R., Bradley, D.W., Houghton, M. (1989). Isolation of a cDNA clone derived from a blood-borne non-A, non-B viral hepatitis genome. *Science*, 244: 359–362.

Chung, R.T., Andersen, J., Volberding, P., Robbins, G.K., Liu, T., Sherman, K.E., Peters, M.G., Koziel, M.J., Bhan, A.K., Alston, B., Colquhoun, D., Nevin, T., Harb, G., van der Horst, C., AIDS Clinical Trials Group A5071 Study Team (2004). Peginterferon alpha-2a plus ribavirin versus interferon alpha-2a plus ribavirin for chronic hepatitis C in HIV-coinfected persons. *New England Journal of Medicine*, 351: 451–459.

Corrao, G., Arico, S. (1998). Independent and combined action of hepatitis C virus infection and alcohol consumption on the risk of symptomatic liver cirrhosis. *Hepatology*, 27: 914–919.

Cramp, M.E., Rossol, S., Chokshi, S., Carucci, P., Williams, R., Naoumov, N.V. (2000). Hepatitis C virus-specific T-cell reactivity during interferon and ribavirin treatment in chronic hepatitis C. *Gastroenterology*, 118: 346–355.

Crosse, K., Umeadi, O.G., Anania, F.A., Laurin, J., Papadimitriou, J., Drachenberg, C., Howell, C.D. (2004). Racial differences in liver inflammation and fibrosis related to chronic hepatitis C. *Clinical Gastroenterology Hepatology*, 2: 463–468.

Crotty, S., Maag, D., Arnold, J.J., Zhong, W., Lau, J.Y., Hong, Z., Andino, R., Cameron, C.E. (2000). The broad-spectrum antiviral ribonucleoside ribavirin is an RNA virus mutagen. *Natural Medicine*, 6: 1357–1379.

Cruzado, J.M., Carrera, M., Torras, J., Grinyo, J.M. (2001). Hepatitis C virus infection and de novo glomerular lesions in renal allografts. *American Journal of Transplants*, 1: 171–178.

Davis, G.L. (2002). Monitoring of viral levels during therapy of hepatitis C. *Hepatology*, 36: S145–S145.

Davis, G.L., Wong, J.B., McHutchison, J.G., Manns, M.P., Harvey, J., Albrecht, J. (2003). Early virologic response to treatment with peginterferon alpha-2b plus ribavirin in patients with chronic hepatitis C. *Hepatology*, 38: 645–652.

Di Bisceglie, A.M., Conjeevaram, H.S., Fried, M.W., Sallie, R., Park, Y., Yurdaydin, C., Swain, M., Kleiner, D.E., Mahaney, K., Hoofnagle, J.H. (1995). Ribavirin as therapy for chronic hepatitis C. A randomized, double-blind, placebo-controlled trial. *Annals of Internal Medicine*, 123: 897–903.

Dienstag, J.L. (2002). The role of liver biopsy in chronic hepatitis C. *Hepatology*, 36: S152–S160.

Dienstag, J.L., McHutchison, J.G. (2006). American Gastroenterological Association medical position statement on the management of hepatitis C. *Gastroenterology*, 130: 225–230.

Dixit, N.M., Layden-Almer, J.E., Layden, T.J., Perelson, A.S. (2004). Modelling how ribavirin improves interferon response rates in hepatitis C virus infection. *Nature*, 432: 922–924.

Everson, G.T. (2005). Should we treat patients with chronic hepatitis C on the waiting list? *Journal of Hepatology*, 42: 456–462.

Fabrizi, F., Poordad, F.F., Martin, P. (2002). Hepatitis C infection and the patient with end-stage renal disease. *Hepatology*, 36: 3–10.

Fabrizi, F., Dulai, G., Dixit, V., Bunnapradist, S., Martin, P. (2003). Meta-analysis: interferon for the treatment of chronic hepatitis C in dialysis patients. *Alimentary Pharmacology Therapy*, 18: 1071–1081.

Frese, M., Schwarzle, V., Barth, K., Krieger, N., Lohmann, V., Mihm, S., Haller, O., Bartenschlager, R. (2002). Interferon-gamma inhibits replication of subgenomic and genomic hepatitis C virus RNAs. *Hepatology*, 35: 694–703.

Fried, M.W., Shiffman, M.L., Reddy, K.R., Smith, C., Marinos, G., Goncales, F.L., Jr., Haussinger, D., Diago, M., Carosi, G., Dhumeaux, D., Craxi, A., Lin, A., Hoffman, J., Yu, J. (2002). Peginterferon alpha-2a plus ribavirin for chronic hepatitis C virus infection. *New England Journal of Medicine*, 347: 975–982.

Furusyo, N., Hayashi, J., Ueno, K., Sawayama, Y., Kawakami, Y., Kishihara, Y., Kashiwagi, S. (1997). Human lymphoblastoid interferon treatment for patients with hepatitis C virus-related cirrhosis. *Clinical Therapy*, 19: 1352–1367.

Furusyo, N., Hayashi, J., Kanamoto-Tanaka, Y., Ariyama, I., Etoh, Y., Shigematsu, M., Kashiwagi, S. (2000a). Liver damage in hemodialysis patients with hepatitis C virus viremia: a prospective 10-year study. *Digestive Diseases and Sciences*, 45: 2221–2228.

Furusyo, N., Hayashi, J., Ariyama, I., Sawayama, Y., Etoh, Y., Shigematsu, M., Kashiwagi, S. (2000b). Maintenance hemodialysis decreases serum hepatitis C virus (HCV) RNA levels in hemodialysis patients with chronic HCV infection. *American Journal of Gastroenterology*, 95: 490–496.

Furusyo, N., Hayashi, J., Kakuda, K., Ariyama, I., Kanamoto-Tanaka, Y., Shimizu, C., Etoh, Y., Shigematsu, M., Kashiwagi, S. (2001). Acute hepatitis C among Japanese hemodialysis patients: a prospective 9-year study. *American Journal of Gastroenterology*, 96: 1592–1600.

Furusyo, N., Hayashi, J., Kashiwagi, K., Nakashima, H., Nabeshima, S., Sawayama, Y., Kinukawa, N., Kashiwagi, S. (2002). Hepatitis C virus (HCV) RNA level determined by second-generation branched-DNA probe assay as predictor of response to interferon treatment in patients with chronic HCV viremia. *Digestive Diseases and Sciences*, 47: 535–542.

Furusyo, N., Kubo, N., Nakashima, H., Kashiwagi, K., Etoh, Y., Hayashi, J. (2004). Confirmation of nosocomial hepatitis C virus infection in a hemodialysis unit. *Infection Control in Hospital Epidemiology*, 25: 584–590.

Furusyo, N., Kubo, N., Toyoda, K., Takeoka, H., Nabeshima, S., Murata, M., Nakamuta, M., Hayashi, J. (2005). Helper T cell cytokine response to ribavirin priming before combined treatment with interferon alpha and ribavirin for patients with chronic hepatitis C. *Antiviral Research*, 67: 46–54.

Furusyo, N., Katoh, M., Tanabe, Y., Kajiwara, E., Maruyama, T., Shimono, J., Sakai, H., Nakamuta, M., Nomura, H., Masumoto, A., Shimoda, S., Takahashi, K., Azuma, K., Hayashi, J. (2006). Interferon alpha plus ribavirin combination treatment of Japanese chronic hepatitis C patients with HCV genotype 2: a project of the Kyushu University Liver Disease Study Group. *World Journal of Gastroenterology*, 12: 784–790.

Gane, E. (2002). Treatment of recurrent hepatitis C. *Liver Transplantation*, 8: S28–S37.

Ghany, M.G., Kleiner, D.E., Alter, H., Doo, E., Khokar, F., Promrat, K., Herion, D., Park, Y., Liang, T.J., Hoofnagle, J.H. (2003). Progression of fibrosis in chronic hepatitis C. *Gastroenterology*, 124: 97–104.

Hadziyannis, S.J., Sette, H., Jr., Morgan, T.R., Balan, V., Diago, M., Marcellin, P., Ramadori, G., Bodenheimer, H., Jr., Bernstein, D., Rizzetto, M., Zeuzem, S., Pockros, P.J., Lin, A., Ackrill, A.M., PEGASYS International Study Group (2004). Peginterferon-alpha2a and ribavirin combination therapy in chronic hepatitis C: a randomized study of treatment duration and ribavirin dose. *Annals of Internal Medicine*, 140: 346–355.

Hagiwara, H., Hayashi, N., Mita, E., Ueda, K., Takehara, T., Kasahara, A., Fusamoto, H., Kamada, T. (1992). Detection of hepatitis C virus RNA in serum of patients with chronic hepatitis C treated with interferon-alpha. *Hepatology*, 15: 37–41.

Hayashi, J., Hirata, M., Nakashima, K., Noguchi, A., Kashiwagi, S., Matsui, M., Ishibashi, H., Maeda, Y. (1991a). Hepatitis C virus is a more likely cause of chronic liver disease in the Japanese population than hepatitis B virus. *Fukuoka Igaku Zasshi*, 82: 648–654.

Hayashi, J., Nakashima, K., Kajiyama, W., Noguchi, A., Morofuji, M., Maeda, Y., Kashiwagi, S. (1991b). Prevalence of antibody to hepatitis C virus in hemodialysis patients. *American Journal of Epidemiology*, 134: 651–657.

Hayashi, J., Ohmiya, M., Kishihara, Y., Tani, Y., Kinukawa, N., Ikematsu, H., Kashiwagi, S. (1994a). A statistical analysis of predictive factors of response to human lymphoblastoid interferon in patients with chronic hepatitis C. *American Journal of Gastroenterology*, 89: 2151–2156.

Hayashi, J., Nakashima, K., Yoshimura, E., Kishihara, Y., Ohmiya, M., Hirata, M., Kashiwagi, S. (1994b). Prevalence and role of hepatitis C viraemia in haemodialysis patients in Japan. *Journal of Infection*, 28: 271–277.

Hayashi, J., Kishihara, Y., Yamaji, K., Furusyo, N., Yamamoto, T., Pae, Y., Etoh, Y., Ikematsu, H., Kashiwagi, S. (1997a). Hepatitis C viral quasispecies and liver damage in patients with chronic hepatitis C virus infection. *Hepatology*, 25: 697–701.

Hayashi, J., Kishihara, Y., Yoshimura, E., Furusyo, N., Yamaji, K., Kawakami, Y., Murakami, H., Kashiwagi, S. (1997b). Correlation between human T cell lymphotropic virus type-1 and Strongyloides stercoralis infections and serum immunoglobulin E responses in residents of Okinawa, Japan. *American Journal of Tropical Medicine Hygiene*, 56: 71–75.

Hayashi, J., Kishihara, Y., Ueno, K., Yamaji, K., Kawakami, Y., Furusyo, N., Sawayama, Y., Kashiwagi, S. (1998a). Age-related response to interferon alfa treatment in women vs men with chronic hepatitis C virus infection. *Archives of Internal Medicine*, 158: 177–181.

Hayashi, J., Kawakami, Y., Nabeshima, A., Kishihara, Y., Furusyo, N., Sawayama, Y., Kinukawa, N., Kashiwagi, S. (1998b). Comparison of HCV RNA levels by branched DNA probe assay and by competitive polymerase chain reaction to predict effectiveness of interferon treatment for patients with chronic hepatitis C virus. *Digestive Diseases and Sciences*, 43: 384–391.

Hayashi, J., Furusyo, N., Ariyama, I., Sawayama, Y., Etoh, Y., Kashiwagi, S. (2000). A relationship between the evolution of hepatitis C virus variants, liver damage, and hepatocellular carcinoma in patients with hepatitis C viremia. *Journal of Infectious Disease*, 181: 1523–1527.

Heathcote, E.J., Shiffman, M.L., Cooksley, W.G., Dusheiko, G.M., Lee, S.S., Balart, L., Reindollar, R., Reddy, R.K., Wright, T.L., Lin, A., Hoffman, J., De Pamphilis, J. (2000). Peginterferon alpha-2a in patients with chronic hepatitis C and cirrhosis. *New England Journal of Medicine*, 343: 1673–1680.

Hoofnagle, J.H., Mullen, K.D., Jones, D.B., Rustgi, V., Di Bisceglie, A., Peters, M., Waggoner, J.G., Park, Y., Jones, E.A. (1986). Treatment of chronic non-A, non-B hepatitis with recombinant human alpha interferon. A preliminary report. *New England Journal of Medicine*, 315: 1575–1578.

Hoofnagle, J.H., Lau, D., Conjeevaram, H., Kleiner, D., Di Bisceglie, A.M. (1996). Prolonged therapy of chronic hepatitis C with ribavirin. *Journal of Viral Hepatology*, 3: 247–252.

Hourigan, L.F., Macdonald, G.A., Purdie, D., Whitehall, V.H., Shorthouse, C., Clouston, A., Powell, E.E. (1999). Fibrosis in chronic hepatitis C correlates significantly with body mass index and steatosis. *Hepatology*, 29: 1215–1219.

Hugle, T., Cerny, A. (2003). Current therapy and new molecular approaches to antiviral treatment and prevention of hepatitis C. *Reviews in Medical Virology*, 13: 361–371.

Jaeckel, E., Cornberg, M., Wedemeyer, H., Santantonio, T., Mayer, J., Zankel, M., Pastore, G., Dietrich, M., Trautwein, C., Manns, M.P., German Acute Hepatitis C Therapy Group (2001). Treatment of acute hepatitis C with interferon alpha-2b. *New England Journal of Medicine*, 345: 1452–1457.

Kasahara, A., Hayashi, N., Mochizuki, K., Takayanagi, M., Yoshioka, K., Kakumu, S., Iijima, A., Urushihara, A., Kiyosawa, K., Okuda, M., Hino, K., Okita, K. (1998). Risk factors for hepatocellular carcinoma and its incidence after interferon treatment in patients with chronic hepatitis C. Osaka Liver Disease Study Group. *Hepatology*, 27: 1394–1402.

Kashiwagi, S., Kajiyama, W., Hayashi, J., Noguchi, A., Nakashima, K., Nomura, H., Ikematsu, H., Sawada, T., Kida, S., Koide, A. (1990). Antibody to p40tax protein of human T cell leukemia virus 1 and infectivity. *Journal of Infectious Diseases*, 161: 426–429.

Kashiwagi, K., Furusyo, N., Kubo, N., Nakashima, H., Nomura, H., Kashiwagi, S., Hayashi, J. (2003). A prospective comparison of the effect of interferon-alpha and interferon-beta treatment in patients with chronic hepatitis C on the incidence of hepatocellular carcinoma development. *Journal of Infection and Chemotherapy*, 9: 333–340.

Kawakami, Y., Nabeshima, S., Furusyo, N., Sawayama, Y., Hayashi, J., Kashiwagi, S. (2000). Increased frequency of interferon-gamma-producing peripheral blood CD4+ T cells in chronic hepatitis C virus infection. *American Journal of Gastroenterology*, 95: 227–232.

Kirchner, H. (1984). Interferons, a group of multiple lymphokines. *Springer Seminar in Immunopathology*, 7: 347–374.

Kishihara, Y., Furusyo, N., Kashiwagi, K., Mitsutake, A., Kashiwagi, S., Hayashi, J. (2001). Human T lymphotropic virus type 1 infection influences hepatitis C virus clearance. *Journal of Infectious Disease*, 184: 1114–1119.

Kottilil, S., Polis, M.A., Kovacs, J.A. (2004). HIV Infection, hepatitis C infection, and HAART: hard clinical choices. *The Journal of the American Medical Association*, 292: 243–250.

Kubo, N., Furusyo, N., Nakashima, H., Kashiwagi, K., Hayashi, J. (2005). Strenuous physical labor is important as a cause of elevated alanine aminotransferase levels in Japanese patients with chronic hepatitis C viremia. *European Journal of Epidemiology*, 20: 251–261.

Lafeuillade, A., Hittinger, G., Chadapaud, S. (2001). Increased mitochondrial toxicity with ribavirin in HIV/HCV coinfection. *Lancet*, 357: 280–281.

Laskus, T., Radkowski, M., Wang, L.F., Jang, S.J., Vargas, H., Rakela, J. (1998). Hepatitis C virus quasispecies in patients infected with HIV-1: correlation with extrahepatic viral replication. *Virology*, 248: 164–171.

Lau, D.T., Kleiner, D.E., Ghany, M.G., Park, Y., Schmid, P., Hoofnagle, J.H. (1998). 10-year follow-up after interferon-alpha therapy for chronic hepatitis C. *Hepatology*, 28: 1121–1127.

Lau, J.Y., Tam, R.C., Liang, T.J., Hong, Z. (2002). Mechanism of action of ribavirin in the combination treatment of chronic HCV infection. *Hepatology*, 35: 1002–1009.

Layden-Almer, J.E., Ribeiro, R.M., Wiley, T., Perelson, A.S., Layden, T.J. (2003). Viral dynamics and response differences in HCV-infected African American and white patients treated with IFN and ribavirin. *Hepatology*, 37: 1343–1350.

Liang, T.J., Rehermann, B., Seeff, L.B., Hoofnagle, J.H. (2000). Pathogenesis, natural history, treatment, and prevention of hepatitis C. *Annals of Internal Medicine*, 132: 296–305.

Lindsay, K.L., Trepo, C., Heintges, T., Shiffman, M.L., Gordon, S.C., Hoefs, J.C., Schiff, E.R., Goodman, Z.D., Laughlin, M., Yao, R., Albrecht, J.K., Hepatitis Interventional Therapy Group (2001). A randomized, double-blind trial comparing pegylated interferon alpha-2b to interferon alpha-2b as initial treatment for chronic hepatitis C. *Hepatology*, 34: 395–403.

Luxon, B.A., Grace, M., Brassard, D., Bordens, R. (2002). Pegylated interferons for the treatment of chronic hepatitis C infection. *Clinical Therapy*, 24: 1363–1383.

Mangia, A., Santoro, R., Minerva, N., Ricci, G.L., Carretta, V., Persico, M., Vinelli, F., Scotto, G., Bacca, D., Annese, M., Romano, M., Zechini, F., Sogari, F., Spirito, F., Andriulli, A. (2005).

Peginterferon alpha-2b and ribavirin for 12 vs. 24 weeks in HCV genotype 2 or 3. *New England Journal of Medicine*, 352: 2609–2617.

Manns, M.P., McHutchison, J.G., Gordon, S.C., Rustgi, V.K., Shiffman, M., Reindollar, R., Goodman, Z.D., Koury, K., Ling, M., Albrecht, J.K. (2001). Peginterferon alpha-2b plus ribavirin compared with interferon alpha-2b plus ribavirin for initial treatment of chronic hepatitis C: a randomised trial. *Lancet*, 358: 958–965.

Marcellin, P., Boyer, N., Degott, C., Martinot-Peignoux, M., Duchatelle, V., Giostra, E., Areias, J., Erlinger, S., Benhamou, J.P. (1994). Long-term histologic and viral changes in patients with chronic hepatitis C who responded to alpha interferon. *Liver*, 14: 302–307.

Marcellin, P., Boyer, N., Gervais, A., Martinot, M., Pouteau, M., Castelnau, C., Kilani, A., Areias, J., Auperin, A., Benhamou, J.P., Degott, C., Erlinger, S. (1997). Long-term histologic improvement and loss of detectable intrahepatic HCV RNA in patients with chronic hepatitis C and sustained response to interferon-alpha therapy. *Annals of Internal Medicine*, 127: 875–881.

McHutchison, J.G., Gordon, S.C., Schiff, E.R., Shiffman, M.L., Lee, W.M., Rustgi, V.K., Goodman, Z.D., Ling, M.H., Cort, S., Albrecht, J.K. (1998). Interferon alpha-2b alone or in combination with ribavirin as initial treatment for chronic hepatitis C. Hepatitis Interventional Therapy Group. *New England Journal of Medicine*, 339: 1485–1492.

Mohsen, A.H., Easterbrook, P.J., Taylor, C., Portmann, B., Kulasegaram, R., Murad, S., Wiselka, M., Norris, S. (2003). Impact of human immunodeficiency virus (HIV) infection on the progression of liver fibrosis in hepatitis C virus infected patients. *Gut*, 52: 1035–1040.

Monga, H.K., Rodriguez-Barradas, M.C., Breaux, K., Khattak, K., Troisi, C.L., Velez, M., Yoffe, B. (2001). Hepatitis C virus infection-related morbidity and mortality among patients with human immunodeficiency virus infection. *Clinical Infectious Disease*, 33: 240–247.

Muir, A.J., Bornstein, J.D., Killenberg, P.G., Atlantic Coast Hepatitis Treatment Group (2004). Peginterferon alpha-2b and ribavirin for the treatment of chronic hepatitis C in blacks and non-Hispanic whites. *New England Journal of Medicine*, 350: 2265–2271.

Murata, M., Nabeshima, S., Maeda, N., Nakashima, H., Kashiwagi, S., Hayashi, J. (2002). Increased frequency of IFN-gamma-producing peripheral CD8+ T cells with memory-phenotype in patients with chronic hepatitis C. *Journal of Medical Virology*, 67: 162–170.

Murata, M., Nabeshima, S., Kikuchi, K., Yamaji, K., Furusyo, N., Hayashi, J. (2006). A comparison of the antitumor effects of interferon-alpha and beta on human hepatocellular carcinoma cell lines. *Cytokine*, 33: 121–128.

National Institutes of Health Consensus Development Conference statement (2002). Management of Hepatitis C: 2002-June 10–12, 2002. *Hepatology*, 36: S3–S20.

Neumann, A.U., Lam, N.P., Dahari, H., Gretch, D.R., Wiley, T.E., Layden, T.J., Perelson, A.S. (1998). Hepatitis C viral dynamics in vivo and the antiviral efficacy of interferon-alpha therapy. *Science*, 282: 103–107.

Niederau, C., Lange, S., Heintges, T., Erhardt, A., Buschkamp, M., Hurter, D., Nawrocki, M., Kruska, L., Hensel, F., Petry, W., Haussinger, D. (1998). Prognosis of chronic hepatitis C: results of a large, prospective cohort study. *Hepatology*, 28: 1687–1695.

Nishiguchi, S., Kuroki, T., Nakatani, S., Morimoto, H., Takeda, T., Nakajima, S., Shiomi, S., Seki, S., Kobayashi, K., Otani, S. (1995). Randomised trial of effects of interferon-alpha on incidence of hepatocellular carcinoma in chronic active hepatitis C with cirrhosis. *Lancet*, 346: 1051–1055.

Nomura, H., Tanimoto, H., Kajiwara, E., Shimono, J., Maruyama, T., Yamashita, N., Nagano, M., Higashi, M., Mukai, T., Matsui, Y., Hayashi, J., Kashiwagi, S., Ishibashi, H. (2004a). Factors contributing to ribavirin-induced anemia. *Journal of Gastroenterology and Hepatology*, 19: 1312–1317.

Nomura, H., Sou, S., Tanimoto, H., Nagahama, T., Kimura, Y., Hayashi, J., Ishibashi, H., Kashiwagi, S. (2004b). Short-term interferon-alpha therapy for acute hepatitis C: a randomized controlled trial. *Hepatology*, 39: 1213–1219.

Okanoue, T., Itoh, Y., Minami, M., Sakamoto, S., Yasui, K., Sakamoto, M., Nishioji, K., Murakami, Y., Kashima K. (1999). Interferon therapy lowers the rate of progression to hepatocellular carcinoma in chronic hepatitis C but not significantly in an advanced stage: a retrospective study in 1148 patients. Viral Hepatitis Therapy Study Group. *Journal of Hepatology*, 30: 653–659.

Pereira, B.J., Natov, S.N., Bouthot, B.A., Murthy, B.V., Ruthazer, R., Schmid, C.H., Levey, A.S. (1998). Effects of hepatitis C infection and renal transplantation on survival in end-stage renal disease. The New England Organ Bank Hepatitis C Study Group. *Kidney International*, 53: 1374–1381.

Perez-Olmeda, M., Nunez, M., Romero, M., Gonzalez, J., Castro, A., Arribas, J.R., Pedreira, J., Barreiro, P., Garcia-Samaniego, J., Martin-Carbonero, L., Jimenez-Nacher, I., Soriano, V. (2003). Pegylated IFN-alpha2b plus ribavirin as therapy for chronic hepatitis C in HIV-infected patients. *AIDS*, 17: 1023–1028.

Poynard, T., Bedossa, P., Chevallier, M., Mathurin, P., Lemonnier, C., Trepo, C., Couzigou, P., Payen, J.L., Sajus, M., Costa, J.M. (1995). A comparison of three interferon alpha-2b regimens for the long-term treatment of chronic non-A, non-B hepatitis. Multicenter Study Group. *New England Journal of Medicine*, 332: 1457–1462.

Poynard, T., Leroy, V., Cohard, M., Thevenot, T., Mathurin, P., Opolon, P., Zarski, J.P. (1996). Meta-analysis of interferon randomized trials in the treatment of viral hepatitis C: effects of dose and duration. *Hepatology*, 24: 778–789.

Poynard, T., Bedossa, P., Opolon, P. (1997). Natural history of liver fibrosis progression in patients with chronic hepatitis C. The OBSVIRC, METAVIR, CLINIVIR, and DOSVIRC groups. *Lancet*, 349: 825–832.

Poynard, T., Marcellin, P., Lee, S.S., Niederau, C., Minuk, G.S., Ideo, G., Bain, V., Heathcote, J., Zeuzem, S., Trepo, C., Albrecht, J. (1998). Randomised trial of interferon alpha2b plus ribavirin for 48 weeks or for 24 weeks versus interferon alpha2b plus placebo for 48 weeks for treatment of chronic infection with hepatitis C virus. International Hepatitis Interventional Therapy Group (IHIT). *Lancet*, 352: 1426–1432.

Reddy, K.R., Wright, T.L., Pockros, P.J., Shiffman, M., Everson, G., Reindollar, R., Fried, M.W., Purdum, P.P. 3rd., Jensen, D., Smith, C., Lee, W.M., Boyer, T.D., Lin, A., Pedder, S., DePamphilis, J. (2001). Efficacy and safety of pegylated (40-kd) interferon alpha-2a compared with interferon alpha-2a in noncirrhotic patients with chronic hepatitis C. *Hepatology*, 33: 433–438.

Robertson, B., Myers, G., Howard, C., Brettin, T., Bukh, J., Gaschen, B., Gojobori, T., Maertens, G., Mizokami, M., Nainan, O., Netesov, S., Nishioka, K., Shini, T., Simmonds, P., Smith, D., Stuyver, L., Weiner, A. (1998). Classification, nomenclature, and database development for hepatitis C virus (HCV) and related viruses: Proposals for standardization. International Committee on Virus Taxonomy. *Archives of Virology*, 143: 2493–2503.

Rodriguez-Luna, H., Khatib, A., Sharma, P., De Petris, G., Williams, J.W., Ortiz, J., Hansen, K., Mulligan, D., Moss, A., Douglas, D.D., Balan, V., Rakela, J., Vargas, H.E. (2004). Treatment of recurrent hepatitis C infection after liver transplantation with combination of pegylated interferon alpha2b and ribavirin: an open-label series. *Transplantation*, 77: 190–194.

Russo, M.W., Goldsweig, C.D., Jacobson, I.M., Brown, R.S., Jr. (2003). Interferon monotherapy for dialysis patients with chronic hepatitis C: an analysis of the literature on efficacy and safety. *American Journal of Gastroenterology*, 98: 1610–1615.

Sanchez-Tapias, J.M., Diago, M., Escartin, P., Enriquez, J., Romero-Gomez, M., Barcena, R., Crespo, J., Andrade, R., Martinez-Bauer, E., Perez, R., Testillano, M., Planas, R., Sola, R., Garcia-Bengoechea, M., Garcia-Samaniego, J., Munoz-Sanchez, M., Moreno-Otero, R., TeraViC-4 Study Group (2006). Peginterferon-alpha2a plus ribavirin for 48 versus 72 weeks in patients with detectable hepatitis C virus RNA at week 4 of treatment. *Gastroenterology*, 131: 451–460.

Scott, L.J., Perry, C.M. (2002). Interferon-alpha-2b plus ribavirin: A review of its use in the management of chronic hepatitis C. *Drugs*, 62: 507–556.

Sheeff, L.B., Hoofnagle, J.H. (2002). National Institutes of Health Consensus Development Conference: management of hepatitis C: 2002. *Hepatology*, 36: S1–S2.

Shiffman, M.L., Di Bisceglie, A.M., Lindsay, K.L., Morishima, C., Wright, E.C., Everson, G.T., Lok, A.S., Morgan, T.R., Bonkovsky, H.L., Lee, W.M., Dienstag, J.L., Ghany, M.G., Goodman, Z.D., Everhart, J.E., Hepatitis C Antiviral Long-Term Treatment Against Cirrhosis Trial

Group (2004). Peginterferon alpha-2a and ribavirin in patients with chronic hepatitis C who have failed prior treatment. *Gastroenterology*, 126: 1015–1023.

Shindo, M., Di Bisceglie, A.M., Cheung, L., Shih, J.W., Cristiano, K., Feinstone, S.M., Hoofnagle, J.H. (1991). Decrease in serum hepatitis C viral RNA during alpha-interferon therapy for chronic hepatitis C. *Annals of Internal Medicine*, 115: 700–704.

Soza, A., Heller, T., Ghany, M., Lutchman, G., Jake Liang, T., Germain, J., Hsu, H.H., Park, Y., Hoofnagle, J.H. (2005). Pilot study of interferon gamma for chronic hepatitis C. *Journal of Hepatology*, 43: 67–71.

Sporea, I., Sirli, R., Golea, O., Totolici, C., Danila, M., Popescu, A. (2004). Peg-interferon alpha 2a (40kDa) in patients on chronic haemodialysis with chronic C hepatitis. Preliminary results. *Roman Journal of Gastroenterology*, 13: 99–102.

Strader, D.B., Wright, T., Thomas, D.L., Seeff, L.B., American Association for the Study of Liver Diseases (2004). Diagnosis, management, and treatment of hepatitis C. *Hepatology*, 39: 1147–1171.

Thomas, D.L., Astemborski, J., Rai, R.M., Anania, F.A., Schaeffer, M., Galai, N., Nolt, K., Nelson, K.E., Strathdee, S.A., Johnson, L., Laeyendecker, O., Boitnott, J., Wilson, L.E., Vlahov, D. (2000). The natural history of hepatitis C virus infection: host, viral, and environmental factors, *The Journal of the American Medical Association*, 284: 450–456.

Tilg, H. (1997). New insights into the mechanisms of interferon alpha: an immunoregulatory and anti-inflammatory cytokine. *Gastroenterology*, 112: 1017–1021.

Torriani, F.J., Rodriguez-Torres, M., Rockstroh, J.K., Lissen, E., Gonzalez-Garcia, J., Lazzarin, A., Carosi, G., Sasadeusz, J., Katlama, C., Montaner, J., Sette, H., Jr., Passe, S., De Pamphilis, J., Duff, F., Schrenk, U.M., Dieterich, D.T., APRICOT Study Group (2004). Peginterferon alpha-2a plus ribavirin for chronic hepatitis C virus infection in HIV-infected patients. *New England Journal of Medicine*, 351: 438–450.

Yamaji, K., Hayashi, J., Kawakami, Y., Furusyo, N., Sawayama, Y., Kishihara, Y., Etoh, Y., Kashiwagi, S. (1998). Hepatitis C viral RNA status at two weeks of therapy predicts the eventual response. *Journal of Clinical Gastroenterology*, 26: 193–199.

Yano, M., Kumada, H., Kage, M., Ikeda, K., Shimamatsu, K., Inoue, O., Hashimoto, E., Lefkowitch, J.H., Ludwig, J., Okuda, K. (1996). The long-term pathological evolution of chronic hepatitis C. *Hepatology*, 23: 1334–1340.

Yoshida, H., Shiratori, Y., Moriyama, M., Arakawa, Y., Ide, T., Sata, M., Inoue, O., Yano, M., Tanaka, M., Fujiyama, S., Nishiguchi, S., Kuroki, T., Imazeki, F., Yokosuka, O., Kinoyama, S, Yamada, G., Omata, M. (1999). Interferon therapy reduces the risk for hepatocellular carcinoma: National surveillance program of cirrhotic and noncirrhotic patients with chronic hepatitis C in Japan. IHIT Study Group. Inhibition of hepatocarcinogenesis by interferon therapy. *Annals of Internal Medicine*, 131: 174–181.

Zeuzem, S., Schmidt, J.M., Lee, J.H., von Wagner, M., Teuber, G., Roth, W.K. (1998). Hepatitis C virus dynamics *in vivo*: effect of ribavirin and interferon alpha on viral turnover. *Hepatology*, 28: 245–252.

Interferon-Based Therapy for Hepatitis C Virus Infections

Ming-Lung Yu* and Wan-Long Chuang

History and Evolution

Evolution of Treatment Regimen for Chronic Hepatitis C

Interferon-alpha (IFN-α) therapy was associated with normalization of alanine aminotransferase (ALT) in some patients diagnosed as having non-A, non-B hepatitis even before the hepatitis C virus (HCV) was identified as the chief etiologic agent in this diagnosis (Hoofnagle et al., 1986). In 1989, the first cases of successful treatment of documented chronic hepatitis C (CHC) with IFN-α were reported, although relapse after the cessation of treatment was common (Davis et al., 1989; Di Bisceglie et al., 1989). The introduction of combination therapy with IFN-α and ribavirin has markedly improved treatment response. However, more than one-half of patients with CHC remain unable to experience a favorable response to the combination therapy (Lai et al., 1996; McHutchison et al., 1998; Poynard et al., 1998). Until recently, the attachment of inert polyethylene glycol to conventional IFN-α—pegylated IFN-α (PegIFN-α)—reduced degradation and clearance, prolonging the half-life of IFN and permitting less frequent, weekly dosing while maintaining higher sustained IFN levels (compared with 3 times weekly for conventional IFN). Now, a PegIFN-α–ribavirin combination treatment has been recommended for all patients infected with HCV. For patients infected with HCV genotype 1 (HCV-1), the recommended treatment duration is 48 weeks, whereas for patients infected with HCV-2 or -3, the recommended treatment duration is 24 weeks (National Institutes of Health 2002; Strader et al., 2004).

Evolution of Assessment of Treatment Response for Hepatitis C

In an earlier study, the primary endpoint for HCV therapies was a biochemical response, defined as the normalization of ALT levels (Davis et al., 1989;

---

* Address for proofs: Hepatobiliary Division, Department of Internal Medicine, Kaohsiung Medical University Hospital, No. 100, Tzyou 1st Road, Kaohsiung 807, Taiwan, Phone: 886-7-312-1101, ext. 7475; Fax: 886-7-323-4553, E-mail: fishya@ms14.hinet.net; fish6069@gmail.com

Di Bisceglie et al., 1989). The introduction of virological assays to detect HCV RNA further allows the assessment of a virological response, defined as polymerase chain reaction (PCR)-seronegative (< 50 IU/ml, or 100 copies/ml) for HCV RNA. The histological response has been assessed in some clinical studies, but there is little indication for posttreatment biopsy in real-world clinical practice.

Three patterns of on-treatment and three patterns of off-treatment virological responses to antiviral therapy for hepatitis C have emerged over the past decade (Figure 1) (Davis & Lau, 1995; Davis et al., 2003; Yu et al., 2006a). They include the following:

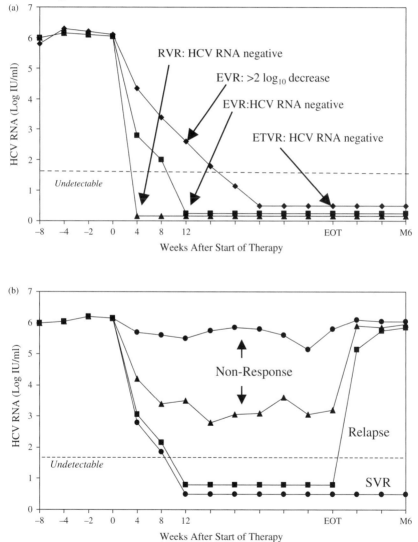

Fig. 1 On-treatment (1a) and off-treatment (1b) virological responses to interferon-based therapy. RVR, rapid virological response; EVR, early virological response; ETVR, end-of-treatment virologic response; SVR, sustained virologic response; EOT, end of treatment; M6, six months after EOT; detection limit of HCV RNA, 50 IU/mL

demonstrated in studies done before the discovery of HCV (Hoofnagle et al., 1986). Initially, a 6-month course of 3 weekly injections of 3 MU IFN-α was approved for treatment of CHC, and a biochemical response, defined as the normalization of ALT levels, was assigned as the primary endpoint (Davis et al., 1989; Di Bisceglie et al., 1989). IFN-α monotherapy suppresses serum HCV RNA to undetectable levels and normalizes the ALT level in 25–40% of CHC patients, usually within the first 2–3 months of treatment. However, these initial responses to IFN-α monotherapy are usually transient, and sustained response is documented in only about 8–9% of patients (Poynard et al., 1996).

Three meta-analyses: (1) 52 RCTs of treatment with IFN-α for 3–6 months (Niederau et al., 1996); (2) 33 RCTs of treatment for a full 6 months (Poynard et al., 1996); (3) 32 RCTs of treatment with at least 2 MU of IFN-α-2b monotherapy 3 times weekly for 24 weeks (Carithers & Emerson, 1997) showed that IFN-α monotherapy resulted in normalization of ALT levels at the end of treatment in 51.2%, 45%, and 47% of patients, respectively, but in only 21.7%, 21%, and 23% of patients, respectively, 3–6 months after the end of treatment. When virological assays for detection of HCV RNA became available, the virological response rates were observed to be lower than those reported with biochemical endpoints. In the meta-analysis of IFN-α monotherapy (Carithers & Emerson, 1997), normalization of ALT levels at the end of treatment and 6 months after stopping treatment was seen in 47% and 23% of treated patients, respectively. ETVR and SVR, however, were observed in only 29% and 8% of treated patients, respectively. Improvement of efficacy on CHC could be achieved with higher doses and/or a longer duration of IFN-α monotherapy. A doubling of the duration of therapy to 12 months increased the frequency of SVRs to approximately 20%. The best efficacy/risk ratio was in favor of 3 MU IFN-α 3 times weekly for at least 12 months in treatment-naïve patients with CHC (Poynard et al., 1996).

IFN-α and Ribavirin Combination Therapy

The introduction of ribavirin in combination with IFN-α was a major breakthrough in the treatment of CHC. Although ribavirin monotherapy was shown to be ineffective (Bodenheimer et al., 1997), the combination of ribavirin and IFN-α has greater antiviral activity than either agent alone in patients with CHC (McHutchison et al., 1998; Poynard et al., 1998). Two small RCTs showed a greater improvement of SVR in CHC patients with IFN–ribavirin combination therapy than in those with IFN monotherapy (Lai et al., 1996; Reichard et al., 1998). The rate of SVRs was 43% and 6% for the IFN-α-2a with and without ribavirin combination (Lai et al., 1996), respectively, and 36% and 18% for the IFN-α-2b with and without ribavirin combination (Reichard et al., 1998). A meta-analysis in 1997 showed that the SVR rate was significantly higher for IFN–ribavirin combination therapy than for IFN or ribavirin monotherapy [odds ratio {OR} IFN–ribavirin vs. IFN = 9.8, 95% confidence interval {CI} = 1.9–50] (Schalm et al., 1997). These reports were so encouraging that it appeared to be of scientific and practical interest to collect the individual data in one large database.

Several landmark studies then followed that consistently demonstrated the dramatically improved responses to combination therapy, especially for HCV-2 or -3 patients. In 1998, two multicenter RCTs (one U.S. study and one international study) totaling 1,744 previously untreated patients with compensated CHC compared 24- and 48-week drug regimens of IFN-α-2b monotherapy (3 MU 3 times weekly) with IFN-α-2b and ribavirin (1,000 mg/day or 1,200 mg/day for patients weighing <75 kg or >75 kg, respectively) combination therapy followed by 24 weeks of off-therapy followup (McHutchison et al., 1998; Poynard et al., 1998). The overall SVR rates for 24 and 48 weeks of therapy were 33% and 41%, respectively, for patients receiving IFN-α-2b-ribavirin, compared with SVR rates of 6% at 24 weeks and 16% at 48 weeks IFN-α-2b monotherapy. In addition to definitively showing the benefit of combination therapy over IFN alone, these studies made several other important clinical points. First, a striking reduction in hepatic inflammation was seen in sustained virological responders. Second, the likelihood of response to treatment was related to pretreatment virus level and genotype. SVRs to 48 or 24 weeks of combination therapy occurred in 29% and 17% of HCV-1 patients, respectively, and in 65% and 66% of HCV-2 or -3 patients. The two studies reinforced the importance of longer-duration therapy for 48 weeks in patients with HCV-1 infection. Similarly, SVRs to 48 or 24 weeks of combination therapy occurred in 38% and 27% of patients with pretreatment HCV RNA levels of 2×10^6 copies/ml or more, respectively, but the SVR rates were no different for those with lower levels (45% and 43%, respectively). A systematic review in 2001 included data from 15 trials in which patients received IFN-α monotherapy or IFN-α–ribavirin combination therapy. Compared with IFN-α monotherapy, combination therapy reduced the nonresponse rate (absence of SVR) by 26% in treatment-naïve patients (relative risk = 0.74, 95% CI = 0.70–0.78).

In 1998, the FDA approved the combination of IFN-α and ribavirin for patients with chronic HCV infection. In 1999, the EASL International Consensus Conference on Hepatitis C (EASL, 1999) recommended that for patients with CHC who have not been previously treated, (1) standard therapy should consist of IFN-α and ribavirin in combination for 24 weeks; and (2) treatment should be extended to 48 weeks in patients with both HCV-1 and HCV RNA levels greater than 2×10^6 copies/ml.

PegIFN-α Monotherapy

Four RCTs compared the efficacy and safety of once-weekly PegIFN-α monotherapy with IFN-α monotherapy three times per week for the treatment of chronic HCV infection in treatment-naïve patients (Heathcote et al., 2000; Lindsay et al., 2001; Reddy et al., 2001; Zeuzem et al., 2000). The initial studies of PegIFN-α evaluated the dose-ranging efficacy of monotherapy. The recommended dose of PegIFN-α-2a monotherapy, administered fixed at 180 μg/week for 48 weeks, achieved higher SVR rates compared with IFN-α-2a monotherapy (30% to 39% vs. 8% to 19%)

(Heathcote et al., 2000; Reddy et al., 2001; Zeuzem et al., 2000); the PegIFN-α-2b monotherapy, administered according to body weight at 1.5 µg/kg/week for 48 weeks, achieved an SVR rate of 23%, compared to 12% with IFN-α-2b monotherapy (Lindsay et al., 2001). In some studies, the PegIFN not only increased the SVR rate compared with nonpegylated ones, but even achieved similar SVR rates compared with the synergic effects of ribavirin added to nonpegylated IFN (Reddy et al., 2001; Zeuzem et al., 2000).

Of note, Heathcote et al. (2000) conducted the first substantive prospective study confined to patients with compensated cirrhosis or advanced fibrosis. Cirrhosis has been a poor predictor of responsiveness and is associated with a high risk of leukopenia and thrombocytopenia (McHutchison et al., 1998; Poynard et al., 1998). This study, however, showed that PegIFN monotherapy was both well tolerated and effective in cirrhotic CHC patients, with an SVR rate of 30%.

Although PegIFN monotherapy has been recommended for patients with contraindications to ribavirin, such as those with renal insufficiency, hemoglobinopathies, and ischemic cardiovascular disease, no clinical trials have been reported to date in these populations. For patients with contraindications to ribavirin but who have indications for antiviral therapy, PegIFN represents the best option of treatment.

PegIFN-α and Ribavirin Combination Therapy

The results of PegIFN-α monotherapy encouraged more clinical trials to go on in the anticipation that combination therapy with PegIFN-α and ribavirin would be even more effective. The earlier two large RCTs were applied with fixed durations of 48 weeks (Fried et al., 2002; Manns et al., 2001). In these trials, PegIFN-α-2b was dosed by weight (1.5 µg/kg was FDA approved) and coupled with 800 mg of ribavirin; PegIFN-α-2a was given at a fixed dose of 180 µg along with a weight-adjusted, higher dose of ribavirin (1,000 mg/day or 1,200 mg/day for patients weighing <75 kg or >75 kg, respectively). The overall response rate in clinical trials was 54–56%. These trials demonstrated that higher SVR rates could be achieved with the combination of PegIFN-α weekly plus oral ribavirin given twice daily than with IFN-α given three times weekly together with ribavirin or PegIFN-α monotherapy. Since the results in an RCT of contemporaneous head-to-head comparisons of the two PegIFN-α compounds combined with similar ribavirin doses have not been reported, no definitive conclusions on their relative efficacies can be drawn.

The issue of the influence of ribavirin dose by body weight on the response rate was first addressed (Manns et al., 2001). In the PegIFN-α-2b study, a post hoc analysis demonstrated that an SVR rate of 61% was achieved in the subgroup whose daily dose of ribavirin exceeded 10.6 mg/kg. Logistic regression analyses observed that the response rates generally increased as ribavirin dose increased up to about 13 mg/kg/day. Although the study was not prospectively designed or sufficiently powered to address the contribution of more optimal ribavirin weight-based dosing (actually, the optimal ribavirin dose has not been defined), other studies highlighted the potential importance of higher doses of ribavirin and adherence to treatment (Lindahl et al., 2005; McHutchison et al., 2002a), and a suboptimal dose of ribavirin may have had an impact on response rates in the original PegIFN-α-2b–ribavirin trial.

Later, the optimal treatment duration and ribavirin dose were investigated in a multicenter RCT in which all CHC patients received PegIFN-α-2a at a dose of 180 µg, while patients in the four arms received either 24 or 48 weeks of ribavirin at a dose of 800 mg or at the higher, weight-based doses of 1,000 or 1,200 mg daily (Hadziyannis et al., 2004). In the subsequent registration trial, a high frequency of SVRs occurred in patients with HCV-2 or -3, regardless of the regimen (79% to 84%), but optimal frequencies of SVRs in HCV-1 (52%) required longer-duration and full-dose ribavirin, independent of the level of HCV RNA. In patients with HCV-1 with a low viral load (<2 $\times$ 10^{6}copies/ml, or 800,000 IU/ml), the SVR was highest in those who had received the higher ribavirin dose and who were treated for 48 weeks (61%). This regimen was also optimal for patients with HCV-1 and a high viral load (SVR rate, 46%). In contrast, in patients with HCV-2 or -3, regardless of the pretreatment viral load, no differences were detected among the four treatment regimens. Another single-arm, open-label, historical-control study of 24 weeks of treatment with PegIFN-α-2b plus ribavirin limited to patients with HCV-2 or -3 demonstrated that 24 weeks of treatment was sufficient in HCV-2 or -3 infected patients, with an overall SVR rate of 81% (Zeuzem et al., 2004). This study supports the current recommendations that patients with HCV-1 require 48 weeks of PegIFN-α therapy with higher doses of ribavirin, while patients with HCV-2 or -3 can be treated for only 24 weeks and with only 800 mg daily of ribavirin (National Institutes of Health, 2002; Strader et al., 2004).

Contraindication and Adverse Events of IFN–Ribavirin and Management

Contraindications and adverse events of IFN–ribavirin therapy are listed in Table 1. Physicians should look specifically for contraindications to antiviral therapy and assess both the therapeutic risk and benefit. Ribavirin is contraindicated in pregnancy, necessitating strict precautions and contraception in women of childbearing age and their sexual partners and in HCV-infected men with female partners of childbearing age. Flu-like side effects of IFN can be managed with acetaminophen or nonsteroidal anti-inflammatory drugs; antidepressants and hypnotics can be used for depression and insomnia, respectively. For management of neutropenia, dose reduction suffices; the addition of granulocyte colony-stimulating factor is generally not recommended, although it may be considered in individual cases of severe neutropenia. Treatment with ribavirin should be avoided in patients with ischemic cardiovascular and cerebrovascular disease and in patients with renal insufficiency. If anemia occurs, options include ribavirin dose reduction or the addition of erythropoietin. Patients with decompensated cirrhosis are at high risk of adverse events and are relatively contraindicated to IFN-ribavirin. However, a recent study

Table 1 Contraindications and Adverse Effects of Hepatitis C Therapy

Contraindications

• **Absolute contraindications**

Major, uncontrolled depressive illness; autoimmune hepatitis or other condition known to be exacerbated by interferon and ribavirin; untreated hyperthyroidism; pregnant or unwilling/unable to comply with adequate contraception; severe concurrent disease such as severe hypertension, heart failure, significant coronary artery disease, poorly controlled diabetes, obstructive pulmonary disease; under 3 years of age; known hypersensitivity to drugs used to treat HCV

• **Relative contraindications**

Decompensated liver disease; solid organ transplantation (except liver); coexisting medical conditions: severe anemia (hemoglobin level < 100 g/L), neutropenia (neutrophil count < 0.75×10^9/L), thrombocytopenia (platelet count < 75×10^9/L), hemoglobinopathy, uncontrolled heart disease (angina, congestive heart failure, significant arrhythmias), cerebrovascular disease, advanced renal failure (creatinine clearance < 50 ml/min)

Adverse effects

• **Interferon or peginterferon**

Flu-like symptoms (fever, fatigue, myalgia, and headaches); mild bone marrow suppression (especially, leucopenia and thrombocytopenia); gastrointestinal manifestation (anorexia, nausea, vomiting, and diarrhea); emotional effects (depression, irritability, difficulty concentrating, memory disturbance, and insomnia); dermatological manifestation (skin irritation, rash, and alopecia); autoimmune disorders (especially thyroid dysfunction); weight loss; tinnitus and hearing loss; retinopathy (usually not clinically significant); hyperglycemia; seizures; renal function impairment; pneumonitis

• **Ribavirin**

Hemolytic anemia (dose-dependent); cough and dyspnea; rash and pruritis; nausea; sinus disorders; teratogenicity

demonstrated that HCV clearance by PegIFN–ribavirin was life-saving and reduced disease progression (Iacobellis et al., 2007).

Patients receiving combination therapy had an increased risk for requiring medication dose reduction (RR = 2.44, 95% CI = 1.58–3.75) or discontinuation (RR = 1.28, 95% CI = 1.07–1.52) compared with those receiving IFN monotherapy (Kjaergard et al., 2001). The rates of IFN dose reduction and discontinuation were similar among subjects receiving PegIFN and conventional IFN (Lindsay et al., 2001; Zeuzem et al., 2000).

Factors Associated with Treatment Efficacy

Although the introduction of new agents and regimens for the treatment of CHC, such as PegIFNs and combination therapy with ribavirin, has resulted in substantial improvements in overall SVR rates, treatment remains a challenge, particularly for certain patient populations. Accurately predicting therapeutic responses is a critical issue in the management of diseases. With the great progress in the management of CHC, predictors for SVR in CHC therapy have been elucidated. All predictors derived from experiences of therapy that can be collected to form clinical data for analyses play an important role in the clinical setting.

Table 2 Factors Associated with Response to Interferon-Based Therapy for Hepatitis C

Baseline predictors
• Virological factors
Hepatitis C virus genotype
Hepatitis C viral loads
Quasispecies
• Host factors
Bridging fibrosis/cirrhosis
Gender
Age
Ethnicity
Insulin resistance
Obesity
Hepatic steatosis
Host genetics
Co-infection with HIV
Nonresponse to previous interferon-based therapy
On-treatment predictors
• Rapid virological response (RVR) at week 4
• Early virological response (EVR) at week 12
• Medical adherence

Clinical factors have been identified as predictors for the efficacy of the IFN-based therapy and divided into two major categories: baseline and on-treatment predictors (Table 2).

Baseline Predictors of Response to IFN-Based Therapy

Virological Factors

The pretreatment variable most strongly predictive of an SVR is the presence of HCV-2 or -3 infection (McHutchison & Poynard, 1999) whether with conventional IFNs or PegIFNs, alone or in combination with ribavirin (Fried et al., 2002; Manns et al., 2001; McHutchison et al., 1998; Poynard et al., 1998). On the basis of variations in the nucleotide sequence of HCV, six genotypes (numbered 1–6) and more than 50 subtypes (identified by lowercase letters, e.g., 1a and 1b) have been identified (Bukh et al., 1995). The HCV genotype and subtype are intrinsic characteristics of a transmitted viral strain and do not change during the course of the infection. Why HCV-1 is harder to treat than other HCV genotypes is not yet fully understood. Several studies demonstrated that there exists a genotype-specific difference among viral kinetics (Neumann et al., 2000; Yu et al., 2006a). The turnover of hepatocytes infected with HCV-1 is slower than that of hepatocytes infected with other HCV genotypes after initiation of IFN-based therapy (Neumann et al., 2000; Zeuzem et al., 2001), implying that HCV-1 might be more resistant to antiviral therapy. Under the current recommendation (Strader et al., 2004), SVR rates were 42–60% for HCV-1 infection with a 48-week PegIFN–ribavirin treatment, compared with

76–95% for HCV-2 or -3 infections with a 24-week regimen (Fried et al., 2002; Hadziyannis et al., 2004; Mangia et al., 2005; Manns et al., 2001; von Wagner et al., 2005; Yu et al., 2007; Zeuzem et al., 2004). Patients with HCV-4, which is common in Egypt, are intermediate in responsiveness to therapy between those infected with HCV-1 and HCV-2 or -3, and it is suggested that they be treated for a full 48 weeks with full-dose ribavirin, like patients with HCV-1 (Di Bisceglie & Hoofnagle, 2002). There is insufficient experience to provide recommendations for the treatment of persons with HCV-5 and -6 so far. Experienced providers need to make treatment judgments on a case-by-case basis. Since HCV genotype is the strongest predictor of responses to IFN-based therapy for CHC, the duration of therapy and the likelihood of response should be determined in all HCV-infected persons prior to treatment (Strader et al., 2004).

Pretreatment HCV RNA level, even less important than HCV genotype, is a predictor of sustained response in IFN-based therapy (Lindsay et al., 2001; Manns et al., 2001; McHutchison et al., 1998; Poynard et al., 1998; Yu et al., 2004). A higher HCV RNA level predicts a lower response rate. The impact of the HCV RNA level on the response to combination therapy was different between patients with different HCV genotype infections. High viral load (with a cutoff value of 2,000,000 copies/ml, or 800,000 IU/ml) influenced the response rate in patients with HCV-1 (41% vs. 56%), but not in patients with HCV-2 or -3 (74% vs. 81%) (Fried et al., 2002). Under the circumstances of a determined HCV genotype for CHC patients, testing HCV RNA levels is beneficial and recommended for HCV-1 patients but seems variable for HCV-2 or -3 patients (Strader et al., 2004).

HCV viral quasispecies evolution is considered another key element determining treatment response (Okada et al., 1992). Higher quasispecies complexity at baseline has been observed in nonresponders than in sustained virological responders (Moribe et al., 1995). An increasing number of mutations within the carboxyl terminal region of the HCV nonstructural 5A protein, named the IFN-sensitivity-determining region (ISDR), was correlated with treatment response in HCV-1 infected patients (Enomoto et al., 1996). Patients infected with the so-called mutant type, defined by four or more amino acid substitutions in the ISDR, showed a more favorable response toward IFN-based therapy in Japan and Taiwan (Enomoto et al., 1996; Hung et al., 2003). However, these findings were not observed in a European study (Pascu et al., 2004).

Host Factors

The presence of bridging fibrosis and cirrhosis has been reported as one of the most unfavorable predictors for IFN-based therapy (McHutchison et al., 1998; McHutchison & Poynard, 1999; Poynard et al., 1998; Schalm et al., 1999; Yu et al., 2005; Zeuzem et al., 2000). Patients with cirrhosis generally respond poorly to conventional IFN monotherapy, with SVR rates of 5–20% (Heathcote et al., 2000; Poynard et al., 1998). Responses are improved when conventional IFNs or PegIFNs are combined with ribavirin, resulting in SVR rates of 33–44% (Fried et al., 2002; Manns et al., 2001; Poynard et al., 1998).

A gender effect on response has been reported. Female gender was a predictor of SVR in studies of conventional IFN-based therapy (McHutchison & Poynard, 1999), but not in the studies of PegIFN–ribavirin (Fried et al., 2002; Hadziyannis et al., 2004; Lindsay et al., 2001). Younger patients (<40 years) had higher SVR rates with PegIFN–ribavirin (Fried et al., 2002; Hadziyannis et al., 2004; Manns et al., 2001). Sustained responders were younger than nonresponders by an average of 5 years (Martinot-Peignoux et al., 1995).

Racial differences in response to the efficacy of IFN exist and have been one of the host factors. A lower response rate to IFN monotherapy was observed among African-American patients compared with white patients (McHutchison et al., 2000; Reddy et al., 1999). A pool analysis of two clinical trials with IFN–ribavirin combination therapy demonstrated that SVRs were highest among Asians (61%), followed by whites (39%), Latinos (23%), and African Americans (14%) (Hepburn et al., 2004). Latinos and African Americans were less likely to respond to PegIFN-α–ribavirin compared to whites (Muir et al., 2004). In studies of Taiwanese CHC patients, the SVR rate was 23.7%, 37.1%, and 63.6% for a 24-week treatment of 3 MU IFN-α 3 times weekly alone, 6 MU 3 times weekly alone, and 3 MU 3 times weekly plus ribavirin, respectively (Lee et al., 2005; Yu et al., 2005). The SVR rate of HCV-1b patients to 24-week PegIFN-α–ribavirin was 48.9–65.8% and could be as high as 80% with a 48-week regimen in Taiwan (Lee et al., 2005; Yu et al., 2006b). The different ethnic response rates may reflect the role of genetics. A relative lower body weight (67–70 kg) in Asian patients compared to U.S. patients (78–81 kg) may also play an important role (Bressler et al., 2003).

Several studies have demonstrated that SVR rates are lower in patients with coexistent insulin resistance and/or hepatic steatosis or steatohepatitis (D'Souza et al., 2005; Wu et al., 2006). In HCV-1 patients treated with PegIFN–ribavirin, a lower SVR rate was observed in patients with insulin resistance (homeostasis model of assessment, HOMA-IR > 2) compared to those without insulin resistance (32.8% vs. 60.5%, $p = 0.007$, OR = 3.12, 95% CI = 1.42–6.89) (Romero-Gomez et al., 2005).

CHC patients with body mass indexes greater than 30 kg/m^2 are more likely to be insulin-resistant, to have more advanced hepatic steatosis or steatohepatitis and fibrosis, and to experience a reduced response to combination therapy (Bressler et al., 2003; Hickman et al., 2003). In addition, other possible mechanisms of the impact of obesity on the therapeutic response include the linear correlation of efficacy and body-weight-based doses of ribavirin (10.6–15 mg/kg/day) (Manns et al., 2001). The most direct approach for formulating more effective treatment regimens is to encourage weight loss and exercise before treatment, which has been associated with a reduction in steatosis fibrosis scores (Hickman et al., 2002).

Excessive alcohol use could reduce the likelihood of a response to therapy (Ohnishi et al., 1996; Tabone et al., 2002). To increase the efficacy of antiviral therapy, it has been suggested that abstinence be recommended before and during treatment for CHC (National Institutes of Health, 2002).

Host genetic variations are probably involved in the efficacy of IFN-based therapies for CHC (Tang & Kaslow, 2004). Genetic polymorphisms of the human leukocyte antigen, CC chemokine receptor 5, cytotoxic T lymphocyte antigen-4, interleukin-10, low molecular mass polypeptide 7, MxA, and transforming growth factor-β1 have been reported to have significant associations with responsiveness (Konishi et al., 2004; Sugimoto et al., 2002; Suzuki et al., 2004; Thursz et al., 1999; Vidigal et al., 2002; Yee et al., 2001, 2003; Yu et al., 2003). In contrast, transporter associated with antigen processing 1 and 2 and IFN-γ have not been associated with treatment outcome (Dai et al., 2005; Sugimoto et al., 2002). Most studies showed that promoter polymorphisms of the tumor necrosis factor α (TNF-α) gene were not related to response to IFN-based therapy (Yee et al., 2001; Yu et al., 2003). However, TNF-α-308 polymorphism was associated with SVRs to IFN–ribavirin in patients with HCV-1b infection and a high viral load (Dai et al., 2006). These results reflect the important role of unique genetic predisposition, at least in part, in the response to IFN-based therapy for CHC. Recent advances in pharmacogenomics have demonstrated the potential applications of genetic single-nucleotide polymorphism and expression patterns in determining treatment responsiveness in CHC (Chen et al., 2005; Hwang et al., 2006). Progress in the technology of the high-throughput genotyping method and bioinformatics will be very helpful in future studies on these issues.

Because of the presumably shared routes of transmission, approximately one-fourth to one-third of all persons infected with HIV are co-infected with HCV (Thomas, 2002). Patients with HIV–HCV co-infection have been shown to respond less favorably to antiviral therapy than patients infected with HCV alone (Perez-Olmeda et al., 2003; Thomas, 2002). Several RCTs recommended 48 weeks of PegIFN–ribavirin for HCV, regardless of HCV genotype, in HCV–HIV co-infected patients (Carrat et al., 2004; Chung et al., 2004).

Dual infections of HCV and hepatitis B virus (HBV) are not uncommon and occur in up to 5% of the general population in HCV-endemic areas (el-Sayed et al., 1997). Combined chronic hepatitis B and C leads to more severe liver disease and an increased risk of HCC (Liaw et al., 2004). Although HBV–HCV dual infection was refractory to conventional IFN monotherapy (Liu et al., 2005), recent studies in Taiwan have demonstrated that conventional IFN–ribavirin combination therapy was effective in HCV clearance among HCV-dominant, HBV/HCV dually infected patients (Chuang et al., 2005; Liu et al., 2003). A large, open-label, comparative, multicenter study is ongoing to evaluate the efficacy of PegIFN–ribavirin for patients with chronic HCV–HBV dual infection in Taiwan.

Nonresponders are more resistant than relapsers to retreatment with subsequent IFN-based therapy (OR = 3.912, 95% CI = 1.459–10.49) (Chuang et al., 2004). Retreatment with PegIFN–ribavirin could achieve an SVR rate of 47–60% for relapsers and 18–23% for nonresponders (Sherman et al., 2006; Shiffman, 2002; Shiffman et al., 2004).

On-Treatment Predictors of Response to IFN-Based Therapy

During IFN-α-based therapy, HCV RNA levels generally fall in a biphasic manner (Neumann et al., 1998). The first, rapid phase of viral suppression, from a few hours after the first IFN-α injection to the end of the first day, is related to an inhibition of viral replication by a direct, nonspecific action of IFN-α. This early initial decline in HCV RNA levels correlates poorly with the eventual response to IFN-based therapy (Layden & Layden, 2001; Neumann et al., 1998). The second, slower phase of viral suppression, beginning on day 2 and gradually leading to seroclearance of HCV RNA, is possibly related to the gradual clearance of infected cells by the patient's immune system, while HCV replication is efficiently inhibited. This phase, less influenced by the dosage of IFN and HCV genotype, exhibits a good response to PegIFN and is an excellent marker of an SVR to the treatment (Neumann et al., 1998; Zeuzem et al., 2001).

An RVR at week 4 could predict an SVR to IFN–ribavirin with a high degree of accuracy in both HCV-1 and -2 patients, with positive predictive values of 78% and 92%, respectively (Yu et al., 2006a). Recent studies have demonstrated that an RVR is the single best predictor of an SVR to PegIFN–ribavirin for HCV-1 (Jensen et al., 2006; Zeuzem et al., 2006) and HCV-2 or -3 patients (Dalgard et al., 2004; Mangia et al., 2005; von Wagner et al., 2005; Yu et al., 2007).

Among patients with an EVR, the likelihood of an SVR is only 72% (Davis et al., 2003). However, as a negative predictor, EVR is an even more robust predictor. The likelihood of an SVR is approximately 0–2% in cases without an EVR (Davis, 2002). The EVR is a significantly negative predictor in HCV-1 patients, but not in HCV-2 patients (Yu et al., 2006a). Thus, it is recommended that patients who do not achieve an EVR at week 12 should discontinue the therapy beyond 12 weeks (Davis et al., 2003; Muir et al., 2004).

Medical adherence is an important factor associated with response to IFN–ribavirin, especially among patients with HCV-1 infection. In a retrospective analysis of data collected in the large registration trials of IFN–ribavirin, SVRs have been reported to be more likely in patients who had taken at least 80% of all projected IFN injections and at least 80% of all ribavirin for at least 80% of the anticipated duration of treatment (McHutchison et al., 2002a).

Individualized Therapy

Individualized therapy has become a major consideration for clinicians. It is desirable to expose CHC patients to the lowest doses and shortest durations of treatment possible to reduce the likelihood of adverse events and to minimize costs, without compromising treatment efficacy. On the other hand, some difficult-to-treat patients have to receive longer and/or higher-dose therapy to achieve responses. Virological predictors, genotype, and baseline viral load are the most widely used factors in tailoring regimens of IFN-based therapy (EASL, 1999; Strader et al., 2004; Yu et al., 2004). Recently, several studies demonstrated that shorter courses of treatment in patients with RVRs at week 4 were as effective as the current recommendations for treatment by HCV genotype (Strader et al., 2004). In one pilot study and three RCTs for HCV-2 or -3 patients, 12–16 weeks of PegIFN–ribavirin were as effective as a standard 24-week treatment in patients with an RVR (SVR rates, 87% to 100% vs. 80% to 92%) (Dalgard et al., 2004; Mangia et al., 2005; von Wagner et al.,

Bressler, B.L., Guindi, M., Tomlinson, G., Heathcote, J. (2003). High body mass index is an independent risk factor for nonresponse to antiviral treatment in chronic hepatitis C. *Hepatology*, 38: 639–644.

Bruchfeld, A., Lindahl, K., Schvarcz, R., Stahle, L. (2002). Dosage of ribavirin in patients with hepatitis C should be based on renal function: a population pharmacokinetic analysis. *Therapeutic Drug Monitoring*, 24: 701–708.

Bukh, J., Miller, R.H., Purcell, R.H. (1995). Genetic heterogeneity of hepatitis C virus: quasispecies and genotypes. *Seminars in Liver Disease*, 15: 41–63.

Carithers, R.L., Jr., Emerson, S.S. (1997). Therapy of hepatitis C: meta-analysis of interferon alpha-2b trials. *Hepatology*, 26: 83S–88S.

Carrat, F., Bani-Sadr, F., Pol, S., Rosenthal, E., Lunel-Fabiani, F., Benzekri, A., Morand, P., Goujard, C., Pialoux, G., Piroth, L., Salmon-Ceron, D., Degott, C., Cacoub, P., Perronne, C. (2004). Pegylated interferon alpha-2b vs standard interferon alpha-2b, plus ribavirin, for chronic hepatitis C in HIV-infected patients: a randomized controlled trial. *Journal of the American Medical Association*, 292: 2839–2848.

Chen, L., Borozan, I., Feld, J., Sun, J., Tannis, L.L., Coltescu, C., Heathcote, J., Edwards, A.M., McGilvray, I.D. (2005). Hepatic gene expression discriminates responders and nonresponders in treatment of chronic hepatitis C viral infection. *Gastroenterology*, 128: 1437–1444.

Chuang, W.L., Dai, C.Y., Chen, S.C., Lee, L.P., Lin, Z.Y., Hsieh, M.Y., Wang, L.Y., Yu, M.L., Chang, W.Y. (2004). Randomized trial of three different regimens for 24 weeks for re-treatment of chronic hepatitis C patients who failed to respond to interferon-alpha monotherapy in Taiwan. *Liver International*, 24: 595–602.

Chuang, W.L., Dai, C.Y., Chang, W.Y., Lee, L.P., Lin, Z.Y., Chen, S.C., Hsieh, M.Y., Wang, L.Y., Yu, M.L. (2005). Viral interaction and responses in chronic hepatitis C and B coinfected patients with interferon-alpha plus ribavirin combination therapy. *Antiviral Therapy*, 10: 125–133.

Chung, R.T., Andersen, J., Volberding, P., Robbins, G.K., Liu, T., Sherman, K.E., Peters, M.G., Koziel, M.J., Bhan, A.K., Alston, B., Colquhoun, D., Nevin, T., Harb, G., van der Horst, C. (2004). Peginterferon alpha-2a plus ribavirin versus interferon alpha-2a plus ribavirin for chronic hepatitis C in HIV-coinfected persons. *New England Journal of Medicine*, 351: 451–459.

Cramp, M.E., Rossol, S., Chokshi, S., Carucci, P., Williams, R., Naoumov, N.V. (2000). Hepatitis C virus-specific T-cell reactivity during interferon and ribavirin treatment in chronic hepatitis C. *Gastroenterology*, 118: 346–355.

Crotty, S., Maag, D., Arnold, J.J., Zhong, W., Lau, J.Y., Hong, Z., Andino, R., Cameron, C.E. (2000). The broad-spectrum antiviral ribonucleoside ribavirin is an RNA virus mutagen. *Nature Medicine*, 6: 1375–1379.

D'Souza, R., Sabin, C.A., Foster, G.R. (2005). Insulin resistance plays a significant role in liver fibrosis in chronic hepatitis C and in the response to antiviral therapy. *American Journal of Gastroenterology*, 100: 1509–1515.

Dai, C.Y., Chuang, W.L., Chang, W.Y., Chen, S.C., Lee, L.P., Hsieh, M.Y., Hou, N.J., Lin, Z.Y., Hsieh, M.Y., Wang, L.Y., Yu, M.L. (2005). Polymorphisms in the interferon-gamma gene at position +874 in patients with chronic hepatitis C treated with high-dose interferon-alpha and ribavirin. *Antiviral Research*, 67: 93–97.

Dai, C.Y., Chuang, W.L., Chang, W.Y., Chen, S.C., Lee, L.P., Hsieh, M.Y., Hou, N.J., Lin, Z.Y., Huang, J.F., Hsieh, M.Y., Wang, L.Y., Yu, M.L. (2006). Tumor necrosis factor- alpha promoter polymorphism at position -308 predicts response to combination therapy in hepatitis C virus infection. *Journal of Infectious Disease*, 193: 98–101.

Dalgard, O., Bjoro, K., Hellum, K.B., Myrvang, B., Ritland, S., Skaug, K., Raknerud, N., Bell, H. (2004). Treatment with pegylated interferon and ribavirin in HCV infection with genotype 2 or 3 for 14 weeks: a pilot study. *Hepatology*, 40: 1260–1265.

Davis, G.L., Balart, L.A., Schiff, E.R., Lindsay, K., Bodenheimer, H.C., Jr., Perrillo, R.P., Carey, W., Jacobson, I.M., Payne, J., Dienstag, J.L., et al. (1989). Treatment of chronic hepatitis C with recombinant interferon alpha. A multicenter randomized, controlled trial. Hepatitis Interventional Therapy Group. *New England Journal of Medicine*, 321: 1501–1506.

viral replication by a direct, nonspecific action of IFN-α. This early initial decline in HCV RNA levels correlates poorly with the eventual response to IFN-based therapy (Layden & Layden, 2001; Neumann et al., 1998). The second, slower phase of viral suppression, beginning on day 2 and gradually leading to seroclearance of HCV RNA, is possibly related to the gradual clearance of infected cells by the patient's immune system, while HCV replication is efficiently inhibited. This phase, less influenced by the dosage of IFN and HCV genotype, exhibits a good response to PegIFN and is an excellent marker of an SVR to the treatment (Neumann et al., 1998; Zeuzem et al., 2001).

An RVR at week 4 could predict an SVR to IFN–ribavirin with a high degree of accuracy in both HCV-1 and -2 patients, with positive predictive values of 78% and 92%, respectively (Yu et al., 2006a). Recent studies have demonstrated that an RVR is the single best predictor of an SVR to PegIFN–ribavirin for HCV-1 (Jensen et al., 2006; Zeuzem et al., 2006) and HCV-2 or -3 patients (Dalgard et al., 2004; Mangia et al., 2005; von Wagner et al., 2005; Yu et al., 2007).

Among patients with an EVR, the likelihood of an SVR is only 72% (Davis et al., 2003). However, as a negative predictor, EVR is an even more robust predictor. The likelihood of an SVR is approximately 0–2% in cases without an EVR (Davis, 2002). The EVR is a significantly negative predictor in HCV-1 patients, but not in HCV-2 patients (Yu et al., 2006a). Thus, it is recommended that patients who do not achieve an EVR at week 12 should discontinue the therapy beyond 12 weeks (Davis et al., 2003; Muir et al., 2004).

Medical adherence is an important factor associated with response to IFN–ribavirin, especially among patients with HCV-1 infection. In a retrospective analysis of data collected in the large registration trials of IFN–ribavirin, SVRs have been reported to be more likely in patients who had taken at least 80% of all projected IFN injections and at least 80% of all ribavirin for at least 80% of the anticipated duration of treatment (McHutchison et al., 2002a).

Individualized Therapy

Individualized therapy has become a major consideration for clinicians. It is desirable to expose CHC patients to the lowest doses and shortest durations of treatment possible to reduce the likelihood of adverse events and to minimize costs, without compromising treatment efficacy. On the other hand, some difficult-to-treat patients have to receive longer and/or higher-dose therapy to achieve responses. Virological predictors, genotype, and baseline viral load are the most widely used factors in tailoring regimens of IFN-based therapy (EASL, 1999; Strader et al., 2004; Yu et al., 2004). Recently, several studies demonstrated that shorter courses of treatment in patients with RVRs at week 4 were as effective as the current recommendations for treatment by HCV genotype (Strader et al., 2004). In one pilot study and three RCTs for HCV-2 or -3 patients, 12–16 weeks of PegIFN–ribavirin were as effective as a standard 24-week treatment in patients with an RVR (SVR rates, 87% to 100% vs. 80% to 92%) (Dalgard et al., 2004; Mangia et al., 2005; von Wagner et al.,

2005; Yu et al., 2007). Another two studies observed that 24 weeks of PegIFN–ribavirin treatment in HCV-1 infected patients resulted in similar SVR rates as a recommended 48-week regimen in patients with an RVR (87% to 89% vs. 73% to 91%) (Jensen et al., 2006; Zeuzem et al., 2006). HCV genotype and on-treatment virological responses will provide information for individualized therapy decision making for CHC patients in the near future.

Acute Hepatitis C

Data exploring the efficacy of treatment for acute HCV infection are very limited since these infections are seldom diagnosed during the acute phase. Given the high rate of progression to chronic infection (>70%) and the relatively limited efficacy of current therapy for chronic infection, the treatment of acute infection has been tried empirically but has not yet proven beneficial. Furthermore, some patients with acute symptomatic HCV infection have high rates of spontaneous clearance and would probably be treated unnecessarily. Nonetheless, the preliminary results of more recent studies suggest that early treatment, even with IFN alone (Jaeckel et al., 2001) or with PegIFN alone (Wiegandet al., 2006) or in combination with ribavirin (Kamal et al., 2002), has a high rate of efficacy (>70%). Initiation of PegIFN treatment <12 weeks after onset resulted in a higher SVR rate than treatment initiated at week 20 (Kamal et al., 2006). In view of these data, early therapy may be advised, but the optimal therapeutic regimen and the best point at which to intervene have not been defined clearly.

Clinical Outcome of IFN-Based Therapy

An important goal of therapy in patients with CHC is to delay disease progression and reduce hepatic complication. Histological response improvement of at least one stage in the fibrosis score was observed in approximately one-half of patients with compensated cirrhosis after treatment with IFN or PegIFN alone or in combination with ribavirin (Poynard et al., 2002). The histological response was observed in treatment-naïve or previously IFN-resistant patients and in patients with or without SVRs (Heathcote et al., 2000; Kjaergard et al., 2001; Poynard et al., 2002). Several earlier studies showed long-term beneficial effects of IFN monotherapy in reducing the progression of cirrhosis (Effect, 1998), hindering HCC development (Effect, 1998; Yoshida et al., 1999), and prolonging survival (Yoshida et al., 2002) among both sustained responders and nonresponders. However, the benefits of preventing disease progression in CHC patients without SVRs no longer existed over a longer observation period (Shiratori et al., 2005; Yoshida, 2004; Yu et al., 2006c). Until recently, a nationwide, multicenter study in Taiwan demonstrated that IFN–ribavirin combination therapy could reduce the risk for HCC and improve survival of CHC patients with SVR (Yu et al., 2006c). However, more than one-third of patients are

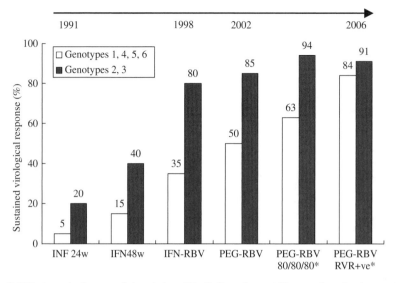

Fig. 2 Milestones in therapy of chronic hepatitis C. Over the past 15 years there has been a dramatic improvement in treatment results for chronic hepatitis C. IFN, interferon; RBV, ribavirin; PEG, pegylated interferon; w, weeks; 80/80/80*, subgroup of treated patients who had taken at least 80% of all projected IFN injections and at least 80% of all ribavirin for at least 80% of the anticipated duration of treatment (McHutchison et al., 2002a); RVR*, subgroup of patients who achieved a rapid virological response, defined as PCR-seronegative of HCV RNA after 4 weeks of therapy

resistant to the recommended antiviral regimens (Fried et al., 2002; Manns et al., 2001). Three large multicenter studies in the United States and Europe are evaluating the long-term effects of 4 years of PegIFN maintenance therapy in a subgroup of nonresponders (Kelleher & Afdhal, 2005; Shiffman, 2004).

Summary

Figure 2 summarizes the progress in treatment of CHC over the past 15 years. The issue of treatment of CHC is in constant flux. There is highly active clinical research in this area, and new information appears with increasing frequency. Future aims should be to develop a treatment beyond IFN with fewer side effects and higher efficacy.

References

Bodenheimer, H.C., Jr., Lindsay, K.L., Davis, G.L., Lewis, J.H., Thung, S.N., Seeff, L.B. (1997). Tolerance and efficacy of oral ribavirin treatment of chronic hepatitis C: a multicenter trial. *Hepatology*, 26: 473–477.

Bressler, B.L., Guindi, M., Tomlinson, G., Heathcote, J. (2003). High body mass index is an independent risk factor for nonresponse to antiviral treatment in chronic hepatitis C. *Hepatology*, 38: 639–644.

Bruchfeld, A., Lindahl, K., Schvarcz, R., Stahle, L. (2002). Dosage of ribavirin in patients with hepatitis C should be based on renal function: a population pharmacokinetic analysis. *Therapeutic Drug Monitoring*, 24: 701–708.

Bukh, J., Miller, R.H., Purcell, R.H. (1995). Genetic heterogeneity of hepatitis C virus: quasispecies and genotypes. *Seminars in Liver Disease*, 15: 41–63.

Carithers, R.L., Jr., Emerson, S.S. (1997). Therapy of hepatitis C: meta-analysis of interferon alpha-2b trials. *Hepatology*, 26: 83S–88S.

Carrat, F., Bani-Sadr, F., Pol, S., Rosenthal, E., Lunel-Fabiani, F., Benzekri, A., Morand, P., Goujard, C., Pialoux, G., Piroth, L., Salmon-Ceron, D., Degott, C., Cacoub, P., Perronne, C. (2004). Pegylated interferon alpha-2b vs standard interferon alpha-2b, plus ribavirin, for chronic hepatitis C in HIV-infected patients: a randomized controlled trial. *Journal of the American Medical Association*, 292: 2839–2848.

Chen, L., Borozan, I., Feld, J., Sun, J., Tannis, L.L., Coltescu, C., Heathcote, J., Edwards, A.M., McGilvray, I.D. (2005). Hepatic gene expression discriminates responders and nonresponders in treatment of chronic hepatitis C viral infection. *Gastroenterology*, 128: 1437–1444.

Chuang, W.L., Dai, C.Y., Chen, S.C., Lee, L.P., Lin, Z.Y., Hsieh, M.Y., Wang, L.Y., Yu, M.L., Chang, W.Y. (2004). Randomized trial of three different regimens for 24 weeks for re-treatment of chronic hepatitis C patients who failed to respond to interferon-alpha monotherapy in Taiwan. *Liver International*, 24: 595–602.

Chuang, W.L., Dai, C.Y., Chang, W.Y., Lee, L.P., Lin, Z.Y., Chen, S.C., Hsieh, M.Y., Wang, L.Y., Yu, M.L. (2005). Viral interaction and responses in chronic hepatitis C and B coinfected patients with interferon-alpha plus ribavirin combination therapy. *Antiviral Therapy*, 10: 125–133.

Chung, R.T., Andersen, J., Volberding, P., Robbins, G.K., Liu, T., Sherman, K.E., Peters, M.G., Koziel, M.J., Bhan, A.K., Alston, B., Colquhoun, D., Nevin, T., Harb, G., van der Horst, C. (2004). Peginterferon alpha-2a plus ribavirin versus interferon alpha-2a plus ribavirin for chronic hepatitis C in HIV-coinfected persons. *New England Journal of Medicine*, 351: 451–459.

Cramp, M.E., Rossol, S., Chokshi, S., Carucci, P., Williams, R., Naoumov, N.V. (2000). Hepatitis C virus-specific T-cell reactivity during interferon and ribavirin treatment in chronic hepatitis C. *Gastroenterology*, 118: 346–355.

Crotty, S., Maag, D., Arnold, J.J., Zhong, W., Lau, J.Y., Hong, Z., Andino, R., Cameron, C.E. (2000). The broad-spectrum antiviral ribonucleoside ribavirin is an RNA virus mutagen. *Nature Medicine*, 6: 1375–1379.

D'Souza, R., Sabin, C.A., Foster, G.R. (2005). Insulin resistance plays a significant role in liver fibrosis in chronic hepatitis C and in the response to antiviral therapy. *American Journal of Gastroenterology*, 100: 1509–1515.

Dai, C.Y., Chuang, W.L., Chang, W.Y., Chen, S.C., Lee, L.P., Hsieh, M.Y., Hou, N.J., Lin, Z.Y., Hsieh, M.Y., Wang, L.Y., Yu, M.L. (2005). Polymorphisms in the interferon-gamma gene at position +874 in patients with chronic hepatitis C treated with high-dose interferon-alpha and ribavirin. *Antiviral Research*, 67: 93–97.

Dai, C.Y., Chuang, W.L., Chang, W.Y., Chen, S.C., Lee, L.P., Hsieh, M.Y., Hou, N.J., Lin, Z.Y., Huang, J.F., Hsieh, M.Y., Wang, L.Y., Yu, M.L. (2006). Tumor necrosis factor- alpha promoter polymorphism at position -308 predicts response to combination therapy in hepatitis C virus infection. *Journal of Infectious Disease*, 193: 98–101.

Dalgard, O., Bjoro, K., Hellum, K.B., Myrvang, B., Ritland, S., Skaug, K., Raknerud, N., Bell, H. (2004). Treatment with pegylated interferon and ribavirin in HCV infection with genotype 2 or 3 for 14 weeks: a pilot study. *Hepatology*, 40: 1260–1265.

Davis, G.L., Balart, L.A., Schiff, E.R., Lindsay, K., Bodenheimer, H.C., Jr., Perrillo, R.P., Carey, W., Jacobson, I.M., Payne, J., Dienstag, J.L., et al. (1989). Treatment of chronic hepatitis C with recombinant interferon alpha. A multicenter randomized, controlled trial. Hepatitis Interventional Therapy Group. *New England Journal of Medicine*, 321: 1501–1506.

Davis, G.L., Lau, J.Y. (1995). Choice of appropriate end points of response to interferon therapy in chronic hepatitis C virus infection. *Journal of Hepatology*, 22: 110–114.

Davis, G. L. (2002). Monitoring of viral levels during therapy of hepatitis C. *Hepatology*, 36: S145–151.

Davis, G.L., Wong, J.B., McHutchison, J.G., Manns, M.P., Harvey, J., Albrecht, J. (2003). Early virologic response to treatment with peginterferon alpha-2b plus ribavirin in patients with chronic hepatitis C. *Hepatology*, 38: 645–652.

Di Bisceglie A.M., Hoofnagle, J.H. (2002). Optimal therapy of hepatitis C. *Hepatology*, 36: S121–127.

Di Bisceglie, A.M., Martin, P., Kassianides, C., Lisker-Melman, M., Murray, L., Waggoner, J., Goodman, Z., Banks, S.M., Hoofnagle, J.H. (1989). Recombinant interferon alpha therapy for chronic hepatitis C. A randomized, double-blind, placebo-controlled trial. *New England Journal of Medicine*, 321: 1506–1510.

Dixit, N.M., Layden-Almer, J.E., Layden, T.J., Perelson, A.S. (2004). Modelling how ribavirin improves interferon response rates in hepatitis C virus infection. *Nature*, 432: 922–924.

EASL International Consensus Conference on Hepatitis C. Paris, 26-27 February 1999. Consensus statement. *Journal of Hepatology*, *31 Suppl. 1*: 3–8.

el-Sayed, H.F., Abaza, S.M., Mehanna, S., Winch, P.J. (1997). The prevalence of hepatitis B and C infections among immigrants to a newly reclaimed area endemic for Schistosoma mansoni in Sinai, Egypt. *Acta Tropica*, 68: 229–237.

Enomoto, N., Sakuma, I., Asahina, Y., Kurosaki, M., Murakami, T., Yamamoto, C., Ogura, Y., Izumi, N., Marumo, F., Sato, C. (1996). Mutations in the nonstructural protein 5A gene and response to interferon in patients with chronic hepatitis C virus 1b infection. *New England Journal of Medicine*, 334: 77–81.

Feld, J.J., Hoofnagle, J.H. (2005). Mechanism of action of interferon and ribavirin in treatment of hepatitis C. *Nature*, 436: 967–972.

Fried, M.W., Shiffman, M.L., Reddy, K.R., Smith, C., Marinos, G., Goncales, F.L., Jr., Haussinger, D., Diago, M., Carosi, G., Dhumeaux, D., Craxi, A., Lin, A., Hoffman, J., Yu, J. (2002). Peginterferon alpha-2a plus ribavirin for chronic hepatitis C virus infection. *New England Journal of Medicine*, 347: 975–982.

Hadziyannis, S.J., Sette, H., Jr., Morgan, T.R., Balan, V., Diago, M., Marcellin, P., Ramadori, G., Bodenheimer, H., Jr., Bernstein, D., Rizzetto, M., Zeuzem, S., Pockros, P.J., Lin, A., Ackrill, A.M. (2004). Peginterferon-alpha2a and ribavirin combination therapy in chronic hepatitis C: a randomized study of treatment duration and ribavirin dose. *Annals of Internal Medicine*, 140: 346–355.

Heathcote, E.J., Shiffman, M.L., Cooksley, W.G., Dusheiko, G.M., Lee, S.S., Balart, L., Reindollar, R., Reddy, R.K., Wright, T.L., Lin, A., Hoffman, J., De Pamphilis, J. (2000). Peginterferon alpha-2a in patients with chronic hepatitis C and cirrhosis. *New England Journal of Medicine*, *343:* 1673–1680.

Hepburn, M.J., Hepburn, L.M., Cantu, N.S., Lapeer, M.G., Lawitz, E.J. (2004). Differences in treatment outcome for hepatitis C among ethnic groups. *American Journal of Medicine*, 117: 163–168.

Hickman, I.J., Clouston, A.D., Macdonald, G.A., Purdie, D.M., Prins, J.B., Ash, S., Jonsson, J.R., Powell, E.E. (2002). Effect of weight reduction on liver histology and biochemistry in patients with chronic hepatitis C. *Gut*, 51: 89–94.

Hickman, I.J., Powell, E.E., Prins, J.B., Clouston, A.D., Ash, S., Purdie, D.M., Jonsson, J.R. (2003). In overweight patients with chronic hepatitis C, circulating insulin is associated with hepatic fibrosis: implications for therapy. *Journal of Hepatology*, 39: 1042–1048.

Hoofnagle, J.H., Mullen, K.D., Jones, D.B., Rustgi, V., Di Bisceglie, A., Peters, M., Waggoner, J.G., Park, Y., Jones, E.A. (1986). Treatment of chronic non-A,non-B hepatitis with recombinant human alpha interferon. A preliminary report. *New England Journal of Medicine*, 315: 1575–1578.

Hung, C.H., Lee, C.M., Lu, S.N., Lee, J.F., Wang, J.H., Tung, H.D., Chen, T.M., Hu, T.H., Chen, W.J., Changchien, C.S. (2003). Mutations in the NS5A and E2-PePHD region of hepatitis

C virus type 1b and correlation with the response to combination therapy with interferon and ribavirin. *Journal of Viral Hepatitis*, 10: 87–94.

Hwang, Y., Chen, E.Y., Gu, Z.J., Chuang, W.L., Yu, M.L., Lai, M.Y., Chao, Y. ., Lee, C. ., Wang, J.H., Dai, C.Y., Shian-Jy Bey, M., Liao, Y.T., Chen, P.J., Chen, D.S. (2006). Genetic predisposition of responsiveness to therapy for chronic hepatitis C. *Pharmacogenomics*, 7: 697–709.

International Interferon-Alpha Hepatocellular Carcinoma Study Group (1998). Effect of interferon-alpha on progression of cirrhosis to hepatocellular carcinoma: a retrospective cohort study. *Lancet*, 351: 1535–1539.

Jaeckel, E., Cornberg, M., Wedemeyer, H., Santantonio, T., Mayer, J., Zankel, M., Pastore, G., Dietrich, M., Trautwein, C., Manns, M.P. (2001). Treatment of acute hepatitis C with interferon alpha-2b. *New England Journal of Medicine*, 345: 1452–1457.

Jensen, D.M., Morgan, T.R., Marcellin, P., Pockros, P.J., Reddy, K.R., Hadziyannis, S.J., Ferenci, P., Ackrill, A.M., Willems, B. (2006). Early identification of HCV genotype 1 patients responding to 24 weeks peginterferon alpha-2a (40 kd)/ribavirin therapy. *Hepatology*, 43: 954–960.

Jian Wu, Y., Shu Chen, L., Gui Qiang, W. (2006). Effects of fatty liver and related factors on the efficacy of combination antiviral therapy in patients with chronic hepatitis C. *Liver International*, 26: 166–172.

Kamal, S.M., Fehr, J., Roesler, B., Peters, T., Rasenack, J.W. (2002). Peginterferon alone or with ribavirin enhances HCV-specific CD4 T-helper 1 responses in patients with chronic hepatitis C. *Gastroenterology*, 123: 1070–1083.

Kamal, S.M., Fouly, A.E., Kamel, R.R., Hockenjos, B., Al Tawil, A., Khalifa, K.E., He, Q., Koziel, M.J., El Naggar, K.M., Rasenack, J., Afdhal, N.H. (2006). Peginterferon alpha-2b therapy in acute hepatitis C: impact of onset of therapy on sustained virologic response. *Gastroenterology*, 130: 632–638.

Kelleher, T. B., Afdhal, N. (2005). Maintenance therapy for chronic hepatitis C. *Current Gastroenterology Reports*, 7: 50–53.

Kjaergard, L.L., Krogsgaard, K., Gluud, C. (2001). Interferon alfa with or without ribavirin for chronic hepatitis C: systematic review of randomised trials. *British Medical Journal*, 323: 1151–1155.

Konishi, I., Horiike, N., Hiasa, Y., Michitaka, K., Onji, M. (2004). CCR5 promoter polymorphism influences the interferon response of patients with chronic hepatitis C in Japan. *Intervirology*, 47: 114–120.

Lai, M.Y., Kao, J.H., Yang, P.M., Wang, J.T., Chen, P.J., Chan, K.W., Chu, J.S., Chen, D.S. (1996). Long-term efficacy of ribavirin plus interferon alfa in the treatment of chronic hepatitis C. *Gastroenterology*, 111: 1307–1312.

Lau, D.T., Kleiner, D.E., Ghany, M.G., Park, Y., Schmid, P., Hoofnagle, J.H. (1998). 10-year follow-up after interferon-alpha therapy for chronic hepatitis C. *Hepatology*, 28: 1121–1127.

Lau, J.Y., Tam, R.C., Liang, T.J., Hong, Z. (2002). Mechanism of action of ribavirin in the combination treatment of chronic HCV infection. *Hepatology*, 35: 1002–1009.

Layden, J.E., Layden, T.J. (2001). How can mathematics help us understand HCV? *Gastroenterology*, 120: 1546–1549.

Lee, S.D., Yu, M.L., Cheng, P.N., Lai, M.Y., Chao, Y.C., Hwang, S.J., Chang, W.Y., Chang, T.T., Hsieh, T.Y., Liu, C.J., Chen, D.S. (2005). Comparison of a 6-month course peginterferon alpha-2b plus ribavirin and interferon alpha-2b plus ribavirin in treating Chinese patients with chronic hepatitis C in Taiwan. *Journal of Viral Hepatitis*, 12: 283–291.

Liaw, Y.F., Chen, Y.C., Sheen, I.S., Chien, R.N., Yeh, C.T., Chu, C.M. (2004). Impact of acute hepatitis C virus superinfection in patients with chronic hepatitis B virus infection. *Gastroenterology*, 126: 1024–1029.

Lindahl, K., Stahle, L., Bruchfeld, A., Schvarcz, R. (2005). High-dose ribavirin in combination with standard dose peginterferon for treatment of patients with chronic hepatitis C. *Hepatology*, 41: 275–279.

Lindsay, K.L., Trepo, C., Heintges, T., Shiffman, M.L., Gordon, S.C., Hoefs, J.C., Schiff, E.R., Goodman, Z.D., Laughlin, M., Yao, R., Albrecht, J.K. (2001). A randomized, double-blind trial comparing pegylated interferon alpha-2b to interferon alpha-2b as initial treatment for chronic hepatitis C. *Hepatology*, 34: 395–403.

Liu, C.J., Chen, P.J., Lai, M.Y., Kao, J.H., Jeng, Y.M., Chen, D.S. (2003). Ribavirin and interferon is effective for hepatitis C virus clearance in hepatitis B and C dually infected patients. *Hepatology*, 37: 568–576.

Liu, C.J., Liou, J.M., Chen, D.S., Chen, P.J. (2005). Natural course and treatment of dual hepatitis B virus and hepatitis C virus infections. *Journal of the Formosan Medical Association*, 104: 783–791.

Mangia, A., Santoro, R., Minerva, N., Ricci, G.L., Carretta, V., Persico, M., Vinelli, F., Scotto, G., Bacca, D., Annese, M., Romano, M., Zechini, F., Sogari, F., Spirito, F., Andriulli, A. (2005). Peginterferon alpha-2b and ribavirin for 12 vs. 24 weeks in HCV genotype 2 or 3. *New England Journal of Medicine*, 352: 2609–2617.

Manns, M.P., McHutchison, J.G., Gordon, S.C., Rustgi, V.K., Shiffman, M., Reindollar, R., Goodman, Z.D., Koury, K., Ling, M., Albrecht, J.K. (2001). Peginterferon alpha-2b plus ribavirin compared with interferon alfa-2b plus ribavirin for initial treatment of chronic hepatitis C: a randomised trial. *Lancet*, 358: 958–965.

Marcellin, P., Boyer, N., Gervais, A., Martinot, M., Pouteau, M., Castelnau, C., Kilani, A., Areias, J., Auperin, A., Benhamou, J.P., Degott, C., Erlinger, S. (1997). Long-term histologic improvement and loss of detectable intrahepatic HCV RNA in patients with chronic hepatitis C and sustained response to interferon-alpha therapy. *Annals of Internal Medicine*, 127: 875–881.

Martinot-Peignoux, M., Marcellin, P., Pouteau, M., Castelnau, C., Boyer, N., Poliquin, M., Degott, C., Descombes, I., Le Breton, V., Milotova, V., et al. (1995). Pretreatment serum hepatitis C virus RNA levels and hepatitis C virus genotype are the main and independent prognostic factors of sustained response to interferon alpha therapy in chronic hepatitis C. *Hepatology*, 22: 1050–1056.

McHutchison, J.G., Gordon, S.C., Schiff, E.R., Shiffman, M.L., Lee, W.M., Rustgi, V.K., Goodman, Z.D., Ling, M.H., Cort, S., Albrecht, J.K. (1998). Interferon alpha-2b alone or in combination with ribavirin as initial treatment for chronic hepatitis C. Hepatitis Interventional Therapy Group. *New England Journal of Medicine*, 339: 1485–1492.

McHutchison, J.G., Poynard, T. (1999). Combination therapy with interferon plus ribavirin for the initial treatment of chronic hepatitis C. *Seminars in Liver Disease, 19 Suppl. 1*: 57–65.

McHutchison, J.G., Poynard, T., Pianko, S., Gordon, S.C., Reid, A.E., Dienstag, J., Morgan, T., Yao, R., Albrecht, J. (2000). The impact of interferon plus ribavirin on response to therapy in black patients with chronic hepatitis C. The International Hepatitis Interventional Therapy Group. *Gastroenterology*, 119: 1317–1323.

McHutchison, J.G., Manns, M., Patel, K., Poynard, T., Lindsay, K.L., Trepo, C., Dienstag, J., Lee, W.M., Mak, C., Garaud, J.J., Albrecht, J.K. (2002a). Adherence to combination therapy enhances sustained response in genotype-1-infected patients with chronic hepatitis C. *Gastroenterology*, 123: 1061–1069.

McHutchison, J.G., Poynard, T., Esteban-Mur, R., Davis, G.L., Goodman, Z.D., Harvey, J., Ling, M.H., Garaud, J.J., Albrecht, J.K., Patel, K., Dienstag, J.L., Morgan, T. (2002b). Hepatic HCV RNA before and after treatment with interferon alone or combined with ribavirin. *Hepatology*, 35: 688–693.

Moribe, T., Hayashi, N., Kanazawa, Y., Mita, E., Fusamoto, H., Negi, M., Kaneshige, T., Igimi, H., Kamada, T., Uchida, K. (1995). Hepatitis C viral complexity detected by single-strand conformation polymorphism and response to interferon therapy. *Gastroenterology*, 108: 789–795.

Muir, A.J., Bornstein, J.D., Killenberg, P.G. (2004). Peginterferon alpha-2b and ribavirin for the treatment of chronic hepatitis C in blacks and non-Hispanic whites. *New England Journal of Medicine*, 350: 2265–2271.

National Institutes of Health Consensus Development Conference Panel (1997). Panel statement: management of hepatitis C. *Hepatology*, 26: 2S–10S.

National Institutes of Health Consensus Development Conference Statement: Management of hepatitis C: 2002—June 10-12, 2002. *Hepatology*, 36: S3–20.

Neumann, A.U., Lam, N.P., Dahari, H., Gretch, D.R., Wiley, T.E., Layden, T.J., Perelson, A.S. (1998). Hepatitis C viral dynamics in vivo and the antiviral efficacy of interferon-alpha therapy. *Science*, 282: 103–107.

Neumann, A.U., Lam, N.P., Dahari, H., Davidian, M., Wiley, T.E., Mika, B.P., Perelson, A.S., Layden, T.J. (2000). Differences in viral dynamics between genotypes 1 and 2 of hepatitis C virus. *Journal of Infectious Disease*, 182: 28–35.

Niederau, C., Heintges, T., Haussinger, D. (1996). Treatment of chronic hepatitis C with a-interferon: an analysis of the literature. *Hepatogastroenterology*, 43: 1544–1556.

Ohnishi, K., Matsuo, S., Matsutani, K., Itahashi, M., Kakihara, K., Suzuki, K., Ito, S., Fujiwara, K. (1996). Interferon therapy for chronic hepatitis C in habitual drinkers: comparison with chronic hepatitis C in infrequent drinkers. *American Journal of Gastroenterology*, 91: 1374–1379.

Okada, S., Akahane, Y., Suzuki, H., Okamoto, H., Mishiro, S. (1992). The degree of variability in the amino terminal region of the E2/NS1 protein of hepatitis C virus correlates with responsiveness to interferon therapy in viremic patients. *Hepatology*, 16: 619–624.

Pascu, M., Martus, P., Hohne, M., Wiedenmann, B., Hopf, U., Schreier, E., Berg, T. (2004). Sustained virological response in hepatitis C virus type 1b infected patients is predicted by the number of mutations within the NS5A-ISDR: a meta-analysis focused on geographical differences. *Gut*, 53: 1345–1351.

Perez-Olmeda, M., Nunez, M., Romero, M., Gonzalez, J., Castro, A., Arribas, J.R., Pedreira, J., Barreiro, P., Garcia-Samaniego, J., Martin-Carbonero, L., Jimenez-Nacher, I., Soriano, V. (2003). Pegylated IFN-alpha2b plus ribavirin as therapy for chronic hepatitis C in HIV-infected patients. *AIDS*, 17: 1023–1028.

Poynard, T., Leroy, V., Cohard, M., Thevenot, T., Mathurin, P., Opolon, P., Zarski, J.P. (1996). Meta-analysis of interferon randomized trials in the treatment of viral hepatitis C: effects of dose and duration. *Hepatology*, 24: 778–789.

Poynard, T., Marcellin, P., Lee, S.S., Niederau, C., Minuk, G.S., Ideo, G., Bain, V., Heathcote, J., Zeuzem, S., Trepo, C., Albrecht, J. (1998). Randomised trial of interferon alpha2b plus ribavirin for 48 weeks or for 24 weeks versus interferon alpha2b plus placebo for 48 weeks for treatment of chronic infection with hepatitis C virus. International Hepatitis Interventional Therapy Group (IHIT). *Lancet*, 352: 1426–1432.

Poynard, T., McHutchison, J., Manns, M., Trepo, C., Lindsay, K., Goodman, Z., Ling, M.H., Albrecht, J. (2002). Impact of pegylated interferon alpha-2b and ribavirin on liver fibrosis in patients with chronic hepatitis C. *Gastroenterology*, 122: 1303–1313.

Reddy, K.R., Hoofnagle, J.H., Tong, M.J., Lee, W.M., Pockros, P., Heathcote, E.J., Albert, D., Joh, T. (1999). Racial differences in responses to therapy with interferon in chronic hepatitis C. Consensus Interferon Study Group. *Hepatology*, 30: 787–793.

Reddy, K.R., Wright, T.L., Pockros, P.J., Shiffman, M., Everson, G., Reindollar, R., Fried, M.W., Purdum, P.P., 3rd, Jensen, D., Smith, C., Lee, W.M., Boyer, T.D., Lin, A., Pedder, S., DePamphilis, J. (2001). Efficacy and safety of pegylated (40-kd) interferon alpha-2a compared with interferon alpha-2a in noncirrhotic patients with chronic hepatitis C. *Hepatology*, 33: 433–438.

Reichard, O., Norkrans, G., Fryden, A., Braconier, J.H., Sonnerborg, A., Weiland, O. (1998). Randomised, double-blind, placebo-controlled trial of interferon alpha-2b with and without ribavirin for chronic hepatitis C. The Swedish Study Group. *Lancet*, 351: 83–87.

Romero-Gomez, M., Del Mar Viloria, M., Andrade, R.J., Salmeron, J., Diago, M., Fernandez-Rodriguez, C.M., Corpas, R., Cruz, M., Grande, L., Vazquez, L., Munoz-De-Rueda, P., Lopez-Serrano, P., Gila, A., Gutierrez, M.L., Perez, C., Ruiz-Extremera, A., Suarez, E., Castillo, J. (2005). Insulin resistance impairs sustained response rate to peginterferon plus ribavirin in chronic hepatitis C patients. *Gastroenterology*, 128: 636–641.

Schalm, S.W., Hansen, B.E., Chemello, L., Bellobuono, A., Brouwer, J.T., Weiland, O., Cavalletto, L., Schvarcz, R., Ideo, G., Alberti, A. (1997). Ribavirin enhances the efficacy but not the adverse effects of interferon in chronic hepatitis C. Meta-analysis of individual patient data from European centers. *Journal of Hepatology*, 26: 961–966.

Schalm, S.W., Weiland, O., Hansen, B.E., Milella, M., Lai, M.Y., Hollander, A., Michielsen, P.P., Bellobuono, A., Chemello, L., Pastore, G., Chen, D.S., Brouwer, J.T. (1999). Interferon-ribavirin for chronic hepatitis C with and without cirrhosis: analysis of individual patient data of six controlled trials. Eurohep Study Group for Viral Hepatitis. *Gastroenterology*, 117: 408–413.

Sen, G.C. (2001). Viruses and interferons. *Annual Review of Microbiology*, 55: 255–281.

Sherman, M., Yoshida, E.M., Deschenes, M., Krajden, M., Bain, V.G., Peltekian, K., Anderson, F., Kaita, K., Simonyi, S., Balshaw, R., Lee, S.S. (2006). Peginterferon alpha-2a (40KD) plus ribavirin in chronic hepatitis C patients who failed previous interferon therapy. *Gut*, 55: 1631–1638.

Shiffman, M.L. (2002). Retreatment of patients with chronic hepatitis C. *Hepatology*, 36: S128–134.

Shiffman, M.L. (2004). Retreatment of patients who do not respond to initial therapy for chronic hepatitis C. *Cleveland Clinic Journal of Medicine*, 71 Suppl. 3: S13–16.

Shiffman, M.L., Di Bisceglie, A.M., Lindsay, K.L., Morishima, C., Wright, E.C., Everson, G.T., Lok, A.S., Morgan, T.R., Bonkovsky, H.L., Lee, W.M., Dienstag, J.L., Ghany, M.G., Goodman, Z.D., Everhart, J.E. (2004). Peginterferon alpha-2a and ribavirin in patients with chronic hepatitis C who have failed prior treatment. *Gastroenterology*, 126: 1015–1023.

Shiratori, Y., Ito, Y., Yokosuka, O., Imazeki, F., Nakata, R., Tanaka, N., Arakawa, Y., Hashimoto, E., Hirota, K., Yoshida, H., Ohashi, Y., Omata, M. (2005). Antiviral therapy for cirrhotic hepatitis C: association with reduced hepatocellular carcinoma development and improved survival. *Annals of Internal Medicine*, 142: 105–114.

Strader, D.B., Wright, T., Thomas, D.L., Seeff, L.B. (2004). Diagnosis, management, and treatment of hepatitis C. *Hepatology*, 39: 1147–1171.

Sugimoto, Y., Kuzushita, N., Takehara, T., Kanto, T., Tatsumi, T., Miyagi, T., Jinushi, M., Ohkawa, K., Horimoto, M., Kasahara, A., Hori, M., Sasaki, Y., Hayashi, N. (2002). A single nucleotide polymorphism of the low molecular mass polypeptide 7 gene influences the interferon response in patients with chronic hepatitis C. *Journal of Viral Hepatitis*, 9: 377–384.

Suzuki, F., Arase, Y., Suzuki, Y., Tsubota, A., Akuta, N., Hosaka, T., Someya, T., Kobayashi, M., Saitoh, S., Ikeda, K., Kobayashi, M., Matsuda, M., Takagi, K., Satoh, J., Kumada, H. (2004). Single nucleotide polymorphism of the MxA gene promoter influences the response to interferon monotherapy in patients with hepatitis C viral infection. *Journal of Viral Hepatitis*, 11: 271–276.

Tabone, M., Sidoli, L., Laudi, C., Pellegrino, S., Rocca, G., Della Monica, P., Fracchia, M., Galatola, G., Molinaro, G.C., Arico, S., Pera, A. (2002). Alcohol abstinence does not offset the strong negative effect of lifetime alcohol consumption on the outcome of interferon therapy. *Journal of Viral Hepatitis*, 9: 288–294.

Tang, J., Kaslow, R.A. (2004). Pharmacogenomic perspectives of chronic hepatitis C virus (HCV) infection. *Pharmacogenomics Journal*, 4: 171–174.

Thomas, D.L. (2002). Hepatitis C and human immunodeficiency virus infection. *Hepatology*, 36: S201–209.

Thursz, M., Yallop, R., Goldin, R., Trepo, C., Thomas, H.C. (1999). Influence of MHC class II genotype on outcome of infection with hepatitis C virus. The HENCORE group. Hepatitis C European Network for Cooperative Research. *Lancet*, 354: 2119–2124.

Vidigal, P.G., Germer, J.J., Zein, N.N. (2002). Polymorphisms in the interleukin-10, tumor necrosis factor-alpha, and transforming growth factor-beta1 genes in chronic hepatitis C patients treated with interferon and ribavirin. *Journal of Hepatology*, 36: 271–277.

von Wagner, M., Huber, M., Berg, T., Hinrichsen, H., Rasenack, J., Heintges, T., Bergk, A., Bernsmeier, C., Haussinger, D., Herrmann, E., Zeuzem, S. (2005). Peginterferon-alpha-2a (40KD) and ribavirin for 16 or 24 weeks in patients with genotype 2 or 3 chronic hepatitis C. *Gastroenterology*, 129: 522–527.

Wiegand, J., Buggisch, P., Boecher, W., Zeuzem, S., Gelbmann, C.M., Berg, T., Kauffmann, W., Kallinowski, B., Cornberg, M., Jaeckel, E., Wedemeyer, H., Manns, M.P. (2006). Early monotherapy with pegylated interferon alpha-2b for acute hepatitis C infection: the HEP-NET acute-HCV-II study. *Hepatology*, 43: 250–256.

Yee, L.J., Tang, J., Gibson, A.W., Kimberly, R., Van Leeuwen, D.J., Kaslow, R.A. (2001). Interleukin 10 polymorphisms as predictors of sustained response in antiviral therapy for chronic hepatitis C infection. *Hepatology*, 33: 708–712.

Yee, L.J., Perez, K A., Tang, J., van Leeuwen, D.J., Kaslow, R.A. (2003). Association of CTLA4 polymorphisms with sustained response to interferon and ribavirin therapy for chronic hepatitis C virus infection. *Journal of Infectious Disease*, 187: 1264–1271.

Yoshida, H., Shiratori, Y., Moriyama, M., Arakawa, Y., Ide, T., Sata, M., Inoue, O., Yano, M., Tanaka, M., Fujiyama, S., Nishiguchi, S., Kuroki, T., Imazeki, F., Yokosuka, O., Kinoyama, S., Yamada, G., Omata, M. (1999). Interferon therapy reduces the risk for hepatocellular carcinoma: national surveillance program of cirrhotic and noncirrhotic patients with chronic hepatitis C in Japan. IHIT Study Group. Inhibition of Hepatocarcinogenesis by Interferon Therapy. *Annals of Internal Medicine*, 131: 174–181.

Yoshida, H., Arakawa, Y., Sata, M., Nishiguchi, S., Yano, M., Fujiyama, S., Yamada, G., Yokosuka, O., Shiratori, Y., Omata, M. (2002). Interferon therapy prolonged life expectancy among chronic hepatitis C patients. *Gastroenterology*, 123: 483–491.

Yoshida, H., Tateishi, R., Arakawa, Y., Sata, M., Fujiyama, S., Nishiguchi, S., Ishibashi, H., Yamada, G., Yokosuka, O., Shiratori, Y., Omata, M. (2004). Benefit of interferon therapy in hepatocellular carcinoma prevention for individual patients with chronic hepatitis C. *Gut*, 53: 425–430.

Yu, M.-L., Dai, C.-Y., Huang, J.-F., Hou, N.-J., Lee, L.-P., Hsieh, M.-Y., Chiu, C.-F., Lin, Z.-Y., Chen, S.-C., Hsieh, M.-Y., Wang, L.-Y., Chang, W.-Y., Chuang, W.-L. (2007). A randomised study of peginterferon and ribavirin for 16 vs 24 weeks in patients with genotype 2 chronic hepatitis C. *Gut*, 56: 553–559.

Yu, M.L., Dai, C.Y., Chen, S.C., Chiu, C.C., Lee, L.P., Lin, Z.Y., Hsieh, M.Y., Wang, L.Y., Chuang, W.L., Chang, W.Y. (2003). Human leukocyte antigen class I and II alleles and response to interferon-alpha treatment, in Taiwanese patients with chronic hepatitis C virus infection. *Journal of Infectious Disease*, 188: 62–65.

Yu, M.L., Dai, C.Y., Chiu, C.C., Lee, L.P., Lin, Z.Y., Chen, S.C., Hsieh, M.Y., Wang, L.Y., Chen, C.J., Chuang, W.L., Chang, W.Y. (2003). Tumor necrosis factor-alpha promoter polymorphisms at position -308 in Taiwanese chronic hepatitis C patients treated with interferon-alpha. *Antiviral Research*, 59: 35–40.

Yu, M.L., Dai, C.Y., Chen, S.C., Lee, L.P., Huang, J.F., Lin, Z.Y., Hsieh, M.Y., Wang, L.Y., Chuang, W.L., Chang, W.Y. (2004). A prospective study on treatment of chronic hepatitis C with tailored and extended interferon-alpha regimens according to pretreatment virological factors. *Antiviral Research*, 63: 25–32.

Yu, M.L., Dai, C.Y., Chen, S.C., Lee, L P., Hsieh, M.Y., Lin, Z.Y., Hsieh, M.Y., Wang, L.Y., Tsai, J.F., Chang, W.Y., Chuang, W.L. (2005). High versus standard doses interferon-alpha in the treatment of naive chronic hepatitis C patients in Taiwan: a 10-year cohort study. *BMC Infectious Disease*, 5: 27.

Yu, M.L., Chuang, W.L., Dai, C.Y., Lee, L.P., Hsieh, M.Y., Lin, Z.Y., Chen, S.C., Hsieh, M.Y., Wang, L.Y., Chang, W.Y., Tsai, S.L., Kuo, H.T. (2006a). Different viral kinetics between hepatitis C virus genotype 1 and 2 as on-treatment predictors of response to a 24-week course of high-dose interferon-alpha plus ribavirin combination therapy. *Translational Research*, 148: 120–127.

Yu, M.L., Dai, C.Y., Lin, Z.Y., Lee, L.P., Hou, N.J., Hsieh, M.Y., Chen, S.C., Hsieh, M.Y., Wang, L.Y., Chang, W.Y., Chuang, W.L. (2006b). A randomized trial of 24- vs. 48-week courses of PEG interferon alpha-2b plus ribavirin for genotype-1b-infected chronic hepatitis C patients: a pilot study in Taiwan. *Liver International*, 26: 73–81.

Yu, M.L., Lin, S.M., Chuang, W.L., Dai, C.Y., Wang, J.H., Lu, S.N., Sheen, I.S., Chang, W.Y., Lee, C.M., Liaw, Y.F. (2006c). A sustained virologic response to interferon or interferon/ribavirin reduces hepatocellular carcinoma and improves survival in chronic hepatitis C: a nation-wide, multicentre study in Taiwan. *Antiviral Therapy*, 11: 985–994.

Zeuzem, S., Feinman, S.V., Rasenack, J., Heathcote, E.J., Lai, M.Y., Gane, E., O'Grady, J., Reichen, J., Diago, M., Lin, A., Hoffman, J., Brunda, M.J. (2000). Peginterferon alpha-2a in patients with chronic hepatitis C. *New England Journal of Medicine*, 343: 1666–1672.

Zeuzem, S., Herrmann, E., Lee, J.H., Fricke, J., Neumann, A.U., Modi, M., Colucci, G., Roth, W.K. (2001). Viral kinetics in patients with chronic hepatitis C treated with standard or peginterferon alpha2a. *Gastroenterology*, 120: 1438–1447.

Zeuzem, S., Hultcrantz, R., Bourliere, M., Goeser, T., Marcellin, P., Sanchez-Tapias, J., Sarrazin, C., Harvey, J., Brass, C., Albrecht, J. (2004). Peginterferon alpha-2b plus ribavirin for treatment of chronic hepatitis C in previously untreated patients infected with HCV genotypes 2 or 3. *Journal of Hepatology*, 40: 993–999.

Zeuzem, S., Buti, M., Ferenci, P., Sperl, J., Horsmans, Y., Cianciara, J., Ibranyi, E., Weiland, O., Noviello, S., Brass, C., Albrecht, J. (2006). Efficacy of 24 weeks treatment with peginterferon alpha-2b plus ribavirin in patients with chronic hepatitis C infected with genotype 1 and low pretreatment viremia. *Journal of Hepatology*, 44: 97–103.

2006). Possible explanations are direct interactions with cellular mechanisms and/or genotype-specific immune modulation (Pawlotsky, 2006).

HCV-RNA load monitoring using serum PCR assays during therapy that allow early stopping rules is currently recommended in order to avoid the dangers and costs associated with the full treatment course in patients with little or no chance of reaching SVR (Davis et al., 2003; Ferenci et al., 2005; Fried et al., 2002). For this purpose, HCV-RNA is quantified at baseline and after 12 weeks of therapy. Treatment is continued in case of early virological response (EVR), defined for patients with at least a 2 $\log_{10}$ decrease from baseline HCV-RNA and/or HCV-RNA undetectability (according to the limit of detection of the HCV-RNA PCR assay used). If HCV-RNA is undetectable at week 24, there is a high likelihood of achieving SVR, and treatment is thus continued until week 48, as established in 2002 by the NIH Consensus Development Conference on Management of Hepatitis C. Among those with an EVR, the probability of reaching SVR is approximately 70%. As a negative predictor, EVR is even more robust, for the lack of EVR is associated to an SVR probability of only 0–2% (Davis, 2002).

The role of viral kinetics even earlier during therapy, such as at week 4, has been also explored in many studies (Davis et al., 2003; Martinot-Peignoux et al., 2006; Moreno et al., 2006). In the analysis performed from pooled data from peg-IFN/RBV pivotal trials (Davis et al., 200), the lack of at least a 1 $\log_{10}$ decrease in baseline HCV-RNA was 91% (83–95%) predictive of treatment failure and showed to be the best cutoff at that time when compared to the lack of negative HCV-RNA (57%) or the lack to achieve at least 3 $\log_{10}$ (73%) or 2 $\log_{10}$ (82%) decreases in HCV-RNA (Davis, 2002; Davis et al., 2003; Ferenci, 2004). A week 4 stopping rule has been proposed using a multivariate model developed on 186 patients with CHC receiving peg-IFN/RBV (Martinot-Peignoux et al., 2006). At this point in time, the model demonstrated a negative predictive value (NPV) of 97% and a positive predictive value (PPV) of 100%, with 95% sensitivity, 89% specificity, and 93% accuracy. In our experience (Moreno et al., 2006), the maximum sensitivity and NPV of 100% as determined by the ROC curve was reached with the cutoff level of a 1 $\log_{10}$ decrease in HCV-RNA levels at week 4, for all subjects with less than a 1 $\log_{10}$ HCV-RNA decrease at that time failed to achieve SVR (Figure 1).

The timing and magnitude of the virological response to antiviral therapy in patients infected with HCV genotype 1 are highly variable. Results of a retrospective trial with peg-IFN-α-2b plus RBV suggested that the duration of therapy after HCV-RNA is suppressed to a level below the limit of detection is critical in maximizing the chance of SVR in genotype 1 patients (Drusano & Preston, 2004). The authors concluded that treatment for 32 to 36 weeks after HCV-RNA becomes undetectable would result in an SVR rate of 80–90%, respectively. It might be speculated that the longer duration of therapy associated with an undetectable virus load allows the virus to be held in check on a long-term basis or that there are sanctuary sites where the penetration of the drugs is blunted and that, at these sites, it simply takes longer to decrease the virus load than in plasma. The results of this study bring up the following question: in the aim of optimizing treatment outcomes in CHC, is treatment beyond 48 weeks ever justified? Two recent large randomized trials have addressed this issue. A prospective study from Germany (Berg et al., 2006)

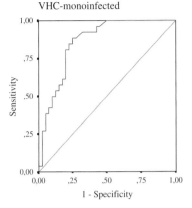

ROC Curve

VHC-monoinfected

Cut-off value predicting failure: 1 log10 decay

ROC Curve Area: 0·86 (95% CI 0·77-0.95)

Se	100%	(100%–100%)
Sp	50%	(35%–66%)
PPV	57%	(42%–71%)
NPV	100%	(100%–100%)

Fig. 1 Cut-off of HCV RNA decrease at week 4 predicting treatment failure at week 72 in HCV monoinfection. Reprinted from *Journal Viral Hepatitis,* 13 (7), Moreno A et al, "Viral kinetics and early prediction of nonresponse to peg-IFN-alpha-2b plus ribavirin in HCV genotypes 1/4 according to HIV serostatus, pages 466-473, Copyright 2006, with permission from Blackwell Publishing

of 459 treatment-naïve genotype 1 patients randomized to 72 versus 48 weeks of 180 μg per week of peg-IFN-α-2a and RBV 800 mg/day failed to find significant differences in the rates of end-of-treatment (ETR), SVR, and relapses: ETR, 63% vs. 71%; SVR, 54% vs. 53%; relapses, 21% vs. 29%. Although subgroup analyses suggested that subjects with persistently detectable HCV-RNA < 6,000 IU/ml at week 12 might benefit from prolonged treatment, the overall SVR rates in the two groups were similar. The frequency of medication dose reductions and adverse events were similar in both groups, but patients on the 72 weeks' arm had significantly higher rates of early treatment discontinuation (41% vs. 24%). Explanations for the poor results in the extended arm include the high rate of early termination, many of them self-discontinuation after week 48, and also the low dose of RBV. Finally, the randomization at baseline may have reduced the ability to see a significant difference in the two arms since the large number of subjects who became HCV-RNA negative at week 4 ($n = 51$) and week 12 ($n = 130$) had a high likelihood of SVR with only 48 weeks of therapy (84% and 81%, respectively). The second study (Sánchez-Tapias et al., 2006) demonstrates a potential benefit of prolonged antiviral therapy only in patients who remain positive HCV-RNA at week 4: 180 μg per week of peg-IFN-α-2a plus RBV 800 mg/day were given to 510 treatment-naïve patients (78% genotype 1 or 4). At week 4, 326 (62%) with persistently detectable HCV-RNA were randomized to either 48 or 72 weeks of therapy. Although the ETR rates were similar (61% in each group), the SVR rate was significantly better in patients allocated to the 72- vs. 48-week arm (45% vs. 32%, $p = 0.01$). The improvement in SVR was most apparent in patients with genotype 1 (44% vs. 28%, $p = 0.003$). As in the Berg study, the frequency of adverse events and that of dose reductions were similar in both groups, but the rate of early discontinuation was significantly higher in the extended arm (36% vs. 18%), mostly due to patients´ preference/withdrawal rather than severe adverse events. These "extended therapy" studies suggest that antiviral therapy beyond 48 weeks may be difficult for patients to adhere to, even when reduced doses of RBV are used. In order to determine if prolonged antiviral therapy is warranted, future studies should only randomize genotype 1 patients with a partial virological response after 4 to 12 weeks of therapy and also include subjects with unfavorable baseline predictors, such as high HCV-RNA levels, advanced fibrosis, or high body mass index. Finally, full-dose RBV (1000–1200 mg/day) should be used, since this will increase the response rate in the control arm. The use of the highly sensitive transcription-mediated amplification assay (TMA, Bayer Healthcare LLC, Tarrytown, NY) for HCV-RNA detection during treatment has also been proposed as another way to identify patients at high risk of virological relapse after discontinuing therapy (Sarrazin et al., 2000).

Another means of increasing the total dose of peg-IFN and/or RBV dose in genotype 1 patients with baseline poor predictors or response is increasing the dose of the drugs, in the so-called induction therapies. Preliminary results of multicenter studies of peg-IFN-α-2b and high-dose RBV (800–1600 mg/day) with or without growth factors demonstrate only a marginal improvement in efficacy compared with fixed doses of RBV in previously untreated genotype 1 patients (Jacobson et al., 2005; Shiffman et al., 2005). Therefore, the lack of improved antiviral efficacy coupled with the increased side effects and costs of high-dose RBV and growth factors do not currently justify these approaches in clinical practice.

Although a prolonged treatment regimen could be more effective for some HCV genotype 1 patients, it is also necessary to identify the subgroup of patients for whom a shorter treatment period would be adequate. The current 12 weeks' EVR rule, when applied to genotype 1 patients, has an NPV approaching 100% (Davis et al., 2003; Ferenci et al., 2005) and is hence able to solve the issue of overtreatment in nonresponders. However, it leaves the field open to the possibility of overtreating some genotype 1 patients with favorable baseline predictors of SVR, such as those with low pretreatment HCV-RNA. This group may represent up to 15–20% of all genotype 1 infected cases (Craxi & Camma, 2006) and, therefore, could hypothetically be cured by shorter and/or less intensive treatment courses, thus reducing costs and improving acceptance and tolerability.

Recent studies (Jensen et al., 2006; Zeuzem et al., 2006) suggest that genotype 1 patients with a low viral load (<600,000 IU/ml) and, especially, fast viral clearance (negative HCV-RNA at week 4) might be efficaciously treated for only 24 weeks. In the Zeuzem study, the rate of SVR was 89% in the subgroup of patients with negative week 4 HCV-RNA (47% of all cases treated for 24 weeks), as compared with 25% and 17% of those who cleared the virus at week 12 or 24, respectively. Among patients treated for 48 weeks, SVR rates were 85%, 93%, and 67% for subjects clearing the virus at week 4, 12, or 24, respectively. The indications of this study cannot, however, be extrapolated to subjects with advanced fibrosis or cirrhosis. In all megatrials (Fried et al., 2002; Hadziyannis et al., 2004; Manns et al., 2001), fibrosis emerges as a significant predictor of resistance to peg-IFN/RBV combinations and, when assessed as a continuous variable (Bruno et al., 2004), a clear-cut inverse relationship with response is observed at all stages. Therefore, in the presence of advanced liver disease, a short treatment schedule would not obtain adequate rates

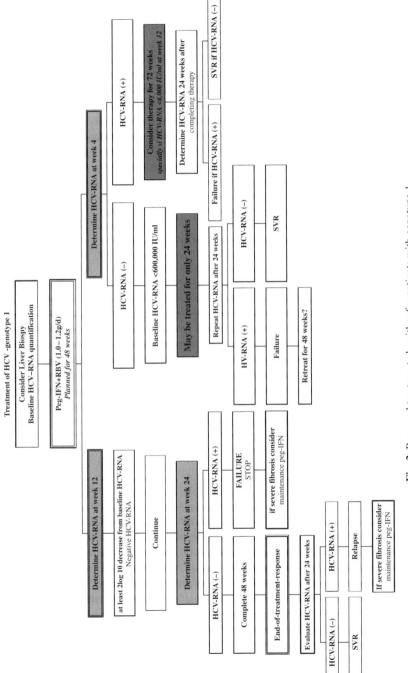

Fig. 2 Proposed treatment algorithm for patients with genotype 1

years, even up to 20 years in long-term followup (Pälsson et al., 2003; Sakata et al., 1996). The restoration of platelets is the result not only of reduced hypersplenism, but also of an increase in liver-synthesized thrombopoietin and a reduction of platelet-associated IgG levels (Rios et al., 2005).

The first studies that specifically have used PSE in HCV cirrhotic patients suffering from cytopenia as a pretreatment for modern combined antiviral therapy with peg-IFN/RBV have recently been published (Foruny et al., 2005; Moreno et al., 2004a). The authors have demonstrated the efficacy of PSE in improving hematological parameters, making it possible not only to start, but more importantly, to maintain full-dose combined therapy in patients who otherwise would never have been considered for therapy. To date, 19 out of 24 patients—89% infected with HCV genotype 1 ($n = 15$) or 4 ($n = 2$)—have complete followup data, with an SVR rate of 26%. Erythropoietin has been used in 5 patients (26%) and G-CSF in 3 (16%). RBV doses were reduced in 4 patients (21%), but full-dose peg-IFN was maintained in all. There were no hepatic decompensation episodes during therapy, and only one patient not receiving G-CSF for neutropenia discontinued therapy prematurely (unpublished personal data). Even with these encouraging treatment results, however, PSE is unlikely to be adopted widely in this setting. The procedure needs a multidisciplinary collaboration, with advanced competence only available in highly specialized centers, and will only be practical in a selected group of HCV-infected patients with hypersplenism. It seems essential to define the patients who will benefit most from this approach, to develop standardized treatment protocols, to perform long-term followup studies, and to find the optimal timing between PSE and the pharmacological therapy and the optimal dosages (Pälsson & Verbaan, 2005).

Although the principal aim of treatment is to achieve viral eradication, in absence of SVR, antiviral therapy may slow the rate of disease progression, decrease the incidence of HCC, and delay the need for transplantation. Ongoing trials are assessing in nonresponders the impact of long-term peg-IFN maintenance therapy on the long-term prognosis of advanced hepatitis C (Curry et al., 2005; Jensen & Marcellin, 2005; Lee et al., 2004).

Patients with HCV Recurrence Following Liver Transplantation

HCV reinfection occurs in almost all patients after LT, with an accelerated course of the disease due to immunosuppressive treatment (Berenguer et al., 2000). Progression to cirrhosis is increased; 20–30% of patients present severe graft damage at 5 years' post-LT (Berenguer, 2005); and the rate of death and allograft failure is increased when compared to other indications (Forman et al., 2002). Various parameters appear to affect the outcome of HCV recurrence, such as HCV-RNA before and after LT, renal function, HCV genotype, the year of transplantation, donor and recipient age, and the nature of immunosuppressive therapy (Terrault & Berenguer, 2006). Overall, it appears that severe hepatitis occurs in approximately 50% of LT subjects, while 15% of them develop cirrhosis and less than 5% require retransplantation (Biggins & Terrault, 2005).

The lack of an efficient strategy to prevent reinfection and the aggressive course of HCV after LT indicate the need for an effective antiviral therapy able to preserve the viability of the graft. Pre-emptive treatment within the first 4 to 6 weeks' post-LT has been disappointing, with SVR rates between 0–33% for different regimens, including IFN-α monotherapy and IFN-α plus RBV (Chalasani et al., 2005; Terrault, 2003). There is more experience on the treatment of established recurrent hepatitis C. The most recent studies using peg-IFN in combination with RBV (Figures 3 and 4) have showed SVR rates similar to those observed among immunocompetent subjects (Dumortier et al., 2004; Moreno et al., 2005; Otón et al., 2006).

Treatment duration should be at least similar to non-transplanted patients, taking into account early viral kinetics and HCV genotype (Otón et al., 2006). Early stopping rules considering week 4 and week 12 HCV-RNA decays seem to be as useful as in immunocompetent patients (Otón et al., 2006): the lack of at least a 1 $\log_{10}$ decrease at week 4 or at least a 2 $\log_{10}$ decrease at week 12 had an NPV of 100% on SVR.

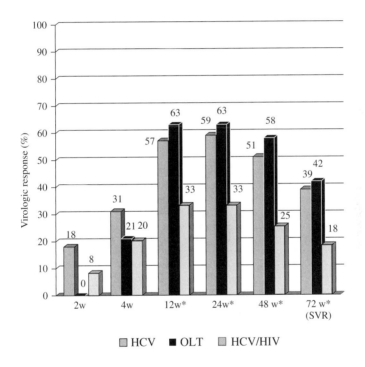

After w 4 p < 0.02* HCV/HIV vs HCV, OLT

HCV: hepatitis C virus; *HIV:* human immunodeficiency virus;
OLT: orthotopic liver transplant

Fig. 3 Viral clearance rates (negative HCV RNA) during therapy (intention-to-treat analysis). Reprinted from *Journal of Hepatology*, 43 (5), Moreno A et al, "HCV clearance and treatment outcome in genotype 1 HCV-monoinfected, HIV-coinfected and liver transplanted patients on peg-IFN-alpha-2b/ribavirin", pages 783–790, Copyright 2005, with permission from The European Association for the Study of the Liver

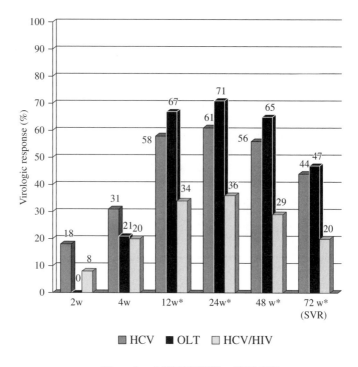

After w 4 p < 0.05* HCV/HIV vs HCV, OLT

HCV: hepatitis C virus; *HIV*: human immunodeficiency virus;
OLT: orthotopic liver transplant

Fig. 4 Viral clearance rates (negative HCV RNA) during therapy (on-treatment-analysis). Reprinted from *Journal of Hepatology*, 43 (5), Moreno A et al, "HCV clearance and treatment outcome in genotype 1 HCV-monoinfected, HIV-coinfected and liver transplanted patients on peg-IFN-alpha-2b/ribavirin", pages 783–790, Copyright 2005, with permission from The European Association for the Study of the Liver

The time between LT and the start of antiviral therapy seems relevant for its impact on virological outcomes and adverse events. A "delayed approach," waiting until chronic hepatitis C is established and immunosuppressive therapy is less intensive, might be the best scenario. In the Otón (2006) study, low baseline HCV-RNA and a length from LT to therapy between 2–4 years were significantly associated to success, probably due to lower HCV-RNA levels than in the immediate postoperative therapy related to less intensive immunosuppression.

The main problem when considering therapy for LT subjects is the high rate and severity of adverse events, with the frequent need for hematopoietic growth factors. In studies of transplant recipients with recurrent hepatitis C, hemolysis was reported in all patients, with 50% experiencing marked anemia (Gane, 2002). Anemia is strongly associated with renal function, a finding consistent with the known renal route of RBV elimination, and RBV dose adjustments and/or erythopoietin use should be considered to prevent serious hematological complications and drug discontinuation (Terrault & Berenguer, 2006). In the largest series on the outcome of peg-IFN plus RBV in LT patients after at least 12 months of surgery in the clinical setting (Otón et al., 2006), the most frequent side effects were neutropenia (76%), anemia (60%), and infectious complications (31%), and toxicity led to peg-IFN withdrawal in 29% of subjects. Erythropoietin was used in 38%, and G-CSF in 14.5%.

The performance of PSE in LT patients with persistent hypersplenism precluding antiviral therapy has shown to be quite useful (Bárcena et al., 2005; Foruny et al., 2006), for it not only allowed full-dose peg-IFN in this setting, but also led to a significant improvement in the graft function even before peg-IFN/RBV, probably related to the reversal of an undiagnosed splenic artery steal syndrome (Bárcena et al., 2006).

Finally, and of note, the risk for IFN-α induced graft rejection seems to be higher if RBV is not used (Bizollon et al., 1997; Manns et al., 2006; Otón et al., 2006), suggesting a potential beneficial effect of RBV on IFN-induced rejection and, thus, the need always to use combined therapy.

Due to the still unsatisfactory rates of response in LT with HCV recurrence, future therapies may include RBV alternatives with lower rates of anemia, such as viramidine, alternative IFN molecules with lower rates of cytopenia, and new antiviral drugs (HCV-protease and polymerase inhibitors) that can be used alone or in combination with either IFN-α or RBV, aimed to enhance SVR rates and improve tolerability (Terrault & Berenguer, 2006).

Treatment of Patients Co-infected with the Human Immunodeficiency Virus (HIV)

Since the introduction of highly active antiretroviral therapy (HAART) in the mid-1990s, liver disease due to HCV has become a leading cause of morbidity and mortality among patients infected with the human immunodeficiency virus (HIV), with estimates that range from 10–45% (Bica et al., 2001; Salmon-Ceron et al., 2005).

The prevalence of anti-HCV antibodies is high in HIV-infected subjects, reaching up to 70–90% among hemophiliacs and intravenous drug users, and is usually associated with active infection as assessed by detectable HCV-RNA (Vallet-Pichard & Pol, 2006). Each virus is able to modify the natural course of the other: HCV infection impairs CD4-cell recovery in HIV co-infected patients receiving potent antiretroviral therapy (Greub et al., 2000), and HIV infection significantly modifies the natural history of HCV infection (Vallet-Pichard & Pol, 2006; Zylberberg & Pol, 1996), increasing the levels of HCV viremia (Bonacini et al., 1999; Cribier et al., 1995) and worsening the histological course of the disease, increasing and accelerating the risk of cirrhosis (Benhamou et al., 1999; Sotó et al., 1997). Indeed, the rate of cirrhosis is increased 2- to 5-fold in HIV/HCV co-infected patients as compared to HCV monoinfected patients, and the mean time elapsed between contamination and cirrhosis is significantly reduced (Vallet-Pichard & Pol, 2006). HAART, and especially protease inhibitors (PI), may decrease the severity of liver disease:

factors of virological non-response to interferon-ribavirin combination therapy for patients infected with hepatitis C virus of genotype 1b and high viral load. *Journal of Medical Virology,* 78: 83–90.

Anonymous. (1999). Global surveillance and control of hepatitis C. Report of a WHO consultation organized in collaboration with the Viral Hepatitis Prevention Board, Antwerp, Belgium. *Journal of Viral Hepatitis,* 6: 35–47.

Ball, L.M., Sulkowski, M.S., Wolf, L, Weiz, K.B., Tice, A.D, Dieterich, D.T. (1999). Filgrastim use for interferon induced neutropenia in patients with hepatitis C (abstract). *Gastroenterology,* 116: A1188.

Bárcena, R., Gil-Grande, L., Moreno, J., Foruny, J.R., Otón, E., García, M., Blázquez, J., Sánchez, J., Moreno, A., Moreno, A. (2005). Partial splenic embolization for the treatment of hypersplenism in liver transplanted patients with hepatitis C virus recurrence before peg-interferon plus ribavirin. *Tranplantation,* 79: 1634–1635.

Bárcena, R., Moreno, A., Foruny, J.R., Moreno, A., Sánchez, J., Gil-Grande, L., Blázquez, J., Nuño, J., Fortún, J., Rodriguez-Gandía, M.A., Otón, E. (2006). Improved graft function in liver-transplanted patients after partial splenic embolization: reversal of splenic artery steal syndrome? *Clinical Transplantation,* 20: 517–522.

Benhamou, Y., Bochet, M., Di Martino, V., Charlotte, F., Azria, F., Coutellier, A., Vidaud, M., Bricaire, F., Opolon, P., Katlama, C., Poynard, T. (1999). Liver fibrosis progression in human immunodeficiency virus and hepatitis C virus coinfected patients. The Multivirc Group. *Hepatology,* 30: 1054–1058.

Benhamou, Y., Di Martino, V., Bochet, M., Colombert, G., Thibault, V., Liou, A., Katlama, C., Poynard, T.; Multivirc Group (2001). Factors affecting liver fibrosis in human immunodeficiency virus and hepatitis C virus-co-infected patients: impact of protease inhibitor therapy. *Hepatology,* 34: 283–287.

Berenguer, M (2005). What determines the natural history of recurrent hepatitis C after liver transplantation? *Journal of Hepatology,* 42: 448–456.

Berenguer, M., Ferrell, L., Watson, J., Prieto, M., Kim, M., Rayon, M., Cordoba, J., Herola, A., Ascher, N., Mir, J., Berenguer, J., Wright, T.L. (2000). HCV-related fibrosis progression following liver transplantation: increase in recent years. *Journal of Hepatology,* 32: 673–684.

Berg, T., von Wagner, M., Nasser, S., Sarrazin, C., Heintges, T., Gerlach, T., Buggisch, P., Goeser, T., Rasenack, J., Pape, G.R., Schmidt, W.E., Kallinowski, B., Klinker, H., Splenger, U., Martus, P., Alshuth, U., Zeuzem, S. (2006). Extended treatment duration for hepatitis C virus type 1: comparing 48 versus 72 weeks of peginterferon-alpha-2a plus ribavirin. *Gastroenterology,* 130: 1086–1097.

Bica, I., McGovern, B., Dhar, R., Stone, D., McGowan, K., Scheib, R., Snydman, D.R. (2001). Increasing mortality due to end-stage liver disease in patients with human immunodeficiency virus infection. *Clinical Infectious Diseases,* 32: 492–497.

Biggins, S.W., Terrault, N.A. (2005). Treatment of recurrent hepatitis C after liver transplantation. *Clinics in Liver Disease,* 9 : 505–523.

Bizollon, T., Palazzo, U., Ducerf, C., Chevallier, M., Elliot, M., Baulieux, J., Pouyet, M., Trepo, C. (1997). Pilot study of the combination of interferon alpha and ribavirin as therapy of recurrent hepatitis C after liver transplantation. *Hepatology,* 26: 500–504.

Bonacini, M., Govindarajan, S., Blatt, L.M., Schmid, P., Conrad, A., Lindsay, K.L. (1999). Patients co-infected with human immunodeficiency virus and hepatitis C virus demonstrate higher levels of hepatic HCV RNA. *Journal of Viral Hepatitis,* 6: 203–208.

Brady, C.W., Muir, A.J. (2005). HIV coinfection shortens the survival of patients with hepatitis C virus-related decompensated cirrhosis. *Hepatology,* 42: 496–497.

Bruno, S., Cammá, C., Di Marco, V., Rumi, M., Vinci, M., Camozzi, M., Rebucci, C., Di Bona, D., Colombo, M., Craxi, A., Mondelli, M.U., Pinzello, G. (2004). Peginterferon alpha-2b plus ribavirin for naïve patients with genotype 1 chronic hepatitis C: a randomized controlled trial. *Journal of Hepatology,* 41: 474–481.

Carrat, F., Bani-Sadr, F., Pol, S., Rosenthal, E., Lunel-Fabiani, F., Benzekri, A., Morand, P., Goujard, C., Pialoux, G., Piroth, L., Salmon-Ceron, D., Degott, C., Cacoub, P., Perrone, C., ANRS HCO2 RIBAVIC Study Team (2004). Pegylated interferon alpha-2b vs standard interferon alpha-2b, plus ribavirin for chronic hepatitis C in HIV-infected patients: a randomized controlled trial. *Journal of the American Medical Association*, 292: 2839–2848.

Chalasani, N., Manzarbeitia, C., Ferenci, P., Vogel, W., Fontana, R.J., Voigt, M., Riely, C., Martin, P., Teperman, L., Jiao, J., Lopez-Talavera, J.C., Pegasys Transplant Study Group (2005). Peginterferon alpha-2a for hepatitis C after liver transplantation: two randomized, controlled trials. *Hepatology*, 41: 289–298.

Choo, Q.L., Kuo, G., Weiner, A.J., Overby, L.R., Bradley, D.W., Houghton, M. (1989). Isolation of a cDNA clone derived from a blood-borne non-A, non-B viral hepatitis genome. *Science*, 244: 359–362.

Chung, R.T., Andersen, J., Volberding, P., Robbins, G.K., Liu, T., Sherman, K.E., Peters, M.G., Koziel, M.J., Bhan, A.K., Alston, B., Colquhoun, D., Nevin, T., Harb, G., van der Horst, C., AIDS Clinical Trials Group A5071 Study Team (2004). Peginterferon alpha-2a plus ribavirin versus interferon alpha-2a plus ribavirin for chronic hepatitis C in HIV-coinfected persons. *New England Journal of Medicine*, 351: 451–459.

Craxi, A, Camma, C. (2006). Treating patients with HCV genotype 1 and low viraemia: more than meets the eye. *Journal of Hepatology*, 44: 4–7.

Cribier, B., Rey, D., Schmitt, C., Lang, J.M., Kirn, A., Stoll-Keller, F. (1995). High hepatitis C viraemia and impaired antibody response in patients coinfected with HIV. *AIDS*, 9: 1131–1136.

Curry, M., Cárdenas, A., Afdhal, N.H. (2005). Effect of maintenance Peg-Interferon therapy on portal hypertension and its complications: results from the COPILOT study. *Journal of Hepatology*, 42: 40 [abstract 95].

Davis, G.L. (2002). Monitoring of viral levels during therapy of hepatitis C. *Hepatology*, 36 (Suppl. 1): S145–151.

Davis, G.L., Wong, J.B., McHutchison, J.G., Manns, M.P., Harvey, J., Albrecht, J. (2003). Early virological response to treatment with peginterferon alpha-2b plus ribavirin in patients with chronic hepatitis C. *Hepatology*, 38: 645–652.

Dieterich, D.T., Spivak J.L. (2003a). Hematologic disorders associated with hepatitis C virus infection and their management. *Clinical Infectious Diseases*, 37 : 533–541.

Dieterich, D.T., Wassermen, R., Bräu, N., Hassanein, T.I., Bini, E.J., Bowers, P.J., Sulkowski, M.S. (2003b). Once-weekly epoietin alpha improves anemia and facilitates maintenance of ribavirin dosing in hepatitis C virus-infected patients receiving ribavirin plus interferon alpha. *American Journal of Gastroenterology*, 98: 2491–2499.

Drusano, G.L., Preston, S.L. (2004). A 48-week duration of therapy with pegylated interferon a2b plus ribavirin may be too short to maximize long-term response among patients infected with genotype-1 hepatitis C virus. *Journal of Infectious Diseases*, 189: 964–970.

Dumortier, J., Scoazec, J.Y., Chevallier, P., Boillot, O. (2004). Treatment of recurrent hepatitis C after liver transplantation: a pilot study of peginterferon alpha-2b and ribavirin combination. *Journal of Hepatology*, 40: 669–674.

Everson, G.T. (2004). Treatment of chronic hepatitis C in patients with decompensated cirrhosis. *Reviews in Gastroenterological Disorders*, 4 (Suppl. 1): S31–S38.

Fattovich, G., Giustina, G., Degos, F., Tremolada, F., Diodati, G., Almasio, P., Nevens, F., Solinas, A., Mura, D., Brouwer, J.T., Thomas, H., Njapoum, C., Casarin, C., Bonetti, P., Fuschi, P., Basho, J., Tocco, A., Bhalla, A., Galassini, R., Noventa, F., Schalm, S.W., Realdi, G. (1997). Morbidity and mortality in compensated cirrhosis type C: a retrospective follow-up study of 384 patients. *Gastroenterology*, 112: 463–472.

Fattovich, G., Pantalena, M., Zagni, I., Realdi, G., Schalm, S.W., Christensen, E. European Concerted Action on Viral Hepatitis (EUROHEP) (2002). Effect of hepatitis B and C virus infections on the natural history of compensated cirrhosis: a cohort study of 297 patients. *American Journal of Gastroenterology*, 97: 2886–2895.

Ferenci, P. (2004). Predicting the therapeutic response in patients with chronic hepatitis C: the role of viral kinetics studies. *Journal of Antimicrobial Chemoterapy*, 43 : 15–18.

Ferenci, P., Fried, M.W., Shiffman, M.L., Smith, C.I., Marinos, G., Goncales, F.L., Jr., Haussinger, D., Diago, M., Carosi, G., Dhumeaux, D., Craxi, A., Chaneac, M., Reddy, K.R. (2005). Predicting sustained virological responses in chronic hepatitis C patients treated with peginterferon alpha-2a (40kd)/ribavirin. *Journal of Hepatology*, 43: 425–433.

Forman, L.M., Lewis, J.D., Berlin, J.A, Feldman, H.I., Lucey, M.R. (2002). The association between hepatitis C infection and survival after orthotopic liver transplantation. *Gastroenterology*, 122: 889–896.

Foruny, J.R., Blázquez, J., Moreno, A., Bárcena, R., Gil-Grande, L., Quereda, C., Pérez-Elías, M.J., Moreno, J., Sánchez, J., Muriel, A., Rodriguez-Sagrado, M.A., Moreno, S. (2005). Safe use of pegylated interferon/ribavirin in hepatitis C virus cirrhotic patients with hypersplenism after partial splenic embolization. *European Journal of Gastroenterology and Hepatology* , 17: 1157–1164.

Foruny, J.R., Bárcena, R., Moreno, A., Blázquez, J., Manzano, R., Gil-Grande, L.A., Moreno, A., Nuño, J. (2006). Benefit of pegylated interferon-alpha-2a/ribavirin in a patient with common variable immunodeficiency and hepatitis C virus cirrhosis after liver transplantation and splenic embolization. *Transplantation*, 82: 289–290.

Fried, M.W. (2002). Side effects of therapy of hepatitis C and their management. *Hepatology* , 36 (Suppl. 1): S237–S244.

Fried, M.W., Shiffman, M.L., Reddy, K.R., Smith, C., Marinos, G., Goncales, F.L., Jr., Haussinger, D., Diago, M., Carosi, G., Dhumeaux, D., Craxi, A., Lin, A., Hoffman, J., Yu, J. (2002). Peginterferon alpha-2a plus ribavirin for chronic hepatitis C virus infection. W2Ti>New</W2Ti> *England Journal of Medicine* , 347: 975–982.

Fuster, D., Planas, R., González, J., Force, L., Cervantes, M., Vilaro, J., Roget, M., García, I., Pedrol, E., Tor, J., Ballesteros, A.L., Salas, A., Sirera, G., Videla, S., Clotet, B., Tural, C. (2006). Results of a study of prolonging treatment with pegylated interferon-alpha-2a plus ribavirin in HIV/HCV-coinfected patients with no early virological response. *Antiviral Therapy*, 11: 473–482.

Gane, E. (2002). Treatment of recurrent hepatitis C. *Liver Transplantation*, 8 (Suppl.): S28–S37.

Greub, G., Ledergerber, B., Battegay, M., Grob, P., Perrin, L., Furrer, H., Burgisser, P., Erb, P., Boggian, K., Piffaretti, J.C., Hirschel, B., Janin, P., Francioli, P., Flepp, M., Telenti, A. (2000). Clinical progression, survival, and immune recovery during antiretroviral therapy in patients with HIV-1 and hepatitis C virus co-infection: the Swiss HIV Cohort Study. *Lancet*, 356: 1800–1805.

Heathcote, E.J. (2003). Treatment considerations in patients with hepatitis C and cirrhosis. *Journal of Clinical Gastroenterology*, 37: 395–398.

Hadziyannis, S.J., Sette, H. Jr., Morgan, T.R., Balan, V., Diago, M., Marcellin, P., Ramadori, G., Bodenheimer, H. Jr., Bernstein, D., Rizzetto, M., Zeuzem, S., Pockros, P.J., Lin, A., Ackrill, A.M., PEGAYS International Study Group (2004). Peginterferon-alpha2a and ribavirin combination therapy in chronic hepatitis C: a randomzed study of treatment duration and ribavirin dose. *Annals of Internal Medicine*, 140: 346–355.

Hayashi, N., Takehara, T. (2006). Antiviral therapy for chronic hepatitis C: past, present, and future. *Journal of Gastroenterology*, 41: 17–27.

Hoofnagle, J.H., Mullen, K.D., Jones, D.B., Rustgi, V., Di Bisceglie, A., Peters, M., Waggoner, J.G., Park, Y., Jones, E.A. (1986). Treatment of chronic non-A, non-B hepatitis with recombinant human alpha interferon. A preliminary report. *New England Journal of Medicine*, 315: 1575–1578.

Höroldt, B., Haydon, G., O'Donnell, K., Dudley, T., Nightingale, P., Mutimer, P. (2006). Results of combination treatment with pegylated interferon and ribavirin in cirrhotic patients with hepatitis C infection. *Liver International*, 26: 650–659.

Jacobson, I.M., Brown, R.S., Freilich, B., Afdhal, N., Kwo, P., Santoro, J., Becker, S., Wakil, A., Pound, D., Godofsky, E., Strauss, R., Bernstein, D., Flamm, S., Bala, N., Araya, V., Davis, M., Monsour, H., Vierling, J., Regenstein, F., Balan, V., Dragutsky, M., Epstein, M., Herring, R.W., Rubin, R., Galler, G., Pauli, M.P., Griffel, L.H., the WIN-RS Group (2005). Weight-based ribavirin dosing increases sustained viral response in patients with chronic hepatitis C: final results of the WIN-R study, a US community based trial. *Hepatology*, 42 (Suppl. 1): Abstract LB03.

Jensen, D.M., Marcellin, P. (2005). The rationale and design of the REPEAT study: a phase III, randomized clinical trial of peginterferon alpha-2a (40kDa) plus ribavirin in non-responders to peginterferon alpha-2b (12kDa) plus ribavirin. *European Journal of Gastroenterology and Hepatology*, 17: 899–904.

Jensen, D.M., Morgan, T.R., Marcellin, P., Pockros, P.J., Reddy, K.R., Hadziyannis, S.J., Ferenci, P., Ackrill, A.M., Willems, B. (2006). Early identification of HCV genotype 1 patients responding to 24 weeks peginterferon a-2a (40kd)/ribavirin therapy. *Hepatology*, 43: 954–960.

Laguno, M., Murillas, J., Blanco, J.L., Martínez, E., Miquel, R., Sánchez-Tapias, J.M., Bargallo, X., García-Criado, A., de Lazzari, E., Larrousse, M., Leon, A., Lonca, M., Milinkovic, A., Gatell, J.M., Mallolas, J. (2004). Peginterferon alpha-2b plus ribavirin compared with interferon alpha-2b plus ribavirin for treatment of HIV/HCV co-infected patients. *AIDS*, 18: F27–F36.

Lee, W.M., Dienstag, J.L., Lindsay, K.L., Lok, A.S., Bonkovsky, H.L., Shiffman, M.L., the HALT-C Trial Group (2004). Evolution of the HALT-C trial: pegylated interferon as maintenance therapy for chronic hepatitis C in previous interferon nonresponders.*Controlled Clinical Trials*, 25: 472–492.

Maddison, F.E. (1973). Embolic therapy for hypersplenism. *Investigative Radiology*, 8: 280–281.

Manns, M.P., McHutchison, J.G., Gordon, S.C., Rustgi, V.K., Shiffman, M., Reindollar, R., Goodman, Z.D., Koury, K., Ling, M., Albrecht, J.K. (2001). Peginterferon alpha-2b plus ribavirin compared with interferon alpha-2b plus ribavirin for initial treatment of chronic hepatitis C: a randomised trial. *Lancet*, 358: 958–965.

Manns, M.P., Wedemeyer, H., Cornberg, M. (2006). Treating viral hepatitis C: efficacy, side effects, and complications. *Gut*, 55: 1350–1359.

Martinot-Peignoux, M., Comanor, L., Minor, J.M., Ripault, M.P., Pham, B.N., Boyer, N., Castelnau, C., Giuily, N., Hendricks, D., Marcellin, P. (2006). Accurate model predicting sustained reponse at week 4 of therapy with pegylated interferon with ribavirin in patients with chronic hepatitis C. *Journal of Viral Hepatitis*, 13: 701–707.

McHutchison, J.G., Gordon, S.C., Schiff, E.R., Shiffman, M.L., Lee, W.M., Rustgi, V.K., Goodman, Z.D., Ling, M.H., Cort, S., Albrecht, J.K. (1998). Interferon alpha-2b alone or in combination with ribavirin as initial treatment for chronic hepatitis C. Hepatitis Interventional Therapy Group. *New England Journal of Medicine*, 339: 1485–1492.

McHutchison, J.G., Manns, M., Patel, K., Poynard, T., Lindsay, K.L., Trepo, C., Dienstag, J., Lee, W.M., Mak, C., Garaud, J.J, Albrecht, J.K., International Hepatitis Interventional Therapy Group (2002). Adherence to combination therapy enhances sustained virological response in genotype-1-infected patients with chronic hepatitis C. *Gastroenterology*, 123: 1061–1069.

McHutchison, J.G., Afdhal, N., Shiffman, M.L., Gordon, S., Mills, P., Sigal, S., Midwinter, D., Campbell, F.M., Williams, D., Theodore, D. (2006). Efficacy and safety of eltrombopag, an oral platelet growth factor, in subjects with HCV associated thrombocytopenia: preliminary results from a phase II dose-ranging study. Program and abstracts of the *41st. Annual Meeting of the European Association for the Study of the Liver*. Abstract 745.

Miró, J.M., Laguno, M., Moreno, A., Rimola, A., the Hospital Clinic OLT in HIV Working Group (2006). Management of end stage liver disease (ESLD): what is the current role of orthotropic liver transplantation (OLT)? *Journal of Hepatology*, 44 (Suppl.): S140–S145.

Miró, J.M., Montejo, M., Vargas, V., Rimola, A., Rafecas, A., Miralles, P., Fortún, J., Blanes, M., de la Torre, J., Pons, J.A., the Spanish LT in HIV-Infected Patients Working Group (2006). Spanish cohort of HIV-infected patients with orthotropic liver transplantation (OLT): evaluation of 50 cases in the highly active antiretroviral (HAART) era (2002-2005). In Program and Abstracts of the *13th Conference on Retroviruses and Opportunistic Infections*, abstract 875.

Moreno, A., Bárcena, R., Blázquez, J., Quereda, C., Gil-Grande, L., Sánchez, J., Moreno, L., Pérez-Elías, M.J., Antela, A., Moreno, J., del Campo, S., Moreno, S. (2004a). Partial splenic embolization for the treatment of hypersplenism in cirrhotic HIV-HCV patients prior to peg-interferon and ribavirin. *Antiviral Therapy*, 9: 1027–1030.

Moreno, A., Quereda, C., Moreno, L., Pérez-Elías, M.J., Muriel, A., Casado, J.L., Antela, A., Dronda, F., Navas, E., Bárcena, R., Moreno, S. (2004b). High rate of didanosine-related mitochondrial toxicity in HIV/HCV co-infected patients receiving ribavirin. *Antiviral Therapy*, 9: 133–138.

Moreno, A., Bárcena, R., García-Garzón, S., Muriel, A., Quereda, C., Moreno, L., Mateos, M.L., Fortún, J., Martín-Dávila, P., García, M., Blesa, C., Otón, E., Moreno, A., Moreno, S. (2005).

HCV clearance and treatment outcome in genotype 1 HCV-monoinfected, HIV-coinfected and liver transplanted patients on peg-INF-α-2b/ribavirin. *Journal of Hepatology*, 43: 783–790.

Moreno, A., Bárcena, R., García-Garzón, S., Moreno, L., Quereda, C., Muriel, A., Zamora, J., Mateos, M.L., Pérez-Elías, M.J., Antela, A., Diz, S., Moreno, A., Moreno, S. (2006). Viral kinetics and early prediction of nonresponse to peg-IFN-α-2b plus ribavirin in HCV genotype 1/4 according to HIV serostatus. *Journal of Viral Hepatitis*, 13: 466–473.

Neumann, A.U., Lam, N.P., Dahari, H., Gretch, D.R., Wiley, T.E., Layden, T.J., Perelson, A.S. (1998). Hepatitis C viral dynamics in vivo and the antiviral efficacy of interferon-alpha therapy. *Science* , 282: 103–107.

N'Kontchou, G., Seror, O., Bourcier, V., Mohand, D., Ajavon, Y., Castera, L., Grando-Lemaire, V.,Ganne-Carrie, N., Sellier, N., Trinchet, J.C., Beaugrand, M. (2005). Partial splenic embolization in patients with cirrhosis: efficacy, tolerance, and long-term outcome in 32 patients. *European Journal of Gastroenterology and Hepatology*, 17: 179–184.

Otón, E., Bárcena, R., Moreno-Planas, J.M., Cuervas-Mons, V., Moreno-Zamora, A., Barrios, C., García-Garzón, S., Moreno, A., Boullosa-Graña, E., Rubio-González, E., García-González, M., Blesa, C., Mateos, M.L. (2006). Hepatitis C recurrence after liver transplantation: viral and histologic response to full-dose peg-interferon and ribavirin. *American Journal of Transplantation*, 6: 2348–2355.

Pälsson, B., Hallen, M., Forsberg-Mandahl, A.M., Alwmark, A. (2003). Partial splenic embolization—long-term outcome. *Langenbecks Archives of Surgery*, 387: 421–426.

Pälsson, B., Verbaan, H. (2005). Partial splenic embolization as pretreatment for antiviral therapy in hepatitis C virus infection. *European Journal of Gastroenterology and Hepatology*, 17: 1153–1155.

Pawlotsky, J.M. (2006). Therapy of hepatitis C: from empiricism to eradication. *Hepatology*, 43: S207–S220.

Pawlotsky, J.M., Hezode, C., Pellegrin, B., Soulier, A., von Wagner, M., Brouwer, J.T., Missale, G,, Germanidis, G., Lurie, Y., Negro, F., Esteban, J., Hellstrand, K., Ferrari, C., Zeuzem, S., Schalm, S.W., Neumann, A.U. (2002). Early HCV genotype 4 replication kinetics during treatment with peginterferon alpha-2a (Pegasys)-ribavirin combination: a comparison with HCV genotypes 1 and 3 kinetics. *Hepatology*, 36 (Suppl.): 219A.

Pineda, J.A., Romero-Gómez, M., Diaz-García, F., Girón-González, J.A., Montero, J.L., Torre-Cisneros, J., Andrade, R.J., Gonzalez-Serrano, M., Aguilar, J., Aguilar-Guisado, M., Navarro, J.M., Salmerón, J., Caballero Granado, F.J., García-García, J.A., Grupo Andaluz para el Estudio de las Enfermedades Infecciosas; Grupo Andaluz para el Estudio del Hígado (2005). HIV coinfection shortens the survival of patients with hepatitis C virus-related decompensated cirrhosis. *Hepatology*, 41: 779–789.

Poynard, T., Marcellin, P., Lee, S.S., Niederau, C., Minuk, G.S., Ideo, G., Bain, V., Heathcote, J., Zeuzem, S., Trepo, C., Albrecht, J. (1998). Randomised trial of interferon alpha-2b plus ribavirin for 48 weeks or for 24 weeks versus interferon alpha-2b plus placebo for 48 weeks for treatment of chronic infection with hepatitis C virus. *Lancet*, 352: 1426–1432.

Rios, R., Sangro, B., Herrero, I., Quiroga, J., Prieto, J. (2005). The role of thrombopoetin in the thrombocytopenia of patients with liver cirrhosis. *American Journal of Gastroenterology* , 100: 1311–1316.

Sakata, K., Hirai, K., Tanikawa, K. (1996). A long-term investigation of transcatheter splenic arterial embolization for hypersplenism. *Hepatogastroenterology*, 43: 309–318.

Salmon-Ceron, D., Lewden, C., Morlat, P., Bevilacqua, S., Jougla, E., Bonnet, F., Heripret, L., Costagliola, D., May, T., Chene, G., The Mortality 2000 Study Group (2005). Liver disease as a major cause of death among HIV infected patients: role of hepatitis C and B viruses and alcohol. *Journal of Hepatology*, 42: 799–805.

Sánchez-Tapias, J.M., Diago, M., Escartín, P., Enríquez, J., Romero-Gómez, M., Bárcena, R., Crespo, J., Andrade, R., Martínez-Bauer, E., Pérez, R., Testillano, M., Planas, R., Solá, R., García-Bengoechea, M., García-Samaniego, J., Muñoz-Sánchez, M., Moreno-Otero, R. (2006). Peginterferon-alpha-2a plus ribavirin for 48 versus 72 weeks in patients with detectable hepatitis C virus RNA at week 4 of treatment. *Gastroenterology*, 131: 451–460.

Sangiovanni, A., Pratti, G.M., Fasani, P., Ronchi, G., Romeo, R., Manini, M., Del Ninno, E., Morabito, A., Colombo, M. (2006). The natural history of compensated cirrhosis due to hepatitis C virus: a 17-year cohort study of 214 patients. *Hepatology*, 43: 1303–1310.

Sangro, B., Bilbao, I., Herrero, I., Corella, C., Longo, J., Beloqui, O., Ruiz, J., Zozaya, J.M., Quiroga, J., Prieto, J. (1993). Partial splenic embolization for the treatment of hypersplenism in cirrhosis. *Hepatology*, 18: 309–314.

Sarrazin, C., Teuber, G., Kokka, R., Rabenau, H., Zeuzem, S. (2000). Detection of residual hepatitis C virus RNA by transcription mediated amplification in patients with complete virologic response according to polymerase chain reaction-based assays. *Hepatology*, 32: 818–823.

Serfaty, L., Aumaitre, H., Chazouilleres, O., Bonnand, A.M., Rosmorduc, O., Poupon, R.E., Poupon, R. (1998). Determinants of outcome of compensated hepatitis C virus-related cirrhosis. *Hepatology*, 27: 1435–1440.

Shiffman, M.L., Price, A., Hubbard, S., Wilson, M., Salvatori, J., Sterling, R.K., Stravitz, R.T., Luketic, V.A., Sanyal, A.J. (2005). Treatment of chronic hepatitis C virus genotype 1 with peginterferon alpha-2b, high weight based dose ribavirin and epoietin-alpha enhances sustained virological response (abstr). *Hepatology*, A55.

Solá, R., Tural, C., Rubio, R., Santín, M., Fuster, D., Moreno, S., Berenguer, J., González, J., Clotet, B., Planas, R. (2005). Lack of benefit of an induction dose of peginterferon alpha2a on early hepatitis C virus kinetics in HIV/HCV coinfected patients: results from the CORAL-1 Pilot, Multicenter Study. In program and abstracts of the *45th Interscience Conference of Antimicrobial Agents and Chemotherapy* (ICAAC): abstract LB H-416b.

Soriano, V. (2006). Treatment of chronic hepatitis C in HIV-positive individuals: selection of candidates. *Journal of Hepatology*, 44 (Suppl.): S44–S48.

Soriano, V., Puoti, M., Sulkowski, M., Mauss, S., Cacoub, P., Cargnel, A., Dieterich, D., Hatzakis, A., Rockstroh, J. (2004). Care of patients with hepatitis C and HIV co-infection: updated recommendations from the HIV-HCV International Panel. *AIDS*, 17: 1–12.

Soto, B., Sánchez-Quijano, A., Rodrigo, L., del Olmo, J.A., García-Bengoechea, M., Hernández-Quero, J., Rey, C., Abad, M.A., Rodriguez, M., Sales Gilabert, M., González, F., Miron, P., Caruz, A., Relimpio, F., Torronteras, R., Leal, M., Lissen, E. (1997). Human immunodeficiency virus infection modifies the natural history of chronic parenterally acquired hepatitis C with an unusually rapid progression to cirrhosis. *Journal of Hepatology*, 26: 1–5.

Strader, D.B., Wright, T., Thomas, D.L., Seef, L.B. (2004). American Association for the Study of Liver Diseases. Diagnosis, management, and treatment of hepatitis C. *Hepatology*, 39: 1147–1171.

Tajiri, T., Onda, M., Yoshida, H., Mamada, Y., Taniai, N., Kumazaki, T. (2002). Long-term hematological and biochemical effects of partial splenic embolization in hepatic cirrhosis. *Hepatogastroenterology*, 49: 1445–1448.

Terrault, N.A. (2003). Prophylactic and preemptive therapies for hepatitis C virus-infected patients undergoing liver transplantation. *Liver Transplantation*, 9 (Suppl.): 95–100. Terrault, N.A., Berenguer, M. (2006). Treating hepatitis C infection in liver transplant recipients. *Liver Transplantation*, 12: 1192–1204.

Torriani, F.J., Rodriguez-Torres, M., Rockstroh, J.K., Lissen, E., González-García, J., Lazzarin, A,, Carosi, G., Sasadeuz, J., Katlama, C., Montaner, J., Sette, H. Jr., Passe, S., De Pamphilis, J., Duff, F., Schrenk, U.M., Dieterich, D.T., APRICOT Study Group (2004). Peginterferon alpha-2a plus ribavirin for chronic hepatitis C virus infection in HIV-infected patients. *New England Journal of Medicine*, 351: 438–450.

Vallet-Pichard, A., Pol, S. (2006). Natural history and predictors of severity of chronic hepatitis C virus (HCV) and human immunodeficiency virus (HIV) co-infection. *Journal of Hepatology*, 44 (Suppl.): S28–S34.

Vujic, I., Lauver, J.W. (1981). Severe complications from partial splenic embolization in patients with liver failure. *British Journal of Radiology*, 54 : 492–495.

Wiesner, R.H., Sorrell, M., Villamil, F. (2003). Report of the First International Liver Transplantation Society Expert Panel Consensus Conference on liver transplantation and hepatitis C. *Liver Transplantation*, 9 (Suppl.): S1–S9.

or to identify the most effective treatment strategy in order to reach liver transplant in optimal conditions.

The aim of the present chapter is to review the evidence available concerning the clinical effect of antiviral treatment, mainly a combined treatment of IFN plus ribavirin, in patients with liver cirrhosis, its efficacy in influencing the natural history of HCV-related disease (i.e., in preventing liver disease progression, liver disease decompensation, and HCC development), and the efficacy of combined treatment in patients with HCV-related disease before and after OLT.

Antiviral Therapy in HCV-Related Cirrhosis

The hepatitis C virus represents one of the main causes of chronic liver disease in the world; in fact, the World Health Organization estimates that there are about 170 million cases of hepatitis C infection worldwide (WHO, 1999). Available epidemiological studies (Davis et al., 2003; Fattovich et al., 1997) have indicated that, although most infected patients develop a mild and long-term disease course, some of them develop, in about 30 years of active infection, liver cirrhosis (Tong et al., 1995) and that these patients are at high risk for disease progression and clinical deterioration (i.e., they will have decompensated cirrhosis). In a European multicenter study (Fattovich et al., 1997) carried out on 384 patients followed up for 5 years, about 18% of the patients developed decompensated cirrhosis (occurrence of ascites, encephalopathy, or variceal bleeding) and 8% developed HCC, with a mortality rate of 1.9%/year and a probability of survival after the onset of the first major complication of the disease of 50% at 5 years. Furthermore, in another study (Davis et al., 2003) in which a mathematical model was used to predict the ongoing complications of HCV-related hepatitis, it was estimated that the proportion of cirrhotic patients will increase from 16% to 32% by 2020 in an untreated population, although a decline of HCV infection by 2040 was predicted.

Given this emerging problem, the importance of finding an effective treatment against HCV for the reduction of cirrhosis development and its complications in the future is obvious.

Fortunately, the development of an antiviral treatment for chronic hepatitis C has been a success story; in fact, in HCV, sustained response rates (SRR) improved from approximately 5% using IFN in monotherapy to up to 60% with the combined and optimized therapy represented by PEGylated interferons (PEG-INFs) plus ribavirin (Manns et al., 2001; Heathcote & Main, 2005).

It is well known that sustained virological response (SVR) is highly influenced by viral genotype and grade of fibrosis. In fact, the role, and consequently the estimated efficacy, of antiviral therapy in patients with high-grade fibrosis or overt cirrhosis is controversial (Everson, 2005): therapy with standard IFN monotherapy was disappointing in terms of both the efficacy and the high frequency of side effects (Janssen et al., 1993; Manns et al., 2001), while the use of PEG-IFN plus ribavirin obtained better results (Heathcote et al., 2005).

The theoretical main goal of therapy in cirrhotic patients is SVR, because virus clearance reduces disease progression toward decompensated cirrhosis; however, at least three other alternative endpoints could be achieved: (1) histological improvement; (2) prevention, or delay, of HCC development; and (3) delay of liver function decompensation.

Although several large, randomized clinical trials of IFN alone, or in combination with ribavirin, have been carried out, only a small percentage of patients are represented by cirrhotic patients (Everson, 2005). In the study by Heathcote and colleagues (Heathcote et al., 2000) that represented the first prospective study focused on patients with bridging fibrosis or cirrhosis, 271 patients were enrolled and were subdivided into three groups with three different treatment regimens. One treatment arm consisted of PEG-IFN-alpha-2a dosed at 90 mcg/weekly; the second was the same drug at 180 mcg/weekly; and the third was IFN-alpha-2a 3 times a week; all treatments lasted 48 months. The results of this study demonstrated that the highest dose of PEG-IFN obtained a 30% SVR vs. 8% standard IFN group; a significant improvement of histological score was more frequent in the arm with the higher doses of PEG-IFN with respect to standard IFN (54% vs. 31%). In a study of 250 patients with bridging fibrosis or cirrhosis (Manns et al., 2001), treatment with standard IFN in combination with ribavirin showed a low SVR (about 38% for 48 weeks), although this percentage was higher compared to that obtained with IFN monotherapy (13%).

There are few data regarding the efficacy of PEG-IFN in combination with ribavirin in cirrhotic patients in terms of SVR; in three large, randomized controlled trials evaluating the efficacy of PEG-IFN in combination with ribavirin using different regimens (Fried et al., 2002; Hadziyannis et al., 2004; Manns et al., 2001) to treat HCV-related chronic hepatitis, in the subgroup of cirrhotic patients, an SVR ranging from 40% for genotype 1 to 70% for genotypes 2 and 3 at a high dose of PEG-IFN (180 mcg/week) was obtained. In a recent open–label, multicenter study carried out in Canada in a "real-world" clinical practice setting, Lee and colleagues (Lee et al., 2004) treated 174 patients having fibrosis F3-F4 with PEG-IFN (180 mcg/week) plus ribavirin 800 mg/day for 48 weeks. The results obtained were comparable to those documented in randomized controlled trials; in fact, the SVR was 34% in genotype 1 and 66% and 44% in genotypes 2 and 3, respectively. In another study, published in abstract form (Gordon et al., 2005), evaluating the efficacy in Child A cirrhotic patients with portal hypertension, an overall SVR of 21% was obtained.

The largest experience in the treatment of patients with HCV-related liver cirrhosis is the Lead-In phase of the HALT-C (Hepatitis C Antiviral Long-Term Treatment to Prevent Cirrhosis) cohort study of the retreatment of nonresponder patients (Everson et al.,2004b); in the second re-evaluation, patients with biopsy-proven cirrhosis and more severe disease (cirrhotic biopsy specimen and platelets $< 125,000/mm^3$) had an SVR ranging from 23% in those with a higher platelet count to 9% in those with a lower platelet count. Thus, this study suggested that the retreatment of previous nonresponder cirrhotic patients is only marginally effective and that extensive fibrosis independently impaired SVR to antiviral therapy.

However, despite the suboptimal response rate of the cirrhotic patients with respect to non-cirrhotic ones, treatment of these patients has recently been poorly tolerated in cirrhotic patients: the presence of hypersplenism with a low platelet count increases the risk of thrombocytopenia during treatment; moreover, renal dysfunction related to the underlying liver disease may increase the risk of ribavirin-related anemia (Jen et al., 2000). The global efficacy of recurrence or prevention of HCV disease does not appear to be superior to that obtained in treating patients after OLT, and the relapse timing seems delayed in some. The use of PEG-IFNs is intriguing for their results in terms of virological response, but in this setting of patients the risk of side effects may be too high.

Prophylactic Therapy Starting at the Time of Transplantation

This approach is based on the following hypotheses:

- treating OLT recipients at the time of transplantation during the early phase of infection could lead to better results in terms of virus eradication: in fact, in immunocompetent patients with acute hepatitis C, clearance of infection occurs in 90% of treated patients (Licata et al., 2003).
- HCV-RNA viral load decreases both during the anhepatic phase of transplantation and immediately after reperfusion of the graft but peaks to the pretransplant levels early after transplant (in some cases as early as 24–48 hours) (Garcia-Retortillo et al., 2002); thus, treating HCV when the viral load is lower could lead to better results. Treatment should be started immediately after surgery.

However, strong immunosuppression carried out in the first post-transplantation week reduces the likelihood of response and produces cytopenia, which could limit the use of IFN. Another important factor limiting the use of IFN during the early phase of transplanted patients is the risk of rejection (Feray et al., 1995). Antiviral actions of IFN-alpha include the induction of several proteins, such as proteinkinase (PKR) and 2',5'-oligoadenylate synthetase, which are important in the suppression of the viral RNA synthesis (Zeuzem et al., 1996); but IFNs induce the expression of major histocompatibility complex (MHC), on both antigen-presenting cells (APC) and hepatocytes, which activates the cytotoxic T cell response and virus-specific lysis of infected cells (Samuel, 2001). Finally, IFNs are involved in the development of autoimmune diseases by the upregulation of MHC in both transplanted (Berardi et al., 2006) and non-transplanted patients (Durelli et al., 2001; Fattovich et al., 1996; Mazzella et al., 1996; Wilson et al., 2002).

IFN or PEG-IFN monotherapy has limited efficacy, with a reported SVR ranging from 0–8% (Chalasani et al., 2005; Sheiner et al., 1995; Shergill et al., 2005; Singh et al., 1998). Moreover, a difference in terms of survival or severity of recurrence has not been shown between treated and untreated patients. Dose reduction is frequently required in these patients, and treatment withdrawal is reported in about 40% of patients despite the use of growth factors (Shergill et al., 2005). The use of combination therapy with ribavirin leads to better results, with reported SVR rates of 18% (Shergill et al., 2005). A study from Japan (Sugawara et al., 2004) reported more encouraging results, with SVR rates of 39% in patients treated with combination therapy within 1 month from OLT, with 57% of patients who required dose modification or treatment discontinuation.

In summary, early antiviral treatment after surgery is a good approach in theory, since in the real world several factors limit the efficacy of the treatment in terms of both tolerability and virological response.

Therapy for Established Severe or Progressive Chronic Hepatitis

In clinical practice, the most common strategy is treating HCV recurrence once there is histological evidence of liver disease (NIH Consensus Statement on Management of Hepatitis C, 2002). This means that therapy is usually delayed months or years after OLT. Data on treatment derive mostly from single centers, which only have experience with a small number of patients rather than with controlled studies (Terrault, 2005); this means that there are no guidelines on treatment duration based on genotypes, the optimal dosage of the drug, the definition of nonresponse, and withdrawal rules for nonresponse.

Even months after OLT, immunosuppression is reduced and patients are stable, IFN tolerability is poor, with high rates of dose reduction and low rates of treatment response. One of the major problems is that ribavirin pharmacokinetics is influenced by renal function (Jen et al., 2000), which is often reduced in patients receiving calcineurin inhibitors. Besides these limitations, the rationale for treating HCV recurrence is based on the fact that SVR is associated with reduced activity, with a rapid decrease in necroinflammatory activity at histology and stable, or reduced, fibrosis at 2 or 5 years after treatment (Alberti et al., 2001; Bizollon et al., 2003, 2005; Burra et al., 2006). The efficacy of antiviral treatment is disappointing, with low rates of SVR, although it seems that using combination therapy with PEG-IFN and ribavirin may lead to a better response.

The efficacy of IFN monotherapy is insignificant in terms of SVR, with rates of response ranging from 0–2.5% (Ahmad et al., 2001; Cotler et al., 2001; Feray et al., 1995; Gane et al., 1998; Wright et al., 1994). The biochemical response is 50–70%.

Better results are reported with combination therapy, with the SVR ranging from 17–30% (Ahmad et al., 2001; Giostra et al., 2004; Samuel et al., 2003). Dose reduction is frequent, with treatment discontinuation present in 22–43% of patients due to adverse events (Table 2).

The use of PEG-IFN in combination with ribavirin gives better results (Table 2): Dumortier and colleagues treated 20 patients with increasing doses of PEG-IFN-alfph-2b (0.5–1 mcg/week) and ribavirin, obtaining an SVR in 45% of the patients (Dumortier et al., 2004). Other studies reported variable rates of response ranging from 24 to 35% with both PEG-IFN–alpha-2b and -alpha-2a plus ribavirin (Castells et al., 2005; Dumortier et al., 2004; Mukherjee, 2005; Neff et al., 2004; Rodriguez-Luna et al., 2004). These controversial results are probably due to the variable dosages of both IFNs and ribavirin and the lack of standard guidelines.

Table 2 Combination Therapy for Established Liver Disease

Author	N	Regimen	SVR	Comments
Samuel, 2003	54	IFN a-2b 3 MU TIW + RBV 1000-1200 mg × 12 months;	21.4%	43% ED
Ahmad, 2001	20	IFNa-2b 3 MU TIW x 1 month then 5MU TIW × 11 months + RBV 600 mg	20%	25% ED
Giostra, 2004	31	RBV 10 mg/Kg/D × 3 months then IFN 3 MU TIW + RBV 10 mg/Kg/d × 12 months	29%	22.6% ED. In SVR decrease of inflammation not fibrosis
Dumortier, 2004	20	PEG-IFN a-2b (0.5 > 1 mcg) + RBV (400 > 1200mg) × 12 months	45%	20% ED; 6% DR; 20% AR
Rodriguez-Luna, 2004	37	PEG-IFN a-2b 0.5–1.5 mcg + RBV 400–1000 mg × 12 month (treatment continued for 1 year if HCV negative)	26%	38% ED. Decrease in inflammation, no change in fibrosis; 1 AR
Mukherjee, 2005	26	PEG-IFN a-2a (180 mg) + RBV (1000–1200 mg) × 6 months (Gen 2) and 12 months (Gen 1)	24%	19% ED; no AR or CR
Castells, 2005	24	PEG-IFN a-2b (1.5 mg) + RBV (400–800 mg) × 6 months	35%	No ED: 58% RBV DR Leukopenia 96%
Neff, 2004	57	PEG-IFN a-2b (1.5 mg) + RBV (400–600 mg) × 12 months	NR	RBV DR 39–45%

IFN: interferon; RBV: ribavirin; MU: millions of units; QD: once daily; TIW: thrice weekly; ETR: end-of-treatment response; SVR: sustained virological response; ED: early discontinuation; DR: dose reduction; AR: acute rejection.

Risk of Rejection and Immunomediated Diseases

The risk of acute rejection (AR) or chronic rejection (CR) during interferon treatment in OLT recipients with chronic hepatitis C is still being debated. However, the absolute risk does not seem to exceed 3% with traditional IFNs but peaks to 14% with PEG-IFNs (Berenguer et al., 2006; Chalasani et al., 2005; Jain et al., 1998; Toniutto et al., 2005; Wang et al., 2006; Yedibela et al., 2005). The overexpression of MHC on cellular membranes induced by IFN, such as immunomodulatory and proinflammatory cytokine, may produce uncontrolled immunological events leading to AR or CR. This may be particularly important in responders to IFN treatment, whose immune systems are so active as to determine the clearance of HCV.

The fear of AR and/or CR has always been a reality during IFN-based treatment of recurrent viral hepatitis (HBV, HCV, and HDV) after liver transplantation; this fear comes from experience with kidney transplantation, in which high rates of graft rejection have been reported (Davis et al., 2003). Dousset and colleagues (Dousset et al., 1994) described two cases of acute vanishing bile duct syndrome and suggested that an increased risk of rejection could be due to the IFN-induced overexpression of MHC class I antigens on liver cells. Feray and colleagues (Feray et al., 1995) reported a notable number of chronic rejections (35%) in a series of 40 transplanted patients treated with 3 MU of IFN-alpha three times a week. More recently, Samuel and colleagues (Samuel et al., 2003) reported no cases of AR and only one case of CR in a series of 28 transplanted patients treated with the same dose of IFN-alpha used in the previous study. This difference might be related to the different immunosuppressive schedules used in recent years.

Few data are available on the risk of AR or CR during PEG-IFN treatment in transplanted patients, even though the improved efficacy of PEG-IFN compared to standard IFN might also expose the transplanted patient to an increased risk of rejection. A recent study reported a relationship between virological response during IFN-based treatment of chronic hepatitis C and AR in a series of transplanted patients (Kugelmas et al., 2003). In this study, the authors investigated the presence of a relationship between viral clearance and immune-suppressant levels. They found that, even though there was no difference between responders and nonresponders in terms of lowest immune-suppressant levels, the time spent at lower immune-suppressant levels was higher among responders. In this study, the authors hypothesized that viral eradication improves microsomal metabolic function, leading to a decrease of immune-suppressant levels, which may predispose patients to allograft rejection.

To minimize the risk of rejection during PEG-IFN, Dumortier and colleagues (Dumortier et al., 2004) progressively increased the dosage of PEG-IFN. Moreover, the level of immune suppressant was carefully maintained at the normal upper limit. In this study, 25% of the patients developed AR; in all cases, this complication was easily controlled only by increasing the immune-suppressant regimen; and no cases of CR were observed. These data may confirm that interferon is capable of increasing the risk of rejection and that maintaining adequate levels of an immune-suppressant drug reduces this risk. However, the study does not mention the response status in those patients who developed AR, so that speculation on the role of immunological activation, and microsomal drug metabolism, is impossible.

In other studies, the incidence of AR during IFN or PEG-IFN treatments in liver-transplanted patients with HCV recurrence varies from 11–35%, while the incidence of CR ranges from 4–9% (Samuel et al., 2003; Stravitz et al., 2004).

Recently, a particular kind of immunomediated graft dysfunction was described in a cohort of transplanted patients receiving PEG-IFN (Berardi et al., 2006). Based on exclusion of other known causes of liver dysfunction, on histology, and on clinical findings, this entity was identified as de novo autoimmune hepatitis (de novo AIH). De novo AIH after OLT is a newly recognized condition affecting patients transplanted for disorders other than AIH (Kerkar et al., 1998; Salcedo et al., 2002). Risk factors and pathogenesis for this disorder remain unknown, and minimum criteria for diagnosis have not been standardized (Czaja, 2002). IFN may trigger the development of autoimmune disorders in virtue of its immunomodulating properties. This pathological entity seems to be very aggressive, since in this study two of the nine patients who developed de novo AIH died, one was reenlisted for OLT, and one had a graft failure.

Conclusions

HCV recurrence after OLT is a major concern, as HCV chronic liver disease is the leading indication for OLT in Western countries, and HCV recurrent liver disease of the graft reduces survival and increases allograft failure (Forman et al., 2002).

IFN-based antiviral treatment is the only one available, but its applicability in the transplant setting is complex. Combination therapy with ribavirin is superior to IFN monotherapy; higher rates of SVR are reported with a combination of PEG-IFNs and ribavirin, but larger studies are needed, and the potential higher risk of severe side effects is unclear.

Pretransplant treatment may be considered in patients with compensated liver disease. Post-transplant treatment is recommended in patients with acute early recurrence and chronic progressive diseases (Wiesner et al., 2003). Tolerability to treatment is poor after transplant; side effects are frequent and sometimes so severe as to lead to graft loss or death. Therefore, the decision to treat should be weighed against the risk of side effects, and disease progression should be assessed by annual biopsies as a protocol (Samuel et al., 2006).

References

Abdelmalek, M.F., Firpi, R.J., Soldevila-Pico, C., Reed, A.I., Hemming, A.W., Liu, C., Crawford, J.M., Davis, G.L., Nelson, D.R. (2004). Sustained viral response to interferon and ribavirin in liver transplant recipients with recurrent hepatitis C. *Liver Transplantation*, 10: 199–207.

Adam, R., McMaster, P., O'Grady, J.G., Castaing, D., Klempnauer, J.L., Jamieson, N., Neuhaus, P., Lerut, J., Salizzoni, M., Pollard, S., Muhlbacher, F., Rogiers, X., Garcia Valdecasas, J.C., Berenguer, J., Jaeck, D., Moreno Gonzalez, E. (2003). Evolution of liver transplantation in Europe: report of the European Liver Transplant Registry. *Liver Transplantation*, 9: 1231–1243.

Afdhal, N., Freilich, B., Levine, R., Black, M., Brown, R., Jr., Mansour, H., O'Brien, M., Brass, C. (2004). Colchicine versus PEG-Interferon long term (COPILOT) trial: interim analysis of clinical outcomes at year 2. *Hepatology*, 40: 239A.

Ahmad, J., Dodson, S.F., Demetris, A.J., Fung, J.J., Shakil, A.O. (2001). Recurrent hepatitis C after liver transplantation: a nonrandomized trial of interferon alpha alone versus interferon alpha and ribavirin. *Liver Transplantation*, 7: 863–869.

Alberti, A.B., Belli, L.S., Airoldi, A., de Carlis, L., Rondinara, G., Minola, E., Vangeli, M., Cernuschi, A., D'Amico, M., Forti, D., Pinzello, G. (2001). Combined therapy with interferon and low-dose ribavirin in posttransplantation recurrent hepatitis C: a pragmatic study. *Liver Transplantation*, 7: 870–876.

Armstrong, G.L., Alter, M.J., McQuillan, G.M., Margolis, H.S. (2000). The past incidence of hepatitis C virus infection: implications for the future burden of chronic liver disease in the United States. *Hepatology*, 31: 777–782.

Azzaroli, F., Accogli, E., Nigro, G., Trere, D., Giovanelli, S., Miracolo, A., Lodato, F., Montagnani, M., Tame, M., Colecchia, A., Mwangemi, C., Festi, D., Roda, E., Derenzini, M., Mazzella, G. (2004). Interferon plus ribavirin and interferon alone in preventing hepatocellular carcinoma: a prospective study on patients with HCV related cirrhosis. *World Journal of Gastroenterology*, 10: 3099–3102.

Benvegnu, L., Chemello, L., Noventa, F., Fattovich, G., Pontisso, P., Alberti, A. (1998). Retrospective analysis of the effect of interferon therapy on the clinical outcome of patients with viral cirrhosis. *Cancer*, 83: 901–909.

Berardi, S., Lodato, F., Gramenzi, A., D'Errico, A., Lenzi, M., Bontadini, A., Morelli, M.C., Tame, M.R., Piscaglia, F., Biselli, M., Sama, C., Mazzella, G., Pinna, A.D., Grazi, G.L., Bernardi, M., Andreone, P. (2006). High incidence of allograft dysfunction in liver transplant patients treated with PEG-Interferon alpha-2b and ribavirin for hepatitis C recurrence: possible de novo autoimmune hepatitis? *Gut*. Jun 23 [Epub ahead of print].

Berenguer, M. (2003). Host and donor risk factors before and after liver transplantation that impact HCV recurrence. *Liver Transplantation*, 9: S44–47.

Berenguer, M., Ferrell, L., Watson, J., Prieto, M., Kim, M., Rayon, M., Cordoba, J., Herola, A., Ascher, N., Mir, J., Berenguer, J., Wright, T.L. (2000). HCV-related fibrosis progression following liver transplantation: increase in recent years. *Journal of Hepatology*, 32: 673–684.

Berenguer, M., Lopez-Labrador, F.X., Wright, T.L. (2001). Hepatitis C and liver transplantation. *Journal of Hepatology*, 35: 666–678.

Berenguer, M., Palau, A., Fernandez, A., Benlloch, S., Aguilera, V., Prieto, M., Rayon, J.M., Berenguer, J. (2006). Efficacy, predictors of response, and potential risks associated with antiviral therapy in liver transplant recipients with recurrent hepatitis C. *Liver Transplantation*, 12: 1067–1076.

Berenguer, M., Prieto, M., Cordoba, J., Rayon, J.M., Carrasco, D., Olaso, V., San-Juan, F., Gobernado, M., Mir, J., Berenguer, J. (1998). Early development of chronic active hepatitis in recurrent hepatitis C virus infection after liver transplantation: association with treatment of rejection. *Journal of Hepatology*, 28: 756–763.

Bizollon, T., Ahmed, S.N., Radenne, S., Chevallier, M., Chevallier, P., Parvaz, P., Guichard, S., Ducerf, C., Baulieux, J., Zoulim, F., Trepo, C. (2003). Long term histological improvement and clearance of intrahepatic hepatitis C virus RNA following sustained response to interferon-ribavirin combination therapy in liver transplanted patients with hepatitis C virus recurrence. *Gut*, 52: 283–287.

Bizollon, T., Pradat, P., Mabrut, J.Y., Chevallier, M., Adham, M., Radenne, S., Souquet, J.C., Ducerf, C., Baulieux, J., Zoulim, F., Trepo, C. (2005). Benefit of sustained virological response to combination therapy on graft survival of liver transplanted patients with recurrent chronic hepatitis C. *American Journal of Transplantation*, 5: 1909–1913.

Brillanti, S., Vivarelli, M., De Ruvo, N., Aden, A.A., Camaggi, V., D'Errico, A., Furlini, G., Bellusci, R., Roda, E., Cavallari, A. (2002). Slowly tapering off steroids protects the graft against hepatitis C recurrence after liver transplantation. *Liver Transplantation,* 8: 884–888.

Bruno, S., Silini, E., Crosignani, A., Borzio, F., Leandro, G., Bono, F., Asti, M., Rossi, S., Larghi, A., Cerino, A., Podda, M., Mondelli, M.U. (1997). Hepatitis C virus genotypes and risk of hepatocellular carcinoma in cirrhosis: a prospective study. *Hepatology,* 25: 754–758.

Burak, K.W., Kremers, W.K., Batts, K.P., Wiesner, R.H., Rosen, C.B., Razonable, R.R., Paya, C.V., Charlton, M.R. (2002). Impact of cytomegalovirus infection, year of transplantation, and donor age on outcomes after liver transplantation for hepatitis C. *Liver Transplantation,* 8: 362–369.

Burra, P., Targhetta, S., Pevere, S., Boninsegna, S., Guido, M., Canova, D., Brolese, A., Masier, A., D'Aloiso, C., Germani, G., Tomat, S., Fagiuoli, S. (2006). Antiviral therapy for hepatitis C virus recurrence following liver transplantation: long-term results from a single center experience. *Transplantation Proceedings,* 38: 1127–1130.

Camma, C., Giunta, M., Andreone, P., Craxi, A. (2001). Interferon and prevention of hepatocellular carcinoma in viral cirrhosis: an evidence-based approach. *Journal of Hepatology,* 34: 593–602.

Castells, L., Vargas, V., Allende, H., Bilbao, I., Luis Lazaro, J., Margarit, C., Esteban, R., Guardia, J. (2005). Combined treatment with PEGylated interferon (alpha-2b) and ribavirin in the acute phase of hepatitis C virus recurrence after liver transplantation. *Journal of Hepatology,* 43: 53–59.

Chalasani, N., Manzarbeitia, C., Ferenci, P., Vogel, W., Fontana, R.J., Voigt, M., Riely, C., Martin, P., Teperman, L., Jiao, J., Lopez-Talavera, J.C. (2005). PEGinterferon alpha-2a for hepatitis C after liver transplantation: two randomized, controlled trials. *Hepatology,* 41: 289–298.

Chiaramonte, M., Stroffolini, T., Vian, A., Stazi, M.A., Floreani, A., Lorenzoni, U., Lobello, S., Farinati, F., Naccarato, R. (1999). Rate of incidence of hepatocellular carcinoma in patients with compensated viral cirrhosis. *Cancer,* 85: 2132–2137.

Colombo, M., de Franchis, R., Del Ninno, E., Sangiovanni, A., De Fazio, C., Tommasini, M., Donato, M.F., Piva, A., Di Carlo, V., Dioguardi, N. (1991). Hepatocellular carcinoma in Italian patients with cirrhosis. *New England Journal of Medicine,* 325: 675–680.

Cotler, S.J., Ganger, D.R., Kaur, S., Rosenblate, H., Jakate, S., Sullivan, D.G., Ng, K.W., Gretch, D.R., Jensen, D.M. (2001). Daily interferon therapy for hepatitis C virus infection in liver transplant recipients. *Transplantation,* 71: 261–266.

Crippin, J.S., McCashland, T., Terrault, N., Sheiner, P., Charlton, M.R. (2002). A pilot study of the tolerability and efficacy of antiviral therapy in hepatitis C virus-infected patients awaiting liver transplantation. *Liver Transplantation,* 8: 350–355.

Czaja, A.J. (2002). Autoimmune hepatitis after liver transplantation and other lessons of self-intolerance. *Liver Transplantation,* 8: 505–513.

Davis, G.L., Albright, J.E., Cook, S.F., Rosenberg, D.M. (2003). Projecting future complications of chronic hepatitis C in the United States. *Liver Transplantation,* 9: 331–338.

De Ruvo, N., Cucchetti, A., Lauro, A., Masetti, M., Cautero, N., Di Benedetto, F., Dazzi, A., Del Gaudio, M., Ravaioli, M., Zanello, M., La Barba, G., di Francesco, F., Risaliti, A., Ramacciato, G., Pinna, A.D. (2005). Preliminary results of immunosuppression with thymoglobuline pretreatment and hepatitis C virus recurrence in liver transplantation. *Transplantation Proceedings,* 37: 2607–2608.

Degos, F., Christidis, C., Ganne-Carrie, N., Farmachidi, J.P., Degott, C., Guettier, C., Trinchet, J.C., Beaugrand, M., Chevret, S. (2000). Hepatitis C virus related cirrhosis: time to occurrence of hepatocellular carcinoma and death. *Gut,* 47: 131–136.

Di Bisceglie, A.M., Hoofnagle, J.H. (2002). Optimal therapy of hepatitis C. *Hepatology,* 36 (5 Suppl. 1): S121–127.

Dousset, B., Conti, F., Houssin, D., Calmus, Y. (1994). Acute vanishing bile duct syndrome after interferon therapy for recurrent HCV infection in liver-transplant recipients. *New England Journal of Medicine,* 330: 1160–1161.

Dumortier, J., Scoazec, J.Y., Chevallier, P., Boillot, O. (2004). Treatment of recurrent hepatitis C after liver transplantation: a pilot study of PEGinterferon alpha-2b and ribavirin combination. *Journal of Hepatology,* 40: 669–674.

Durelli, L., Ferrero, B., Oggero, A., Verdun, E., Ghezzi, A., Montanari, E., Zaffaroni, M. (2001). Thyroid function and autoimmunity during interferon beta-1b treatment: a multicenter prospective study. *Journal of Clinical Endocrinology & Metabolism*, 86: 3525–3532.

Eason, J.D., Nair, S., Cohen, A.J., Blazek, J.L., Loss, G.E., Jr. (2003). Steroid-free liver transplantation using rabbit antithymocyte globulin and early tacrolimus monotherapy. *Transplantation*, 75: 1396–1399.

El-Serag, H.B. (2001). Epidemiology of hepatocellular carcinoma. *Clinical Liver Disease*, 5: 87–107, vi.

Everson, G.T. (2005). Treatment of hepatitis C in the patient with decompensated cirrhosis. *Clinical Gastroenterology and Hepatology*, 3 (10 Suppl. 2): S106–112.

Everson, G.T., Heathcote, J., Pappas, S.C. (2004a). Histologic benefit of Peg-Interferon alpha 2a (40 KD) (PEGASYS) monotherapy in patients with advanced fibrosis or cirrhosis due to chronic hepatitis C. *Hepatology*, 40: 316A.

Everson, G.T., Hoefs, J.C., Malet, P. (2004b). Impaired virologic response in patients with advanced liver disease due to chronic hepatitis C is independently linked to severity of disease: results from the HALT C trial. *Hepatology*, 40: 180A.

Fattovich, G., Giustina, G., Degos, F., Tremolada, F., Diodati, G., Almasio, P., Nevens, F., Solinas, A., Mura, D., Brouwer, J.T., Thomas, H., Njapoum, C., Casarin, C., Bonetti, P., Fuschi, P., Basho, J., Tocco, A., Bhalla, A., Galassini, R., Noventa, F., Schalm, S.W., Realdi, G. (1997). Morbidity and mortality in compensated cirrhosis type C: a retrospective follow-up study of 384 patients. *Gastroenterology*, 112: 463–472.

Fattovich, G., Giustina, G., Favarato, S., Ruol, A. (1996). A survey of adverse events in 11,241 patients with chronic viral hepatitis treated with alpha interferon. *Journal of Hepatology*, 24: 38–47.

Fattovich, G., Pantalena, M., Zagni, I., Realdi, G., Schalm, S.W., Christensen, E. (2002). Effect of hepatitis B and C virus infections on the natural history of compensated cirrhosis: a cohort study of 297 patients. *American Journal of Gastroenterology*, 97: 2886–2895.

Fattovich, G., Stroffolini, T., Zagni, I., Donato, F. (2004). Hepatocellular carcinoma in cirrhosis: incidence and risk factors. *Gastroenterology*, 127 (5 Suppl. 1): S35–50.

Feray, C., Samuel, D., Gigou, M., Paradis, V., David, M.F., Lemonnier, C., Reynes, M., Bismuth, H. (1995). An open trial of interferon alpha recombinant for hepatitis C after liver transplantation: antiviral effects and risk of rejection. *Hepatology*, 22 (4 Pt. 1): 1084–1089.

Forman, L.M., Lewis, J.D., Berlin, J.A., Feldman, H.I., Lucey, M.R. (2002). The association between hepatitis C infection and survival after orthotopic liver transplantation. *Gastroenterology*, 122: 889–896.

Forns, X., Ampurdanes, S., Sanchez-Tapias, J.M., Guilera, M., Sans, M., Sanchez-Fueyo, A., Quinto, L., Joya, P., Bruguera, M., Rodes, J. (2001). Long-term follow-up of chronic hepatitis C in patients diagnosed at a tertiary-care center. *Journal of Hepatology*, 35: 265–271.

Fried, M.W., Shiffman, M.L., Reddy, K.R., Smith, C., Marinos, G., Goncales, F.L., Jr., Haussinger, D., Diago, M., Carosi, G., Dhumeaux, D., Craxi, A., Lin, A., Hoffman, J., Yu, J. (2002). PEGinterferon alfa-2a plus ribavirin for chronic hepatitis C virus infection. *New England Journal of Medicine*, 347: 975–982.

Gane, E.J., Lo, S.K., Riordan, S.M., Portmann, B.C., Lau, J.Y., Naoumov, N.V., Williams, R. (1998). A randomized study comparing ribavirin and interferon alpha monotherapy for hepatitis C recurrence after liver transplantation. *Hepatology*, 27: 1403–1407.

Gane, E.J., Naoumov, N.V., Qian, K.P., Mondelli, M.U., Maertens, G., Portmann, B.C., Lau, J.Y., Williams, R. (1996). A longitudinal analysis of hepatitis C virus replication following liver transplantation. *Gastroenterology*, 110: 167–177.

Garcia-Retortillo, M., Forns, X., Feliu, A., Moitinho, E., Costa, J., Navasa, M., Rimola, A., Rodes, J. (2002). Hepatitis C virus kinetics during and immediately after liver transplantation. *Hepatology*, 35: 680–687.

Ghobrial, R.M., Farmer, D.G., Baquerizo, A., Colquhoun, S., Rosen, H.R., Yersiz, H., Markmann, J.F., Drazan, K.E., Holt, C., Imagawa, D., Goldstein, L.I., Martin, P., Busuttil, R.W. (1999). Orthotopic liver transplantation for hepatitis C: outcome, effect of immunosuppression, and causes of retransplantation during an 8-year single-center experience. *Annals of Surgery,* 229: 824–831; discussion 831–823.

Giostra, E., Kullak-Ublick, G.A., Keller, W., Fried, R., Vanlemmens, C., Kraehenbuhl,S., Locher, S., Egger, H.P., Clavien, P.A., Hadengue, A., Mentha, G., Morel, P., Negro, F. (2004). Ribavirin/interferon-alpha sequential treatment of recurrent hepatitis C after liver transplantation. *Transplant International,* 17: 169–176.

Gordon, A., Mitchell, J.L., Morales, B.M., Thomson, K., Pedersen J.S., McLean, C., Angus, P.W., Roberts, S.K. (2005). Sustained virological response is associated with reduction of HVPG in chronic hepatitis C related cirrhosis. *Journal of Hepatology,* 42: 205A.

Gordon, S.C., Bayati, N., Silverman, A.L. (1998). Clinical outcome of hepatitis C as a function of mode of transmission. *Hepatology,* 28: 562–567.

Gramenzi, A., Andreone, P., Fiorino, S., Camma, C., Giunta, M., Magalotti, D., Cursaro, C., Calabrese, C., Arienti, V., Rossi, C., Di Febo, G., Zoli, M., Craxi, A., Gasbarrini, G., Bernardi, M. (2001). Impact of interferon therapy on the natural history of hepatitis C virus related cirrhosis. *Gut,* 48: 843–848.

Guerrero, R.B., Batts, K.P., Burgart, L.J., Barrett, S.L., Germer, J.J., Poterucha, J.J., Wiesner, R.H., Charlton, M.R., Persing, D.H. (2000). Early detection of hepatitis C allograft reinfection after orthotopic liver transplantation: a molecular and histologic study. *Modern Pathology,* 13: 229–237.

Hadziyannis, S.J., Sette, H., Jr., Morgan, T.R., Balan, V., Diago, M., Marcellin, P., Ramadori, G., Bodenheimer, H., Jr., Bernstein, D., Rizzetto, M., Zeuzem, S., Pockros, P.J., Lin, A., Ackrill, A.M. (2004). PEGinterferon-alpha2a and ribavirin combination therapy in chronic hepatitis C: a randomized study of treatment duration and ribavirin dose. *Annals of Internal Medicine,* 140: 346–355.

Heathcote, J., Main, J. (2005). Treatment of hepatitis C. *Journal of Viral Hepatitis,* 12: 223–235.

Heathcote, E.J., Shiffman, M.L., Cooksley, W.G., Dusheiko, G.M., Lee, S.S., Balart, L., Reindollar, R., Reddy, R.K., Wright, T.L., Lin, A., Hoffman, J., De Pamphilis, J. (2000). PEGinterferon alpha-2a in patients with chronic hepatitis C and cirrhosis. *New England Juornal of Medicine,* 343: 1673–1680.

Jain, A., Demetris, A.J., Manez, R., Tsamanadas, A.C., Van Thiel, D., Rakela, J., Starzl, T.E., Fung, J.J. (1998). Incidence and severity of acute allograft rejection in liver transplant recipients treated with alfa interferon. *Liver Transplantation and Surgery,* 4: 197–203.

Janssen, H.L., Brouwer, J.T., Nevens, F., Sanchez-Tapias, J.M., Craxi, A., Hadziyannis, S. (1993). Fatal hepatic decompensation associated with interferon alpha. European concerted action on viral hepatitis (Eurohep). *British Medical Journal,* 306: 107–108.

Jen, J.F., Glue, P., Gupta, S., Zambas, D., Hajian, G. (2000). Population pharmacokinetic and pharmacodynamic analysis of ribavirin in patients with chronic hepatitis C. *Therapeutic Drug Monitoring,* 22: 555–565.

Kerkar, N., Hadzic, N., Davies, E.T., Portmann, B., Donaldson, P.T., Rela, M., Heaton, N.D., Vergani, D., Mieli-Vergani, G. (1998). De-novo autoimmune hepatitis after liver transplantation. *Lancet,* 351: 409–413.

Kim, W.R. (2002). The burden of hepatitis C in the United States. *Hepatology,* 36 (5 Suppl. 1): S30–34.

Kowdley, K.V. (2005). Hematologic side effects of interferon and ribavirin therapy. *Jouranl of Clinical Gastroenterology,* 39 (1 Suppl.): S3–8.

Kugelmas, M., Osgood, M.J., Trotter, J.F., Bak, T., Wachs, M., Forman, L., Kam, I., Everson, G.T. (2003). Hepatitis C virus therapy, hepatocyte drug metabolism, and risk for acute cellular rejection. *Liver Transplantation,* 9: 1159–1165.

Lake, J.R., Shorr, J.S., Steffen, B.J., Chu, A.H., Gordon, R.D., Wiesner, R.H. (2005). Differential effects of donor age in liver transplant recipients infected with hepatitis B, hepatitis C and without viral hepatitis. *American Journal of Transplantation,* 5: 549–557.

Lee, S.S., Bain, V.G., Peltekian, K., Krajden, M., Yoshida, E.M., Deschenes, M., Heathcote, J., Bailey, R.J., Simonyi, S., Sherman, M. (2006). Treating chronic hepatitis C with PEGylated interferon alfa-2a (40 KD) and ribavirin in clinical practice. *Alimentary Pharmacology & Therapeutics,* 23: 397–408.

Lee, W.M., Dienstag, J.L., Lindsay, K.L., Lok, A.S., Bonkovsky, H.L., Shiffman, M.L., Everson, G.T., Di Bisceglie, A.M., Morgan, T.R., Ghany, M.G., Morishima, C., Wright, E.C., Everhart, J.E. (2004). Evolution of the HALT-C trial: PEGylated interferon as maintenance therapy for chronic hepatitis C in previous interferon nonresponders. *Controlled Clinical Trials,* 25: 472–492.

Licata, A., Di Bona, D., Schepis, F., Shahied, L., Craxi, A., Camma, C. (2003). When and how to treat acute hepatitis C? *Journal of Hepatology,* 39: 1056–1062.

Llovet, J.M., Burroughs, A., Bruix, J. (2003). Hepatocellular carcinoma. *Lancet,* 362: 1907–1917.

Manns, M.P., McHutchison, J.G., Gordon, S.C., Rustgi, V.K., Shiffman, M., Reindollar, R., Goodman, Z.D., Koury, K., Ling, M., Albrecht, J.K. (2001). PEGinterferon alpha-2b plus ribavirin compared with interferon alpha-2b plus ribavirin for initial treatment of chronic hepatitis C: a randomised trial. *Lancet,* 358: 958–965.

Manns, M.P., Wedemeyer, H., Cornberg, M. (2006). Treating viral hepatitis C: efficacy, side effects, and complications. *Gut,* 55: 1350–1359.

Marcos, A., Eghtesad, B., Fung, J.J., Fontes, P., Patel, K., Devera, M., Marsh, W., Gayowski, T., Demetris, A.J., Gray, E.A., Flynn, B., Zeevi, A., Murase, N., Starzl, T.E. (2004). Use of alemtuzumab and tacrolimus monotherapy for cadaveric liver transplantation: with particular reference to hepatitis C virus. *Transplantation,* 78: 966–971.

Marrache, F., Consigny, Y., Ripault, M.P., Cazals-Hatem, D., Martinot, M., Boyer, N., Degott, C., Valla, D., Marcellin, P. (2005). Safety and efficacy of PEGinterferon plus ribavirin in patients with chronic hepatitis C and bridging fibrosis or cirrhosis. *Journal of Viral Hepatitis,* 12: 421–428.

Mazzella, G., Accogli, E., Sottili, S., Festi, D., Orsini, M., Salzetta, A., Novelli, V., Cipolla, A., Fabbri, C., Pezzoli, A., Roda, E. (1996). Alpha interferon treatment may prevent hepatocellular carcinoma in HCV-related liver cirrhosis. *Juornal of Hepatology,* 24: 141–147.

Mazziotti, G., Sorvillo, F., Morisco, F., Carbone, A., Rotondi, M., Stornaiuolo, G., Precone, D.F., Cioffi, M., Gaeta, G.B., Caporaso, N., Carella, C. (2002). Serum insulin-like growth factor I evaluation as a useful tool for predicting the risk of developing hepatocellular carcinoma in patients with hepatitis C virus-related cirrhosis: a prospective study. *Cancer,* 95: 2539–2545.

McCaughan, G.W., Zekry, A. (2002). Pathogenesis of hepatitis C virus recurrence in the liver allograft. *Liver Transplantation,* 8 (10 Suppl. 1): S7–S13.

McHutchison, J.G., Gordon, S.C., Schiff, E.R., Shiffman, M.L., Lee, W.M., Rustgi, V.K., Goodman, Z.D., Ling, M.H., Cort, S., Albrecht, J.K. (1998). Interferon alpha-2b alone or in combination with ribavirin as initial treatment for chronic hepatitis C. Hepatitis Interventional Therapy Group. *New England Journal of Medicine,* 339: 1485–1492.

Mukherjee, S. (2005). PEGylated interferon alpha-2a and ribavirin for recurrent hepatitis C after liver transplantation. *Transplantation Proceedings,* 37: 4403–4405.

Nakagawa, M., Sakamoto, N., Enomoto, N., Tanabe, Y., Kanazawa, N., Koyama, T., Kurosaki, M., Maekawa, S., Yamashiro, T., Chen, C.H., Itsui, Y., Kakinuma, S., Watanabe, M. (2004). Specific inhibition of hepatitis C virus replication by cyclosporin A. *Biochemical and Biophysical Research Communications,* 313: 42–47.

Nakagawa, M., Sakamoto, N., Tanabe, Y., Koyama, T., Itsui, Y., Takeda, Y., Chen, C.H., Kakinuma, S., Oooka, S., Maekawa, S., Enomoto, N., Watanabe, M. (2005). Suppression of hepatitis C virus replication by cyclosporin a is mediated by blockade of cyclophilins. *Gastroenterology,* 129: 1031–1041.

Neff, G.W., Montalbano, M., O'Brien, C.B., Nishida, S., Safdar, K., Bejarano, P.A., Khaled, A.S., Ruiz, P., Slapak-Green, G., Lee, M., Nery, J., De Medina, M., Tzakis, A., Schiff, E.R. (2004). Treatment of established recurrent hepatitis C in liver-transplant recipients with PEGylated interferon-alpha-2b and ribavirin therapy. *Transplantation,* 78: 1303–1307.

Neumann, U.P., Berg, T., Bahra, M., Seehofer, D., Langrehr, J.M., Neuhaus, R., Radke, C., Neuhaus, P. (2004). Fibrosis progression after liver transplantation in patients with recurrent hepatitis C. *Journal of Hepatology,* 41: 830–836.

Niederau, C., Heintges, T., Lange, S., Goldmann, G., Niederau, C.M., Mohr, L., Haussinger, D. (1996). Long-term follow-up of HBeAg-positive patients treated with interferon alpha for chronic hepatitis B. *New England Journal of Medicine,* 334: 1422–1427.

Niederau, C., Lange, S., Heintges, T., Erhardt, A., Buschkamp, M., Hurter, D., Nawrocki, M., Kruska, L., Hensel, F., Petry, W., Haussinger, D. (1998). Prognosis of chronic hepatitis C: results of a large, prospective cohort study. *Hepatology*, 28: 1687–1695.

NIH Consensus Statement on Management of Hepatitis C: 2002. (2002). *NIH Consensus and State of the Science Statements*, 19: 1–46.

Papatheodoridis, G.V., Manesis, E., Hadziyannis, S.J. (2001). The long-term outcome of interferon-alpha treated and untreated patients with HBeAg-negative chronic hepatitis B. *Journal of Hepatology*, 34: 306–313.

Parkin, D.M., Bray, F., Ferlay, J., Pisani, P. (2001). Estimating the world cancer burden: Globocan 2000. *International Journal of Cancer*, 94: 153–156.

Pelletier, S.J., Raymond, D.P., Crabtree, T.D., Berg, C.L., Iezzoni, J.C., Hahn, Y.S., Sawyer, R.G., Pruett, T.L. (2000). Hepatitis C-induced hepatic allograft injury is associated with a pretransplantation elevated viral replication rate. *Hepatology*, 32: 418–426.

Pelletier, S.J., Raymond, D.P., Crabtree, T.D., Iezzoni, J.C., Sawyer, R.G., Hahn, Y.S., Pruett, T.L. (2000). Pretransplantation hepatitis C virus quasispecies may be predictive of outcome after liver transplantation. *Hepatology*, 32: 375–381.

Planas, R., Balleste, B., Alvarez, M.A., Rivera, M., Montoliu, S., Galeras, J.A., Santos, J., Coll, S., Morillas, R.M., Sola, R. (2004). Natural history of decompensated hepatitis C virus-related cirrhosis. A study of 200 patients. *Journal of Hepatology*, 40: 823–830.

Poynard, T., McHutchison, J., Manns, M., Trepo, C., Lindsay, K., Goodman, Z., Ling, M.H., Albrecht, J. (2002). Impact of PEGylated interferon alpha-2b and ribavirin on liver fibrosis in patients with chronic hepatitis C. *Gastroenterology*, 122: 1303–1313.

Poynard, T., Schiff, E., Terg, R., Goncales, F., Diago, M., Reichen, J., Moreno, R, Bedossa, P., Burroughs, M., Albrecht, J. (2005). Sustained virologic response (SVR) in EPIC3 trial: week twelve virology predicts SVR in previous interferon/ribavirin treatment failures receiving PEG-Intron/Rebetol (PR) weight based dosing (WBD). *Journal of Hepatology*, 42: 40A.

Prieto, M., Berenguer, M., Rayon, J.M., Cordoba, J., Arguello, L., Carrasco, D., Garcia-Herola, A., Olaso, V., De Juan, M., Gobernado, M., Mir, J., Berenguer, J. (1999). High incidence of allograft cirrhosis in hepatitis C virus genotype 1b infection following transplantation: relationship with rejection episodes. *Hepatology*, 29: 250–256.

Rifai, K., Sebagh, M., Karam, V., Saliba, F., Azoulay, D., Adam, R., Castaing, D., Bismuth, H., Reynes, M., Samuel, D., Feray, C. (2004). Donor age influences 10-year liver graft histology independently of hepatitis C virus infection. *Journal of Hepatology*, 41: 446–453.

Rodriguez-Luna, H., Khatib, A., Sharma, P., De Petris, G., Williams, J.W., Ortiz, J., Hansen, K., Mulligan, D., Moss, A., Douglas, D.D., Balan, V., Rakela, J., Vargas, H.E. (2004). Treatment of recurrent hepatitis C infection after liver transplantation with combination of PEGylated interferon alpha2b and ribavirin: an open-label series. *Transplantation*, 77: 190–194.

Rosen, H.R., Shackleton, C.R., Higa, L., Gralnek, I.M., Farmer, D.A., McDiarmid, S.V., Holt, C., Lewin, K.J., Busuttil, R.W., Martin, P. (1997). Use of OKT3 is associated with early and severe recurrence of hepatitis C after liver transplantation. *American Journal of Gastroenterology*, 92: 1453–1457.

Saito, Y., Saito, H., Tada, S., Nakamoto, N., Horikawa, H., Kurita, S., Kitamura, K., Ebinuma, H., Ishii, H., Hibi, T. (2005). Effect of long-term interferon therapy for refractory chronic hepatitis C: preventive effect on hepatocarcinogenesis. *Hepatogastroenterology*, 52: 1491–1496.

Salcedo, M., Vaquero, J., Banares, R., Rodriguez-Mahou, M., Alvarez, E., Vicario, J.L., Hernandez-Albujar, A., Tiscar, J.L., Rincon, D., Alonso, S., De Diego, A., Clemente, G. (2002). Response to steroids in de novo autoimmune hepatitis after liver transplantation. *Hepatology*, 35: 349–356.

Samuel, C.E. (2001). Antiviral actions of interferons. *Clinical Microbiology Reviews*, 14: 778–809, table of contents.

Samuel, D., Bizollon, T., Feray, C., Roche, B., Ahmed, S.N., Lemonnier, C., Cohard, M., Reynes, M., Chevallier, M., Ducerf, C., Baulieux, J., Geffner, M., Albrecht, J.K., Bismuth, H., Trepo, C. (2003). Interferon-alpha 2b plus ribavirin in patients with chronic hepatitis C after liver transplantation: a randomized study. *Gastroenterology*, 124: 642–650.

Samuel, D., Forns, X., Berenguer, M., Trautwein, C., Burroughs, A., Rizzetto, M., Trepo, C. (2006). Report of the monothematic EASL conference on liver transplantation for viral hepatitis (Paris, France, January 12-14, 2006). *Journal of Hepatology,* 45: 127–143.

Sanchez-Fueyo, A., Restrepo, J.C., Quinto, L., Bruguera, M., Grande, L., Sanchez-Tapias, J.M., Rodes, J., Rimola, A. (2002). Impact of the recurrence of hepatitis C virus infection after liver transplantation on the long-term viability of the graft. *Transplantation,* 73: 56–63.

Serfaty, L., Aumaitre, H., Chazouilleres, O., Bonnand, A.M., Rosmorduc, O., Poupon, R.E., Poupon, R. (1998). Determinants of outcome of compensated hepatitis C virus-related cirrhosis. *Hepatology,* 27: 1435–1440.

Sheiner, P.A., Schwartz, M.E., Mor, E., Schluger, L.K., Theise, N., Kishikawa, K., Kolesnikov, V., Bodenheimer, H., Emre, S., Miller, C.M. (1995). Severe or multiple rejection episodes are associated with early recurrence of hepatitis C after orthotropic liver transplantation. *Hepatology,* 21: 30–34.

Shergill, A.K., Khalili, M., Straley, S., Bollinger, K., Roberts, J.P., Ascher, N.A., Terrault, N.A. (2005). Applicability, tolerability and efficacy of preemptive antiviral therapy in hepatitis C-infected patients undergoing liver transplantation. *American Journal of Transplantation,* 5: 118–124.

Shiffman, M.L., Hofmann, C.M., Thompson, E.B., Ferreira-Gonzalez, A., Contos, M.J., Koshy, A., Luketic, V.A., Sanyal, A.J., Mills, A.S., Garrett, C.T. (1997). Relationship between biochemical, virological, and histological response during interferon treatment of chronic hepatitis C. *Hepatology,* 26: 780–785.

Shiratori, Y., Ito, Y., Yokosuka, O., Imazeki, F., Nakata, R., Tanaka, N., Arakawa, Y., Hashimoto, E., Hirota, K., Yoshida, H., Ohashi, Y., Omata, M. (2005). Antiviral therapy for cirrhotic hepatitis C: association with reduced hepatocellular carcinoma development and improved survival. *Annals of Internal Medicine,* 142: 105–114.

Singh, N., Gayowski, T., Wannstedt, C.F., Shakil, A.O., Wagener, M.M., Fung, J.J., Marino, I.R. (1998). Interferon-alpha for prophylaxis of recurrent viral hepatitis C in liver transplant recipients: a prospective, randomized, controlled trial. *Transplantation,* 65: 82–86.

Stravitz, R.T., Shiffman, M.L., Sanyal, A.J., Luketic, V.A., Sterling, R.K., Heuman, D.M., Ashworth, A., Mills, A.S., Contos, M., Cotterell, A.H., Maluf, D., Posner, M.P., Fisher, R.A. (2004). Effects of interferon treatment on liver histology and allograft rejection in patients with recurrent hepatitis C following liver transplantation. *Liver Transplantation,* 10: 850–858.

Sugawara, Y., Makuuchi, M., Matsui, Y., Kishi, Y., Akamatsu, N., Kaneko, J., Kokudo, N. (2004). Preemptive therapy for hepatitis C virus after living-donor liver transplantation. *Transplantation,* 78: 1308–1311.

Terrault, N.A. (2005). Treatment of recurrent hepatitis C in liver transplant recipients. *Clinical Gastroenterology and Hepatology,* 3 (10 Suppl. 2): S125–131.

Thomas, R.M., Brems, J.J., Guzman-Hartman, G., Yong, S., Cavaliere, P., Van Thiel, D.H. (2003). Infection with chronic hepatitis C virus and liver transplantation: a role for interferon therapy before transplantation. *Liver Transplantation,* 9: 905–915.

Tong, M.J., el-Farra, N.S., Reikes, A.R., Co, R.L. (1995). Clinical outcomes after transfusion-associated hepatitis C. *New England Journal of Medicine,* 332: 1463–1466.

Toniutto, P., Fabris, C., Fumo, E., Apollonio, L., Caldato, M., Avellini, C., Minisini, R., Pirisi, M. (2005). PEGylated versus standard interferon-alpha in antiviral regimens for post-transplant recurrent hepatitis C: comparison of tolerability and efficacy. *Journal of Gastroenterology and Hepatology,* 20: 577–582.

Tsukuma, H., Hiyama, T., Tanaka, S., Nakao, M., Yabuuchi, T., Kitamura, T., Nakanishi, K., Fujimoto, I., Inoue, A., Yamazaki, H., et al. (1993). Risk factors for hepatocellular carcinoma among patients with chronic liver disease. *New England Journal of Medicine,* 328: 1797–1801.

Valla, D.C., Chevallier, M., Marcellin, P., Payen, J.L., Trepo, C., Fonck, M., Bourliere, M., Boucher, E., Miguet, J.P., Parlier, D., Lemonnier, C., Opolon, P. (1999). Treatment of hepatitis C virus-related cirrhosis: a randomized, controlled trial of interferon alfa-2b versus no treatment. *Hepatology,* 29: 1870–1875.

Velazquez, R.F., Rodriguez, M., Navascues, C.A., Linares, A., Perez, R., Sotorrios, N.G., Martinez, I., Rodrigo, L. (2003). Prospective analysis of risk factors for hepatocellular carcinoma in patients with liver cirrhosis. *Hepatology,* 37: 520–527.

Wang, C.S., Ko, H.H., Yoshida, E.M., Marra, C.A., Richardson, K. (2006). Interferon-based combination anti-viral therapy for hepatitis C virus after liver transplantation: a review and quantitative analysis. *American Journal of Transplantation,* 6: 1586–1599.

Watashi, K., Hijikata, M., Hosaka, M., Yamaji, M., Shimotohno, K. (2003). Cyclosporin A suppresses replication of hepatitis C virus genome in cultured hepatocytes. *Hepatology,* 38: 1282–1288.

WHO (1999). World Health Organization. Weekly Epidemiological Record, pp. 421–428.

Wiesner, R.H., Sorrell, M., Villamil, F. (2003). Report of the first International Liver Transplantation Society expert panel consensus conference on liver transplantation and hepatitis C. *Liver Transplantation,* 9: S1–9.

Wilson, L.E., Widman, D., Dikman, S.H., Gorevic, P.D. (2002). Autoimmune disease complicating antiviral therapy for hepatitis C virus infection. *Seminars in Arthritis and Rheumatism,* 32: 163–173.

Wright, T.L., Combs, C., Kim, M., Ferrell, L., Bacchetti, P., Ascher, N., Roberts, J., Wilber, J., Sheridan, P., Urdea, M. (1994). Interferon-alpha therapy for hepatitis C virus infection after liver transplantation. *Hepatology,* 20 (4 Pt. 1): 773–779.

Yano, M., Kumada, H., Kage, M., Ikeda, K., Shimamatsu, K., Inoue, O., Hashimoto, E., Lefkowitch, J.H., Ludwig, J., Okuda, K. (1996). The long-term pathological evolution of chronic hepatitis C. *Hepatology,* 23: 1334–1340.

Yedibela, S., Schuppan, D., Muller, V., Schellerer, V., Tannapfel, A., Hohenberger, W., Meyer, T. (2005). Successful treatment of hepatitis C reinfection with interferon-alpha2b and ribavirin after liver transplantation. *Liver International,* 25: 717–722.

Yoshida, H., Shiratori, Y., Moriyama, M., Arakawa, Y., Ide, T., Sata, M., Inoue, O., Yano, M., Tanaka, M., Fujiyama, S., Nishiguchi, S., Kuroki, T., Imazeki, F., Yokosuka, O., Kinoyama, S., Yamada, G., Omata, M. (1999). Interferon therapy reduces the risk for hepatocellular carcinoma: national surveillance program of cirrhotic and noncirrhotic patients with chronic hepatitis C in Japan. IHIT Study Group. Inhibition of Hepatocarcinogenesis by Interferon Therapy. *Annals of Internal Medicine,* 131:174–181.

Yoshida, H., Tateishi, R., Arakawa, Y., Sata, M., Fujiyama, S., Nishiguchi, S., Ishibashi, H., Yamada, G., Yokosuka, O., Shiratori, Y., Omata, M. (2004). Benefit of interferon therapy in hepatocellular carcinoma prevention for individual patients with chronic hepatitis C. *Gut,* 53: 425–430.

Zeuzem, S., Schmidt, J.M., Lee, J.H., Ruster, B., Roth, W.K. (1996). Effect of interferon alpha on the dynamics of hepatitis C virus turnover in vivo. *Hepatology,* 23: 366–371.

Pegylated Interferons: Clinical Applications in the Management of Hepatitis C Infection

S. James Matthews and Christopher McCoy

Introduction

Currently, there are two pegylated interferon (PEG-IFN) products available for the treatment of hepatitis C virus (HCV) infection [(pegylated interferon, PEG-IFN alpha-2a (PEGASYS®), and pegylated interferon, PEG-IFN alpha-2b (PEG-INTRON®)]. Pegylation is a process by which the interferon-alpha is bound to a polyethylene glycol moiety (Kozlowski & Harris, 2001). The PEG-IFN-alpha-2a product is bound to a single-branched bis-monomethoxy polyethylene glycol (PEG) chain (40,000 daltons) for a final molecular weight of 60,000 daltons or 60 kDa (kilodaltons). Four major positional isomers exist for this compound (Bailon et al., 2001) In contrast, PEG-IFN-alpha-2b is formed by attaching a single chain of PEG (12 kDa mono-methoxy PEG) to interferon-alpha-2b via an ester linkage. The PEG moiety is conjugated to the His[34] amino acid residue, forming 12 positional isomers (Wang et al., 2000). The combined molecular weight of PEG-IFN-alpha-2b is smaller, about 31 kDa. The chemical structure and linkages of PEG-IFNs (alpha-2a and alpha-2b) are shown in Figure 1.

Pegylation does change the pharmacokinetic properties of unmodified interferon-alpha demonstrably (Luxon et al., 2002; Kozlowski & Harris, 2001). These properties allow for once-weekly dosing, more stable interferon-alpha blood concentrations throughout the dosing interval, and improved efficacy (Luxon et al., 2002; Pockros et al., 2004; Lindsay et al., 2001).

Currently, the National Institutes of Health identifies the use of PEG-IFN-alpha plus ribavirin (RBV) as the preferred therapy for the treatment of chronic HCV infection. While an improvement over rates of treatment success with unmodified interferons, the critical endpoint, a sustained virological response (SVR), defined as an undetectable level of HCV RNA 6 months after the completion of therapy, remains low particularly for genotype 1, at 42–52%. Higher rates are observed in patients with genotype 2 or 3, at 76–84% (Fried et al., 2002; Hadziyannis et al., 2004; Manns et al., 2001; National Institutes of Health Consensus Development Conference Statement: Management of Hepatitis C, 2002). Efforts to develop more effective dosing regimens, especially for subjects with genotype 1 and patients who do not respond or relapse after completion of therapy, are under study. This chapter will review the pharmacokinetics, efficacy in phase II and III clinical studies,

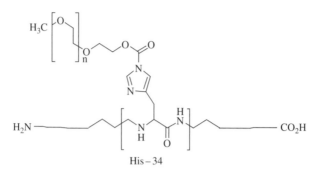

His – 34

Pegylated IFN – alpha 2b

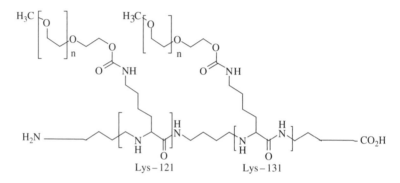

Lys – 121   Lys – 131

Pegylated IFN – alpha 2a

Fig. 1 Pegylated IFN-alpha 2a-Primary sites of pegylation are on amino terminus of the side chains of lysines-121 and 131
Pegylated IFN-alpha 2b-Primary site of pegylation is on imidazole side chain of histidine-34

and therapeutic options for the use of the PEG-IFNs in difficult-to-treat conditions. Safety and drug interaction considerations will also be reviewed.

Pharmacokinetics

This review will involve comparisons among unmodified interferon-alpha, PEG-INF-alpha-2a, and PEG-INF-alpha-2b. The pharmacokinetic parameters of the two PEG-IFNs are quite different and vary markedly from the corresponding unmodified interferon (Table 1). There are no significant differences in core parameters between male and female subjects, or among Black, Latino, and Caucasian patients with chronic HCV infection (PEGASYS®; Hoffmann-La Roche, Basel, Switzerland, 2004; PEG-Intron®; Schering Corporation, Kenilworth, NJ, 2005; Brennan et al., 2005).

PEG-IFN products are absorbed slowly. The absorption half-life for unmodified interferon-alpha and PEG-IFN-alpha-2a and -alpha-2b are 2.3, 50, and 4.6 hours,

Table 1 Pharmacokinetics of Unmodified and Pegylated Interferons

Peg-interferon	Peg-interferon alpha-2a	Interferon alpha	Peg-interferon alpha-2b	Reference
Absorption $t_{1/2}$ (h)	50	2.3	4.6	Glue et al. (2000a); Harris et al. (2001)
T_{max} (h)	45–81	7.3–12	16–44	Algranati et al. (1999); Glue et al. (2000a); Martin et al. (2000a, 2000b); Modi et al. (2000a); Wills (1990); Chatelut et al. (1999)
$T_{1/2}$ (h)	61–110	4–16	30–58	Algranati et al. (1999); Glue et al. (2000a, 2000b); Gupta et al. (2003); Martin et al. (2000b); Wills et al. (1990)
AUC	3334 ng · h/mL[a]	17,600pg · h/mL[d]	86,400pg · h/mL[b]	Glue et al. (2000a); Modi et al. (2000a); Wills et al. (1984)
V_d (L)	6–14	31–98	69.3[c]	Glue et al. (2000a); Lamb & Martin (2002); Harris et al. (2001)
Cl/F	60–118 mL/h	4.9–21 L/h	22 mL/h/kg	Algranati et al. (1999); Glue et al. (2000a); Martin et al. (2000b); Modi et al. (2000a)

[a] 180-μg/week multiple doses (subcutaneous injection);
[b] 1.5-μg/kg/week (subcutaneous injection) measured at week 4;
[c] based on a 70-kg person (0.99 L/kg);
[d] 36×10^6 interferon alpha-2a (subcutaneous injection).

respectively (Harris et al., 2001; Glue et al., 2000). The time to achieve the maximum serum (T_{max}) concentration is markedly increased by pegylation. The T_{max} for unmodified interferon-alpha varies from 7.3 to 12 hours versus approximately 45 to 81 and 16 to 44 hours for PEG-IFN-alpha-2a and -alpha-2b in healthy adults and patients with compensated chronic HCV infection (Algranati et al., 1999; Glue et al., 2000; Wills, 1990; Wills et al., 1984; Chatelut et al., 1999; Martin et al., 2000, 2000a; Modi et al., 2000). Near-dose proportional maximum serum concentrations (C_{max}) of PEG-IFN-alpha-2a are obtained approximately 80 hours after the first dose (Heathcote et al., 1999). Dose-related but not dose-proportional increases between the dose and C_{max} are noted with PEG-IFN-alpha-2b (Glue et al., 2000, 2000a).

Absorption of PEG-IFN-alpha-2a has been shown to be particularly delayed in elderly males (mean T_{max} = 116 hours) versus a mean of 81 hours in young males (Martin et al., 2000). Age does not affect the absorption of PEG-IFN-alpha-2b (Gupta et al., 2003). For children, limited data exist on the pharmacokinetics of PEG-IFN-alpha. Schwarz (2003) presented data on the absorption of PEG-IFN-alpha-2a after subcutaneous administration in 14 treatment-naïve HCV-infected children. The children received PEG-IFN-alpha-2a [(180 µg/1.73 m^2) × patient body surface area] for 48 weeks. Rapid and sustained absorption was noted after the first dose, with mean concentrations 24 and 96 hours after a dose of 22.3 ng/ml and 19.0 ng/ml, respectively. Steady-state concentrations were reached by week 12 of therapy.

The mean apparent volume of distribution (V_d) of unmodified interferon-alpha-2a and PEG-IFN-alpha-2a are markedly different, at 31 to 73 L and approximately 6 to 14 L, respectively (Harris et al., 2001; Lamb & Martin, 2002) (Table 1). In contrast, the mean apparent V_d for unmodified interferon-alpha-2b is approximately 98 L (1.4 liters/kg) versus 69.3 L (0.99 liters/kg) for PEG-IFN-alpha-2b for a 70-kg person (Glue et al., 2000). The volume of distribution of PEG-IFN-alpha-2a is also lower than unmodified and PEG-IFN-alpha-2b. The lower V_d for PEG-IFN-alpha-2a limits the distribution to well-perfused organs, allowing for a fixed-dose regimen (irrespective of body weight). On the other hand, the distribution of PEG-IFN-alpha-2b is closer to that of unmodified interferon.

Unmodified interferon-alpha is eliminated by glomerular filtration with reabsorption occurring in the proximal tubules (Wills, 1990). The liver plays a small part in the elimination of unmodified interferon-alpha. Additional catabolism may occur via interactions with cellular interferon receptors (Glue et al., 2000). In contrast, animal studies demonstrate that PEG-IFN-alpha-2a is metabolized mainly by the liver (Modi et al., 2000a). Non-renal clearance of PEG-IFN-alpha-2b accounts for approximately two-thirds of the total clearance (Gupta et al., 2002).

Unmodified interferon-alpha is rapidly eliminated from the body, with an average half-life between 4 and 16 hours (Glue et al., 2000; Wills, 1990). The mean apparent clearance varies from 4.9 to 21 liters/hour (Table 1) (Wills, 1990). Elimination half-lives for the PEG-IFN products are far longer. The mean half-life for PEG-IFN-alpha-2a varies from 61 to 110 hours, and the mean apparent clearance varies from 60 to 118 mL/hour in subjects with stable renal function. The mean half-life for PEG-IFN-alpha-2b is 30 to 58 hours (Table 1) (Glue et al., 2000, 2000a).

Steady-state concentrations of PEG-IFN-alpha-2a occur after 5–8 weeks of weekly administration (Modi et al., 2000).

Renal dysfunction affects the pharmacokinetics of PEG-IFNs differently (Martin et al., 2000a; Lamb et al., 2001; Gupta et al., 2002). For PEG-IFN-alpha-2a, the C_{max}, distribution, and apparent total body clearance in subjects with stable chronic renal impairment were comparable to patients with normal renal function. Clearance (Cl/F) of PEG-IFN-alpha-2a was 25–45% lower in patients on hemodialysis, however, compared with subjects with normal kidney function (PEGASYS® package insert, 2004; Martin et al., 2000; Lamb et al., 2001). Dosage adjustment is recommended in patients with end-stage kidney disease on hemodialysis (PEGASYS® package insert, 2004).

In contrast, the C_{max} and AUC (area under the concentration-time curve) increased up to two times in patients with decreased renal function receiving PEG-IFN-alpha-2b compared with patients with normal baseline renal function (Gupta et al., 2002). The mean half-life increased and the mean apparent body clearance decreased by 40% and 45%, respectively. Renal clearance accounts for approximately 30% of the total body clearance of PEG-IFN-alpha-2b. Patients with a calculated creatinine clearance less than 50 mL/min should receive PEG-IFN-alpha-2b therapy only after assessing the risks and benefits (PEG-Intron package insert, 2005).

As to removal of PEG-IFNs by hemodialysis, Barril (2004) performed an *in vitro* study comparing the effect of permeability and pore size of hemodialysis membranes on the blood levels of unmodified interferon-alpha-2a versus PEG-IFN-alpha-2a and PEG-IFN-alpha-2b. Pore size was more important than permeability in the removal of interferon products from the blood. Unmodified interferon and PEG-IFN-alpha-2b, but not PEG-IFN-alpha-2a, were cleared appreciably by membranes with middle to middle-large pore size (43–60 Å) high-flux dialysers. Given that PEG-IFN-alpha-2a is not as dependent upon renal clearance for elimination, it is the preferred choice for patients with renal dysfunction. In the case that a patient requires hemodialysis, PEG-IFN-alpha-2b may require supplemental dosing depending on the type of dialyser being used.

Concern about the effect of fixed dosing of PEG-IFN-alpha-2a in patients at the extremes of body weight has been raised. Although the drug has a low volume of distribution, morbidly obese patients may not achieve therapeutic interferon concentrations, while lighter patients may experience an increase in adverse events due to supratherapeutic levels. Obesity has been identified as an independent risk factor for nonresponse to unmodified and PEG-IFNs (Fried et al., 2002; Hadziyannis et al., 2004; Manns et al., 2001). It is unknown whether the decreased efficacy can be attributed to lower interferon levels, intrinsic resistance to the effects of interferon, or the presence of negative prognostic factors in obese patients (Swain et al., 2005). It has been reported that the C_{max} and AUC of unmodified interferon are 35% and 25% lower in obese subjects versus non-obese patients (Lam et al., 1997). Bressler (2005) noted a 21% reduction in trough levels of PEG-IFN-alpha-2a in obese versus non-obese subjects receiving the 180-μg dose. The trough concentrations after PEG-IFN-alpha-2a at 270 μg/week were similar to those noted in the 180-μg/week dose in non-obese subjects. In a single-dose study, the AUC for PEG-IFN-alpha-2a and -2b was not related to body weight. However, the number of obese patients was small (Bruno et al., 2005). The significance of this finding will need to be determined in clinical trials. The manufacturer of PEG-IFN-alpha-2a is undertaking a study to determine the effectiveness of higher doses of PEG-IFN-alpha-2a in patients with HCV infection and who are over 85 kg (ClinicalTrials.gov Identifier: NCT00077649). Swain (2005) reported a higher incidence of serious adverse events in lighter patients ($\leq$75.5 kg) receiving fixed-dose PEG-IFN-alpha-2a, but the number of withdrawals due to adverse events was not increased.

The pharmacokinetics of the two available PEG-IFNs vary. The decision to treat with one product over the other should take into consideration renal function and body size.

Drug Interactions

Drug-drug interactions for the PEG-IFNs are infrequent but are marked by consequence. Additive side effects are the most notable mechanism of interaction. Agents with similar adverse drug event profiles (e.g., hepatotoxins, or drugs that cause anemia) run highest on the list. As the metabolism and elimination of PEG-IFNs are not dependent on extensive oxidative metabolism, through any of the major cytochrome P450 enzyme systems, these agents are not subject to the inductive or inhibitory activity of other drugs. Additionally, neither PEG-IFN product exerts an effect on the majority of P450 enzyme systems, including 2C9, 2C19, 2D6, and 3A4 (Package insert, FDA Briefing Document, 2002). PEG-IFN-alpha-2a demonstrates mild inhibition of cytochrome P450 1A2, the enzyme responsible for metabolism of drugs such as theophylline, risperidone, clozapine, tricyclic antidepressants, and caffeine (FDA Briefing Document, 2002).

Treatment with PEG-IFN-alpha-2a once weekly for four weeks in healthy subjects taking theophylline was associated with a 25% increase in the total theophylline area under the concentration-time curve. The resultant effects were adverse effects of theophylline, including nausea, vomiting, palpitations, and seizures. The effect could be offset by a 25% reduction in the dose of theophylline. PEG-IFN-alpha-2a does not uniformly affect the disposition of methadone metabolism (Sulkowski et al., 2005a). Methadone exposure was increased by a mean of 10–15% in patients co-treated for four weeks. The levels of methadone doubled in two patients. No subject had clinical signs or symptoms of intoxication or withdrawal related to methadone. Mauss (2004) examined the effect of PEG-IFN-alpha-2b on 50 patients on methadone maintenance. While there was a high rate of discontinuation of therapy among this group of patients, there was no increase in adverse effects. RBV does not affect the disposition of PEG-IFN-alpha but contributes significantly to the untoward effects on hematological counts, particularly red blood cells (RBC).

Directly hepatotoxic agents used in combination with PEG-IFN can elicit additive drug toxicity. Concomitant use of the nonnucleoside reverse transcriptase inhibitor nevirapine, a known hepatotoxin, has been associated with a higher incidence of advanced liver fibrosis (Macias et al., 2004). Numerous case reports and retrospective reviews identify didanosine, when used in combination with PEG-IFN products with ribavirin, as a prominent risk factor for cases of fatal hepatic failure, peripheral neuropathy, pancreatitis, and lactic acidosis (Bani-Sadr et al., 2005). Use of ritonavir or ritonavir plus saquinavir in HIV patients infected with the hepatitis C or B virus has also been associated with increased rates of severe hepatotoxicity.

PEG-IFNs, in combination with RBV, may additionally antagonize stavudine and zidovudine antiretroviral activity, secondary to inhibition of intracellular phosphorylation. Use of alternate antiretroviral agents is recommended. For all co-infected patients (HCV and HIV infection), increased vigilance for signs of hepatotoxicity is highly recommended.

Mechanism of Action

PEG-IFNs were developed in order to achieve a more sustained antiviral and immunodulatory effect than unmodified interferon (Kamal et al., 2002). Their mechanism of action is the same as for unmodified interferon. The chapter by Dr. Dash in this manuscript provides extensive insight into the mechanism of interferon action and resistance.

PEG-IFN-alpha combined with RBV is the recommended treatment for the management of patients with chronic HCV infection (National Institutes of Health Consensus Development Conference Statement: Management of Hepatitis C, 2002). When combined with PEG-IFN, RBV significantly improves the possibility of achieving an SVR in patients with chronic HCV infection when compared with an interferon-alpha by itself, therapy with RBV, or PEG-IFN-alpha alone (Fried et al., 2002; Manns et al., 2001). When used as monotherapy in patients with chronic HCV infection, RBV has been shown to improve serum alanine aminotransferase (ALT) concentrations (Bodenheimer et al., 1997; Dusheiko et al., 1996). No significant decrease in HCV RNA concentrations were noted during therapy. On the other hand, Pawlotsky (2004) noted a moderate (-0.5 to -1.6 $\log_{10}$) but transient (days 2 and 3 and disappearing by day 4) decrease in HCV RNA concentrations in 4 of 7 subjects receiving monotherapy with RBV.

The exact mechanism of action of RBV when combined with interferon-alpha therapy is unknown. The chemical structure of RBV is depicted in Figure 2. Several hypotheses have been proposed. RBV may act by inhibiting DNA, RNA, and protein synthesis, leading to a decrease in production of pro-inflammatory cytokines such

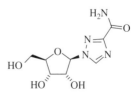

Fig. 2 Ribivirin

as interferon-γ and may induce apoptosis of cells in the inflammatory infiltrate in the infected liver (Meier et al., 2003). Modulation of the T helper 1/Th2 cytokine-mediated immune response with an emphasis on the type 1 cytokine profile is consistent with the results of combined therapy with PEG-IFN-alpha-2a (Kamal et al., 2002; Tam et al., 1999; Hultgren et al., 1998). The mutagenic activity of RBV on HCV is also a possible mechanism (Asahina et al., 2005).

Pharmacodynamics

Administration of unmodified interferon-alpha or PEG-IFN-alpha products results in a multiphase decline of HCV RNA concentrations. A study by Zeuzem (2001) noted a rapid first-phase decline in HCV RNA concentrations in 79% and 76% of patients receiving unmodified interferon and PEG-IFN-alpha-2a, respectively. This response was noted within 48 hours of initiation of interferon therapy. The first phase of viral decline is thought to reflect an interferon-induced blockade of virus production or release, as well as the degradation rate of free virus (Neumann et al., 1998). After about 24 to 48 hours, the viral decline slows and a second phase of HCV RNA decay is noted. The second-phase decline is thought to reflect the degradation rate of infected cells or the elimination of virus from infected hepatocytes (Neumann et al., 1998; Lutchman & Hoofnagle, 2003). The decrease in HCV RNA concentrations was faster in both phases in patients infected with a non-1 HCV genotype (Zeuzem et al., 2001). The degree of second-phase decline of HCV RNA was predictive of SVR (Zeuzem et al., 2001). A similar pattern of HCV RNA decline has been noted in patients who are infected with HCV genotype 1 and receive PEG-IFN-alpha-2b (Buti et al., 2002).

Herrmann (2003) studied the effect of the addition of RBV on the viral kinetics of unmodified interferon-alpha-2b and PEG-IFN-alpha-2a in patients with genotype 1 chronic hepatitis C infection. In addition to the two phases reported by Zeuzem (2001), they noted a third phase of viral decay in 57%, 56%, and 89% of patients receiving unmodified interferon-alpha-2b with RBV and PEG-IFN-alpha-2a without and with RBV, respectively. The third phase of viral decline began at day 7 to day 28 after initiation of therapy. The third-phase kinetic parameters were predictive of an end-of-treatment response (ETR—undetectable HCV RNA concentrations at the end of 48 weeks of therapy) ($P = 0.001$) but not SVR ($P = 0.11$). This third phase was attributed to a treatment-enhanced loss of HCV-infected cells. Pawlotsky (2002) and Neumann (2002) found that a triphasic HCV RNA decay pattern was more common in HCV genotypes 4 and 1 (38%) versus genotypes 2 and 3 (3%). All genotype 1 patients who had a bi-phasic response to PEG-IFN-alpha-2a plus RBV had a rapid viral response (RVR) (a second slope greater than 0.3 IU/ml/week), whereas only 62% of those with a tri-phasic response had an RVR.

Zeuzem (2005) investigated whether individualizing therapy for chronic HCV-infected patients based on viral responses to treatment increased the likelihood of achieving an SVR. An SVR was defined as an undetectable serum HCV RNA (<50

IU/mL) 24 weeks after the end of therapy. Study results indicated that individualized therapy was no better than standard therapy in the management of patients.

Therapeutic Efficacy

Determination of genotype is the first step in the consideration of treatment with either of the PEG-IFNs. This will help determine the length of therapy and the likelihood of treatment success, aiding the patient and provider in the benefit versus risk analysis. Patients infected with genotype 2 or 3 respond better to therapy than subjects infected with genotype 1 or 4 (National Institutes of Health Consensus Development Conference Statement: Management of Hepatitis C, 2002; Sherman et al., 2004)

Monotherapy with PEG-IFN-Alpha

Combination therapy with PEG-IFN and RBV is the therapy of choice for the management of HCV infection. However, in those situations where the patient is intolerant or has a contraindication to the use of RBV, monotherapy with PEG-IFN is indicated. Four randomized controlled trials in interferon-naïve subjects with chronic HCV infection demonstrated the superiority of treatment with PEG-IFN-alpha-2a over unmodified interferon-alpha-2a (Reddy et al., 2001; Heathcote et al., 2000; Zeuzem et al., 2000; Pockros et al., 2004). A sustained biochemical response, end–of-treatment virological response, sustained virological response, and histological response were therapeutic endpoints. A sustained biochemical response was defined as a normal serum ALT at week 72, and a histological response was defined as a $\geq$2-point decrease in the total histological activity index (HAI) in liver biopsy results between baseline and week 72. An end-of-treatment virological response and a sustained virological response were defined as an undetectable plasma HCV RNA after 48 weeks of therapy and an undetectable HCV RNA concentration (<100 copies/mL) at week 72, respectively.

The specific relative rates of response are summarized in Table 2. For all therapeutic endpoints, a statistically significant benefit was seen with treatment with PEG-IFN-alpha-2a over unmodified interferon. The greatest relative benefit for all parameters was noted with treatment at the highest dose, of 180 μg/week. The 135-μg/week dose achieved an identical SVR as the 180-μg/week dosing regimen, but the histological response was inferior and no different from that of unmodified interferon-alpha-2a (Pockros et al., 2004). The histological response correlates to SVR (Reddy et al., 2001; Heathcote et al., 2000; Zeuzem et al., 2000; Cammà et al., 2004). A histological response, however, was found in some patients who failed to respond to therapy or who relapsed after an end-of-treatment response (Heathcote et al., 2000). These data indicate that therapy with PEG-IFN can have a beneficial effect even without an SVR.

Table 2 Monotherapy with Unmodified Versus Peg-Interferon-Alpha in Patients with Chronic HCV Infection

Medication	Sustained Biochemical Response (%)	P value	End of Treatment Virologic Response (%)	P value	Sustained Virologic Response (%)	P value
Unmodified interferon alpha-2a[a,b]	9	P = 0.004	12	P = 0.0002	3	P = 0.0006
Peginterferon alpha-2a 180 μg/week[c]	38		60		36	
Unmodified interferon alpha-2a[b,c,d]	15	P = 0.004	14	P = 0.0010	8	P = 0.0010
Peginterferon alpha-2a 180 μg/week[c]	34		44		30	
Unmodified interferon alpha-2a[c,e,f]	25	P = 0.001	28	P = 0.0010	19	P = 0.0010
Peg-interferon alpha-2a 180 μg/week[c]	45		69		39	
Unmodified interferon alpha-2a[b,c,g]	18	P = 0.001	22	P = 0.0010	11	P = 0.0010
Peg-interferon alpha-2a 135 μg/week[c]	32		53		28	
Peg-interferon alpha-2a 180 μg/week[c]	32		55		28	
Unmodified interferon alpha-2b[b,h]	NA	NA	24	P < 0.0010	12	P < 0.0010
Peg-interferon alpha-2b 1.0 μg/kg/week[c]	NA		41		25	
Peg-interferon alpha-2b 1.5 μg/kg/week[c]	NA		49		23	

[a] Reddy et al. (2001);
[b] 3 million units three times weekly for 48 weeks;
[c] 48 weeks of therapy;
[d] Heathcote et al. (2000);
[e] 6 million units for 12 weeks then 3 million units for 36 weeks;
[f] Zeuzem et al. (2000);
[g] Pockros et al. (2004);
[h] Lindsay et al. (2001).

The likelihood of achieving an SVR was lower in patients infected with HCV genotype 1 when compared to non-1 genotypes. The chance of achieving an SVR in patients infected with HCV genotype 1 varied from 12–31% (180 µg PEG-IFN-alpha-2a weekly for 48 weeks) versus unmodified interferon-alpha-2a (2–7%) (Reddy et al., 2001; Heathcote et al., 2000; Pockros et al., 2004; Zeuzem et al., 2000). In patients infected with HCV non-1 or unknown genotype, the response was approximately 41–51% in the 180-µg/week PEG-IFN-alpha-2a groups versus 0–19% for unmodified interferon-alpha-2a (Reddy et al., 2001; Heathcote et al., 2000; Pockros et al., 2004).

Lindsay (2001) performed a randomized, double-blind study (for PEG-IFN-alpha-2b doses) comparing PEG-IFN-alpha-2b with unmodified interferon-alpha-2b in interferon-naïve patients with chronic HCV infection (Table 2). An end-of-treatment virological response was recorded in 24%, 41%, and 49% of patients receiving unmodified interferon, PEG-IFN-alpha-2b (1.0 µg/kg/week), and PEG-IFN-alpha-2b (1.5 µg/kg/week), respectively. The end-of-treatment virological response was statistically greater for all of the PEG-IFN dosing groups when compared with unmodified interferon ($P < 0.001$). The SVR for the 1.0-µg/kg/week (25%) and 1.5-µg/kg/week (23%) doses of PEG-IFN-alpha-2b were identical and significantly different from unmodified interferon-alpha-2b (12%) ($P < 0.001$ for both PEG-IFN doses).

Combination Therapy with PEG-IFNAlpha-2a and RBV

Major randomized controlled trials have been used to develop consensus guidelines for the use of PEG-IFN combined with RBV for the management of patients with chronic HCV infection (Manns et al., 2001; Fried et al, 2002; Hadziyannis et al., 2004). Like monotherapy, the treatment duration and the likelihood of achieving an SVR are determined by HCV genotype. In this section, we will review the available information concerning the treatment of chronic HCV genotypes 1–6.

Overall SVR and SVR by genotype results are listed in Table 3. Combination therapy with PEG-IFN plus RBV generally exceeds monotherapy and combination therapy with unmodified IFN in terms of the SVR rates ($P = 0.01$, Manns et al., 2001; $P < 0.001$, Fried et al., 2002). The overall SVR in patients receiving PEG-IFN-alpha plus RBV for 48 weeks varied in these trials from 54–63%. The dose of RBV studied was from 800 mg daily in the Manns study (2001) to between 1–1.2 g daily based on body weight in other major studies (Fried et al., 2002; Hadziyannis et al., 2004).

Manns (2001) suggest that higher doses of RBV can maximize response. In their randomized, open-label trial, they noted that the likelihood of achieving an SVR was higher for all genotype groups in patients receiving higher RBV doses, between 10.6–15 mg/kg per day. From this finding and the desire to determine if the results from controlled randomized studies could be replicated in a community setting, the WIN-R study (Jacobson et al., 2005) was designed. This study was a U.S. community-based trial enrolling 4,913 patients. The relative benefit of combining 1.5 µg/kg/week of PEG-IFN-alpha-2b with either a fixed dose (FD 800 mg/day) or

Table 3 Efficacy of (Combination Therapy) PEG-IFN Plus RBV in Subjects with Chronic HCV Infection

Regimen	SVR Overall (%)	SVR Genotype 1(%)	SVR Genotype 2 or 3(%)
Interferon alpha-2b plus RBV (1–1.2 g/day) (48 weeks)[a]	47	33	79
PEG-IFN alpha-2b 1.5 µg/kg/week plus RBV 800 mg daily(48 weeks)	54	42	82
PEG-IFN alpha-2b1.5 µg/kg/week then 0.5 µg/kg/week plus RBV 1–2 g daily (48 weeks)[b]	47	34	80
Interferon alpha-2b plus RBV 1–1.2 g/day (48 weeks)[c]	44	36	61
PEG-IFN alpha-2a 180 µg/week plus placebo (48 weeks)	29	21	45
PEG-IFN alpha-2a 180 µg/week plus RBV 1–1.2 g/day (48 weeks)	56	46	76
PEG-IFN alpha-2a 180 µg/week plus RBV 800 mg/d (24 weeks)[d]	NA	29	84
PEG-IFN alpha-2a 180 µg/week plus RBV 1–1.2 g/d (24 weeks)	NA	41	81
PEG-IFN alpha-2a 180 µg/week plus RBV 800 mg/d (48 weeks)	NA	40	79
PEG-IFN alpha-2a 180 µg/week plus RBV 1–1.2 g/d (48 weeks)	63	52	76

[a] Manns et al. (2001);
[b] 1.5 µg/kg/week PEG-IFN alpha-2b for 4 weeks followed by 0.5 µg/kg/week for 44 weeks;
[c] Fried et al. (2002);
[d] Hadziyannis et al. (2004), PEGASYS®; Hoffmann-La Roche, Basel, Switzerland,2004.

a weight-based dose (WBD 800–1400 mg/day) of RBV was examined. Treatment-naïve patients infected with HCV genotype 1 received 48 weeks of therapy, and patients with HCV genotype 2 or 3 received either 24 or 48 weeks of therapy. The overall SVR in the WBD and FD groups was 44% and 41%, respectively ($P = 0.02$). These results were lower than those reported by Manns (2001) (54%) utilizing a similar protocol with fixed-dose RBV. This may have been a reflection of the greater need for dose reductions of RBV in the WBD group due to anemia. Afdhal (2006) investigated the influence of RBV dose and the presence of liver fibrosis and cirrhosis on SVR in the WIN-R trial. With advanced histological changes, specifically patients classified at Metavir stage 3 to 4, SVR rates were significantly better for patients receiving the weight-based dose versus the fixed dose of RBV (43% versus 37%, $P = 0.02$). Patients with cirrhosis had the lowest overall SVR (34%) when compared with subjects with stages 0–3 (44–46%) ($P < 0.0001$).

Genotype 1

When analyzing the response in HCV genotype 1 infected patients receiving combination therapy with PEG-IFN-alpha plus RBV for 48 weeks, response rates varied from 42–52% (Manns et al., 2001; Fried et al., 2002; Hadziyannis et al., 2004).

Combination PEG-IFN plus RBV regimens are significantly better than unmodified interferon/RBV combinations ($P = 0.02$, Manns et al., 2001; $P = 0.01$, Fried et al., 2002). An even higher rate of response was noted by Sakai (2006), at 61% in treatment-naïve Japanese patients infected with HCV genotype 1b. Post hoc subgroup analyses, by Fried (2002) and Hadziyannis (2004) and Zarski (2005), demonstrate a significantly greater chance of achieving an SVR in patients infected with HCV genotype 1b (53%) versus 1a (45%) ($P = 0.004$).

Even lower rates of treatment success were reported in the community-based trial, WIN-R, wherein SVRs in patients infected with HCV genotype 1 were 34% (WBD) and 29% (FD) ($P = 0.004$) (Jacobson et al., 2005). Subjects with a low viral load (< 2 million copies/mL) had a greater likelihood of experiencing an SVR than patients with a high viral load ($\geq$2 million copies/mL), 35% versus 31%, $P = 0.006$ (Jacobson et al., 2006).

Hadziyannis (2004) studied the effect of treatment duration and RBV dose on the likelihood of achieving an SVR in HCV genotype 1 infected patients. They found a significant correlation with SVR in patients treated for 48 weeks versus those treated for 24 weeks ($P < 0.001$). They also reported a significant effect of weight-based RBV versus low-dose RBV (800 mg daily) on the achievement of an SVR ($P = 0.005$). Subjects with bridging fibrosis or cirrhosis responded best to weight-based RBV and 48 weeks of therapy.

Genotype 2 or 3

The SVR for patients infected with genotype 2 or 3 varied between 76–82% in patients receiving PEG-IFN-alpha/RBV for 48 weeks (Table 3) (Manns et al., 2001; Fried et al., 2002; PEGASYS®; Hoffmann-La Roche, Basel, Switzerland, 2004). Hadziyannis (2004) found no difference in the achievement of an SVR in patients infected with genotype 2 or 3 based on length of therapy or RBV dose ($p > 0.2$ for both). In fact, the 24-week regimen (800-mg dose of RBV) had the highest SVR (84%). Manns (2001) reported no differences in SVR between PEG-IFN-alpha-2b/RBV study groups and unmodified interferon/RBV in patients infected with genotype 2 or 3 treated for 48 weeks.

In the community-based WIN-R study, fixed-dose RBV (800 mg/day) was as equally effective as WBD RBV (800–1400 mg daily) in inducing an SVR in patients with chronic HCV genotype 2 or 3 infection (60% versus 62%, $P = 0.26$) (Jacobson et al., 2005). Patients with genotype 2 had a higher SVR than those with genotype 3 (72% versus 63%) ([Brown et al., 2006). The authors suggested that patients infected with HCV genotype 3 may benefit from WBD of RBV.

Zeuzem (2004) performed a phase 4 study to investigate the safety and efficacy of treatment for 24 weeks with 1.5 μg/kg/week of PEG-IFN-alpha-2b with weight-based RBV (800–1,400mg/day) in patients infected with HCV genotype 2 or 3. The overall SVR was 81%, and the SVR for HCV-2 and HCV-3 was 93% and 79%, respectively. The reduced SVR in patients infected with HCV genotype 3 was attributed to the presence of a high baseline viral load and high-grade hepatic steatosis. The authors raised the question of whether patients infected with HCV-3 with poor prognostic factors should be treated for longer than 24 weeks. Controlled studies will be necessary to answer this question.

Genotype 4

Hepatitis C virus genotype 4 is more common in Africa and eastern Mediterranean countries and has a prevalence of 60% in Saudi Arabia and up to 90% in Egypt. Fried (2001) and Hadziyannis (2004) noted SVR rates in HCV genotype 4 infected patients receiving PEG-IFN-alpha-2a plus weight-based RBV for 48 weeks at 77–82%. The study size was small, however. Khuroo (2004) performed a meta-analysis of available trials of PEG-IFN-alpha plus RBV for the management of chronic HCV genotype 4 infection. They reported an overall predicted SVR rate of 72% in patients treated for 48 weeks with PEG-IFN-alpha (2a: 180 μg/week; or 2b: 1.5 μg/kg/week) and RBV (1–1.2 g/day based on body weight). Trials utilizing these guidelines have reported SVR values between 61–69% (Elmakhzangy et al., 2005; Hasan et al., 2004; Kamal et al., 2005). Kamal (2005) investigated the effect of the duration of therapy on the achievement of an SVR in patients infected with HCV genotype 4. Patients were randomized to receive PEG-IFN-alpha-2b (1.5 μg/kg/weeky) with RBV (1–1.2 g/day) for 24, 36, or 48 weeks. An SVR occurred in 24%, 66%, and 69% of patients treated for 24, 36, and 48 weeks, respectively. No statistical difference was found between 36 or 48 weeks of therapy ($P = 0.3$). The 36- and 48-week regimens were statistically different when compared to the 24-week arm of the study ($p < 0.001$, for both). An early virological response at 12 weeks was predictive of an SVR as long as therapy was at least 36 weeks in duration. The authors concluded, however, that patients with high baseline HCV RNA viral loads may do better with 48 weeks of therapy.

Genotype 5 or 6

Much less data exist on the management of patients infected with chronic HCV genotype 5 or 6. In a retrospective review of patients with HCV genotype 5 infection, D'Heygere (2005) reported an SVR of 55%. Patients were treated for 48 weeks with 180 μg/week of PEG-IFN-alpha-2a and RBV (1–1.2 g/day). Bonny (2005) reported an SVR similar to that of unmodified interferon (64%), in treatment-naïve patients infected with HCV genotype 5, receiving PEG-IFN-alpha-2b or PEG-IFN-alpha-2a plus RBV (60%). Both regimens were administered for 48 weeks. Nguyen (2003) reported an SVR of 54% in patients treated with peginterferon/ribavirin for 24 weeks. The SVR rate in two small studies of HCV genotype 6 infected individuals utilizing unmodified interferon/ribavirin were 79% and 62.5%, respectively (Dev et al., 2002; Hui et al., 2003). Therapy duration was 52 and 48 weeks, respectively. A study to determine the ideal length of therapy in patients infected with

HCV genotype 6 receiving peginterferon-alpha-2b/ribavirin therapy is under way (ClinicalTrials.gov Identifier: NCT00255008).

Comparative Efficacy Between Pegylated Products

To date, there have been no large head-to-head trials evaluating the safety and/or efficacy between the two products. As of preparation of this chapter, the manufacturer of PEG-IFN-alpha-2b (Schering-Plough) is funding a head-to-head safety and efficacy trial of the two PEG-IFN products entitled the "Individualized Dosing Efficacy Versus Flat Dosing to Assess Optimal Pegylated Therapy (IDEAL) Trial." Treatment-naïve patients infected with HCV genotype 1 are being randomized to receive PEG-IFN-alpha-2b at a dose of 1.0 or 1.5 µg/kg/week plus weight-based RBV (800–1,400 mg/day) or 180 µg/week PEG-IFN-alpha-2a with RBV (1–1.2 g/day) for a total of 48 weeks. The primary endpoint of the study will be the SVR rate. One concern about the design of the study is the way the protocol deals with alterations in the RBV dose. Patients in the PEG-IFN-alpha-2b arms will have a two-step dose reduction process. If necessary, the dose of RBV will be decreased to between 600 and 1,000 mg depending on the starting dose. If necessary, the dose may be reduced by another 200 mg. In contrast, for the PEG-IFN-alpha-2a group, the dose is reduced to 600 mg/day. Since it is known that the RBV dose is extremely important to the achievement of an SVR, the study design could reflect a bias in favor of PEG-IFN-alpha-2b.

Importance of Completing the Therapeutic Regimen

Ferenci (2005) investigated the importance of completing the prescribed regimen with PEG-IFN-alpha-2a and RBV on the likelihood of achieving an SVR in patients infected with HCV genotype 1. Patients who received $\geq 80\%$ of the prescribed RBV dose experienced a greater chance of having an SVR than those who received <80%. The SVR rate was also lower in patients who received $\geq 80\%$ of the prescribed PEG-IFN-alpha dose but less than 80% of the RBV dose when compared with subjects who completed $\geq 80\%$ of the RBV dose (48% vs. 69%, $P = 0.014$). Patients who completed $\geq 80\%$ of both PEG-IFN and ribavirin had the lowest SVR (29%). Ribavirin dose changes after 24 weeks in patients infected with HCV genotype 2 or 3 had no effect on SVR rates. Shiffman (2004) found similar results in a retreatment study in patients who had failed previous unmodified interferon-based therapy. In this trial, the reduction of the PEG-IFN-alpha-2a dose during the first 20 weeks of therapy did not greatly affect SVR. However, the reduction of the RBV dose from >80% to <60% of the target dose during the first 20 weeks did significantly reduce the SVR rate (21% vs. 11%, $P = 0.03$). Decreases in dosages of either medication had little effect on SVR after 20 weeks of therapy.

respectively). In patients with HCV genotype 1 infection, the SVR was 28% (Group A) and 44% (Group B) ($P = 0.003$). As in the study by Berg (2006), the dropout rate was higher in the 72-week versus the 48-week group. Both studies by Berg (2006) and Sanchez-Tapias (2006) indicated that it is difficult for patients to complete 72 weeks of therapy. . The RBV dose used in these studies was low when compared to current standards, and this may have influenced response to therapy. More study will be necessary to determine the value of prolonging PEG-IFN-alpha therapy utilizing weight-based ribavirin (i.e., 1–1.2 g/daily) in HCV-infected patients with a slow virological response to therapy.

Use of Rapid Virological Response at 4 Weeks to Determine Shorter Duration of Therapy n Select Groups of Patients with Chronic HCV Infection

Investigations into the possibility of treating select patients for a shorter period of time have been undertaken. Shorter duration of therapy with equal efficacy has the advantage of increased cost efficiency and potentially reduced adverse drug events. Zeuzem (2006) performed a study to investigate whether 24 weeks of PEG-IFN-alpha-2b plus RBV (weight-based) would be as effective as 48 weeks (historical control group) in patients infected with chronic HCV genotype 1 and low baseline HCV RNA concentrations ($\leq$600,000 IU/mL). When SVR was compared, the 24-week regimen (50%) was inferior to the 48-week regimen (69%). Subgroup analysis, however, showed that those subjects with undetectable serum HCV RNA after 4 weeks of therapy (rapid virological response, RVR) had an SVR rate of 89%. If the serum HCV RNA was undetectable after both 4 and 24 weeks of therapy, the SVR was 92%. Liver biopsy revealed the presence of mild inflammation and an average Knodell fibrosis score of 1.2. Adverse events that led to the discontinuation of therapy or dose reduction were less in the 24-week study compared to rates during the initial 24 weeks and 48 weeks in the historical control group.

Jensen (2006) performed a retrospective analysis reviewing the association between an RVR at 4 weeks and SVR in patients receiving PEG-IFN-alpha-2a with RBV, infected with HCV genotype 1. They utilized data from the Hadziyannis study (2004). The primary focus of the analysis was on those patients who achieved an SVR after 24 weeks of therapy and had an RVR. Of the patients treated for 24 weeks, 24% achieved an RVR. Eighty-nine percent of patients with an RVR at 4 weeks achieved an SVR after 24 weeks of therapy versus 19% without an RVR. The authors compared the length of therapy and noted no difference in SVR in patients with an RVR at 4 weeks treated for 24 or 48 weeks. Multiple logistic analyses revealed an RVR at 4 weeks and baseline HCV RNA concentration (<2,000,000 IU/mL versus > 600,000 IU/mL, $P < 0.026$) correlated with an SVR. It is important to note that only 7% and 18% of subjects in this study had cirrhosis or bridging fibrosis on liver biopsy, respectively. Whether these results can be extrapolated to patients with more severe liver disease is unknown. Ferenci (2006) reported a 75%

SVR in patients (HCV genotype 1 or 4) treated with PEG-IFN-alpha-2a and RBV for 24 weeks and an RVR at 4 weeks of therapy. In this prospective study, patients with an SVR had lower baseline HCV RNA concentrations than those who relapsed after an end-of-therapy response.

The European Medicines Agency (EMEA) has approved short-course therapy (24 weeks) of combined PEG-IFN-alpha-2b (1.5 µg/kg/week) plus RBV (800–1,200 mg/daily) in patients with chronic HCV genotype infection with low viral HCV load. Patients must also have an undetectable serum HCV RNA at 4 and 24 weeks of therapy.

Trials to investigate the efficacy of a shorter duration of therapy for the patients with HCV genotype 2 or 3 who experience an RVR after 4 weeks of PEG-IFN-alpha-2b or -2a combined with RBV have been performed. In the study by Mangia (2005), patients received 1.0 µg/kg/week PEG-IFN-alpha-2b plus RBV 1–1.2 g daily via a randomization to one of two groups. Group 1 received therapy for 24 weeks. Patients who experienced an RVR at 4 weeks (Group 2) received therapy for 12 weeks, and subjects who did not have an RVR were treated for 24 weeks. Overall, patients in Group 1 and Group 2 that had an RVR had an SVR rate of 91% and 85%, respectively. Patients with an RVR infected with genotype 2 had an SVR rate of 89% and 87% in Groups 1 and 2, respectively. Although the SVR rate in patients infected with HCV genotype 3 who achieved an RVR was not statistically different, the response was 100% in Group 1 ($n = 10$) and 77% in Group 2 ($n = 24$).

Von Wagner (2005) performed a similarly designed study in patients infected with chronic HCV genotype 2 or 3 who received PEG-IFN-alpha-2a (180 µg/week) plus RBV (800–1,200 mg daily). Patients were treated for 4 weeks, and those subjects with an undetectable HCV RNA (RVR) were allocated to receive a total of 16 (Group A) or 24 weeks (Group B) of therapy. Subjects with detectable HCV RNA concentrations at 4 weeks received 24 weeks of therapy (Group C). An SVR was achieved in 82% and 80% of patients in Groups A and B, respectively. Interestingly, subjects without an RVR (Group C) had an SVR of 36% versus 81% in Group B even though both groups received therapy for 24 weeks. As in the study above, the SVR for subjects infected with HCV genotype 2 was greater than in those with HCV genotype 3, 92% versus 73%. Baseline viral load did not affect the SVR rate for patients with HCV genotype 2. However, a baseline viral load >800,000 IU/mL significantly affected the SVR rate in patients infected with genotype 3 compared with lower viral loads ($P = 0.003$). Patients infected with HCV genotype 3 and a baseline viral load >800,000 IU/mL experienced a higher SVR rate when treated for 24 weeks than 16 weeks, although the difference was not statistically different ($P > 0.2$). These results strongly favor a shorter duration of therapy in patients infected with HCV genotype 2 who experience an RVR to therapy. The results in patients infected with HCV genotype 3 raise the question of whether 24 weeks or longer of therapy may be necessary to assure maximum SVR rates. This may be especially true in subjects with high baseline HCV RNA viral loads. The Genotype 3 Extended Treatment for Hepatitis C (GET-C) Study is designed to determine the efficacy of 24 or 48 weeks of therapy with PEG-IFN-alpha-2b plus RBV in patients infected with HCV genotype 3 and high baseline viral loads (ClinicalTrials.gov identifier NCT00255034).

Health-Related Quality-of-Life and Cost-Effectiveness Analysis of Peginterferon-Alpha with or Without Ribavirin

Current treatment regimens for the management of chronic HCV infection can affect quality of life and are costly. Rasenack and colleagues (2003) have compared the effect of monotherapy with PEG-IFN-alpha-2a or unmodified interferon on the health-related quality of life in treatment-naïve subjects with chronic HCV infection. The Fatigue Severity Scale (FSS), including the visual analogue scale (VAS), was used to measure fatigue severity. Health-related quality of life measures were assessed using the Short-Form health survey (SF-36). Therapy duration in both groups was 48 weeks. The PEG-IFN-alpha-2a patients had better mean FSS total scores and FSS VAS scores during the first 24 weeks and at week 48, respectively, when compared with subjects receiving unmodified interferon. The SF-36 domain scores were better at week 12 but not at week 24 or 48 in patients in the PEG-IFN-alpha-2a versus the unmodified interferon group. Overall, subjects receiving PEG-IFN-alpha-2a reported less fatigue and less body pain and performed better on a daily basis than patients receiving unmodified interferon. The mean FSS and SF-36 scores were improved in patients who achieved an SVR versus those subjects who did not experience an SVR.

The question as to whether monotherapy with PEG-IFN-alpha is cost-beneficial has been studied in a cost-benefit analysis of treatment-naïve subjects with chronic HCV infection. Outcome data were reviewed from published studies of PEG-IFN-alpha monotherapy (Shepherd et al., 2004). The authors concluded that the use of peginterferon-alpha was overall cost–effective, with an increase in this benefit in subjects infected with genotypes 2 and 3.

Hassanein (2004) performed a health-related quality-of-life analysis utilizing data from a large multinational study of patients with chronic HCV infection. Patients were treated with PEG-IFN-alpha-2a with or without RBV or unmodified interferon-alpha-2b plus RBV. The Fatigue Severity Scale (FSS) and the SF-36 Health Survey were used to assess the effect of therapy on health-related quality of life. Patients receiving PEG-IFN-alpha-2a plus placebo fared better on health-related quality-of-life measures (SF-36 and FSS) versus subjects in the PEG-IFN-alpha-2a/RBV and unmodified interferon/RBV groups. Patients receiving PEG-IFN-alpha-2a/RBV reported better health-related quality-of-life measures on both SF-36 and FSS surveys over the 48 weeks of therapy than those receiving unmodified interferon/RBV. Significant differences were reported in vitality, social functioning, body pain, and total fatigue and fatigue severity. Improvements in health-related quality-of-life scores in favor of the PEG-IFN/RBV group were observed as early as week 2 of therapy. Patients who achieved an SVR had improved quality of life, with the greatest improvement over baseline in role-emotional, vitality, general health, and role limitation physical domains in the SFG-36 survey. Fatigue severity decreased significantly in subjects experiencing an SVR versus those who did not. Studies investigating the cost-effectiveness of treatment of chronic HCV infection utilizing PEG-IFN-alpha-2b/RBV compared with unmodified interferon/RBV have been published (Bernfort et al., 2006; Buti et al., 2005; Siebert et al., 2003).

Although the studies varied in design, the PEG-IFN-alpha-2b/RBV regimens were cost-effective when compared with unmodified interferon/RBV regimens in the Swedish, Spanish, and German healthcare systems.

Sullivan (2006) analyzed the cost-effectiveness of the combination of PEG-IFN-alpha-2a/RBV compared with unmodified interferon-alpha-2b/RBV. The authors concluded that the combination of PEG-IFN-alpha-2a plus RBV prolongs life expectancy and saves medical costs when compared with unmodified interferon-alpha-2b with RBV in the U.S. healthcare system. Hornberger (2005) reported similar findings with PEG-IFN-alpha-2a/RBV compared with no therapy in subjects with mild chronic HCV infection. The authors utilized United Kingdom treatment patterns in the design of their study.

Improvement in the quality of life, especially early in therapy, with PEG-IFN with or without RBV may reduce the need to alter doses based on subjective findings, encourage patients to complete the duration of therapy, and increase the likelihood of achieving an SVR. Therapy of chronic HCV infection with PEG-IFN with or without RBV is cost-effective when compared with unmodified interferon therapy in a variety of healthcare systems.

Special Patient Populations

Acute HCV Infection

Another avenue to decrease the impact of prolonged HCV infection and its complications is to treat patients upon presentation with acute HCV infection. Acute HCV infection may be difficult to identify and is often asymptomatic. Infection is acquired through intravenous drug abuse, needle-stick injury in healthcare workers, medical procedures utilizing inappropriately sterilized equipment, and sexual activity. Spontaneous viral clearance may occur in 14–46% of patients and may occur within 8 to 14 weeks of the onset of disease (Seeff, 2002; Kamal et al., 2006). The optimal timing to initiate therapy and the length of treatment have not been definitively determined. Kamal and colleagues (2006, 2006a) have investigated these two issues in randomized controlled clinical trials. In the first study, Kamal (2006) enrolled 129 patients with acute HCV infection to determine the optimal time to initiate therapy. Patients were divided into three groups and therapy was begun 8 weeks, 12 weeks, and 20 weeks after onset of disease. Patients in each group received PEG-IFN-alpha-2b (1.5 µg/kg/week) for a total of 12 weeks. Patients who did not want therapy were followed as a control group. Approximately 30% of subjects in the control group had a spontaneous clearance of HCV. The overall SVR was 88%. The SVR rates for the weeks of initiation of therapy were 95% (week 8), 93% (week 12), and 76% (week 20). The overall SVR rates by genotype were 72% (genotype 1), 100% (genotype 2), 93% (genotype 3), and 84% (genotype 4). Predictors of an SVR were non-1 genotype and a low baseline HCV RNA viral load. Kamal (2006a) performed a study to determine the optimal duration of therapy for the management

are at risk for disease progression. Zeuzem (2004a) reported the results of an international, multicenter, randomized trial of the efficacy of PEG-IFN-alpha-2a (180 μg/week) plus RBV (800 mg daily) in chronic HCV-infected patients with PNALT levels. Patients were treated for 24 weeks (Group A) or 48 weeks (Group B). Patients with cirrhosis, other liver diseases, or co-infection with the human immunodeficiency virus (HIV) were excluded. An SVR was achieved in 30% (Group A) and 52% (Group B) of patients ($P < 0.001$). For HCV genotype 1 infected patients, an SVR occurred in 13% (Group A) and 40% (Group B) ($P < 0.001$). Patients infected with genotype 2 or 3 achieved an SVR 72% (Group A) and 78% (Group B) of the time ($P = 0.45$). Thirteen percent and 56% of subjects infected with HCV genotype 4 achieved an SVR in Group A and B, respectively. Treatment duration (48 weeks) and low baseline HCV load correlated with the likelihood of achieving an SVR in patients with genotype 1 or 4 but not for patients infected with genotype 2 or 3. Patients younger than 40 years of age had a higher chance of achieving an SVR than subjects over 40 years of age. Arora (2006) assessed the effect of achieving an SVR on overall quality-of-life measures in subjects with chronic HCV infection and PNALT. Patients who experienced an SVR had improved quality of life and lower fatigue than patients who did not respond to therapy.

Zehnter (2006) compared the value of PEG-IFN-alpha-2a (180 μg/week) combined with RBV (1–1.2 g/day) in 911 patients infected with HCV genotype 1 and elevated ALT concentrations and 265 subjects with PNALT levels. Patients received therapy for 48 weeks with a 24-week followup off therapy. Patient groups were well matched; however, subjects with PNALT levels tended to be younger. The SVR rates were 74% and 51% in subjects with PNALT levels versus those with elevated ALT concentrations, respectively. The authors surmised that the higher SVR in the patients with PNALT levels could be attributed to the younger age in this group. Overall, patients with PNALT concentrations should be considered for treatment based on factors similar to those in patients with elevated ALT levels. Therapy choices appear to be the same as in patients with chronic HCV infection and elevated ALT concentrations.

Nonresponders and Relapsers to Unmodified Interferon with or Without Ribavirin

Patients who do not achieve an SVR to interferon with or without RBV are designated as nonresponders. Null responders are nonresponders who fail to achieve less than a 2 $\log_{10}$ decline in HCV RNA concentrations from baseline 12 weeks after therapy initiation. These patients rarely respond to additional therapy, and the most appropriate therapy for these patients is best determined in controlled trials (DiBisceglie et al., 2006). Evidence from Tang (2005) that rapid clearance of HCV may be important in enhancing host response to the virus may provide clues to the design of new treatment combinations for these difficult-to-treat patients. Although patients who relapse are not technically nonresponders, they will be discussed here.

The lead-in component of the HALT-C trial is investigating the value of retreatment of patients with chronic HCV infection and advanced fibrosis or cirrhosis who failed therapy with unmodified interferon with or without ribavirin (Shiffman et al., 2004). Subjects received PEG-IFN-alpha-2a at 180 μg/week with RBV (1–1.2 g/day). Patients with undetectable HCV RNA levels at week 20 were treated for a total of 48 weeks and followed for SVR at 24 weeks' post-therapy. Patients with detectable HCV RNA concentrations at 20 weeks were entered into the maintenance phase of the study. Although 32% of subjects were HCV RNA negative at week 48, the relapse rate was high and only 18% achieved an SVR. Factors associated with an SVR were absence of cirrhosis, prior treatment with interferon monotherapy, genotype 2 or 3, HCV RNA concentrations <1.5 million IU/mL, and an AST:ALT ratio <1.0.

Diago (2006) performed an induction dose study with PEG-IFN-alpha-2a plus RBV in HCV genotype 1 infected patients who had not responded to ≥ 22 weeks of unmodified interferon/RBV. No patients had cirrhosis. Patients were randomized to 180 μg/week, 270 μg/week, or 360 μg/week of PEG-IFN-alpha-2a with RBV (1–1.2 g/day) for 12 weeks followed by PEG-IFN at 180 μg/week with the same RBV dose for an additional 36 weeks. An SVR was achieved in 18% (180 μg/week), 30% (270 μg/week), and 38% (360 μg/week) in the respective dosing regimen arms. Izumi (2006) investigated the effect of PEG-IFN-alpha-2a (180 μg/week) plus RBV (600–1,000 mg/day) in Japanese patients with chronic HCV infection who had not responded to prior unmodified interferon monotherapy. A much higher rate of SVR was noted in this population than was found in the HALT-C trial or by Diago (2006). Forty-eight percent of patients had an SVR overall, and 50% of subjects infected with HCV genotype 1b experienced an SVR after 48 weeks of therapy.

Similar retreatment studies in subjects who did not respond to unmodified interferon with or without RBV have been performed by Taliani (2006), Jacobson (2005a), and Poynard (2005) utilizing PEG-IFN-alpha-2b with RBV. Although study population, trial design, and dosing regimens were different, the SVR rates are similar to those found with PEG-IFN-alpha-2a (180 μg/week/RBV) (15–21%). Gaglio (2005) found that fixed-dose RBV was equally as effective as weight-based dosing in patients who failed prior unmodified interferon-based therapy. Jacobson (2005a) noted an SVR of only 8% in subjects who were nonresponders to unmodified interferon/RBV therapy versus an SVR rate of 21% in subjects with no response to unmodified interferon monotherapy. Factors associated with an SVR in these studies were low baseline HCV RNA levels, low γ-glutamyltransferase (γ-GT), weight >75 kg, genotype non-1, and nonresponse to interferon monotherapy.

The "RENEW" study was designed to compare the benefits of 1.5 μg/kg/week versus 3.0 μg/kg/week of PEG-IFN-alpha-2b with RBV (800–1,400 mg daily) in the management of patients who had failed previous HCV therapy with unmodified interferon/RBV (Gross et al., 2005). Ninety-one percent of subjects had HCV genotype 1, 40% had a Metavir score 3 or 4 (F3/4), and 16% were African American (AA). Treatment duration was 48 weeks with a 24-week followup. An SVR occurred in 12% and 17% in the 1.5- and the 3.0-μg/kg/week dosing groups, respectively ($P = 0.03$). Overall, patients with F3/4 and African Americans had lower SVR rates; however, subjects in these groups receiving the higher dose of

PEG-IFN achieved comparable SVR rates to the other study members. The SVR rate for the 1.5- and 3.0-μg/kg/week PEG-IFN-alpha-2b dosing regimens for AA subjects were 2.0% and 14%, respectively (Gross et al., 2005a). Therapy discontinuations and dose changes were comparable between the two dosing groups. Despite differences in study design, patient population, and dosing, the SVR rates with both PEG-IFN products in nonresponders to unmodified interferon-based regimens were remarkably similar (15–21%). Induction dosing with PEG-IFN-alpha-2a appeared to improve the chance of achieving an SVR, but further studies with larger numbers of patients will be necessary to validate these findings.

The likelihood of achieving an SVR is greater in patients who relapse after unmodified interferon monotherapy or combination therapy with RBV. Relapse is defined as a detectable HCV RNA in serum after a patient has had an end-of-therapy response (undetectable HCV RNA). Yoshida (2005) conducted a study including 119 patients who had relapsed after interferon monotherapy or combination therapy with RBV. Patients were assigned to receive either 24 or 48 weeks of PEG-IFN 180 μg/week with 800 mg/day of RBV. Forty-seven percent of patients had advanced fibrosis. Overall, the SVR was 40%. Thirty-five percent and 51% of patients with genotype 1/2 or 3 experienced an SVR, respectively. Nevens (2005) found an overall SVR of 43% in subjects receiving 180μg/week of PEG-IFN-alpha-2a and RBV (1–1.2 g/day). Patients had relapsed after receiving unmodified interferon with or without RBV.

In the EPIC trial reported by Poynard (2005), patients who relapsed after therapy with unmodified interferon with RBV were treated for 48 weeks with 1.5 μg/kg/week of PEG-IFN-alpha-2b with weight-based RBV. Nonresponders to this therapy were placed into the maintenance phase of the study. An overall SVR of 39% was achieved in relapse patients. The SVR rate was higher in the patients infected with HCV genotype 2 or 3 than in those with genotype 1 (58% vs. 29%). Similarly, Jacobson (2005a) reported an overall SVR of 42% in subjects who had relapsed after combination interferon/ribavirin therapy. Even though the medication regimens and patient populations were different, the overall SVR for patients who relapsed after unmodified interferon with or without ribavirin were comparable between the two peginterferons.

Herrine (2005) investigated the value of combining PEG-IFN-alpha-2a 180 μg/week with or without RBV with amantadine (AMD) or mycophenolate mofetil (MMF) in patients who had an increase in HCV RNA levels during therapy (breakthrough) or relapse during or after receiving unmodified interferon/RBV therapy. Subjects were assigned to one of four treatment groups. Group A received PEG-IFN-alpha-2a plus RBV (800–1,000 mg daily), Group B received PEG-IFN-alpha-2a plus 1 g of MMF orally twice daily, Group C received PEG-IFN-alpha-2a and AMD 200 mg daily, and Group D received PEG-IFN-alpha-2a plus AMD 200 mg daily plus ribavirin (800–1,000 mg daily). Treatment duration was 48 weeks with a 24-week followup period. An SVR was achieved in 37.5 % (Group A), 17.2% (Group B), 9.7% (Group C), and 45.2% (Group D) of subjects. Only Groups D and C were significantly different when compared, $P = 0.02$. The study revealed no clear advantage of combining amantadine to PEG-IFN/RBV therapy. The combination of AMD with MMF and PEG-IFN alone offered no advantage over RBV-containing regimens. More study will be necessary with combination therapy (Group A versus Group D) plus standard doses of ribavirin (1–1.2 g/day) to determine efficacy in relapse or breakthrough situations.

The ideal therapy with which to treat patients who do not respond or relapse after therapy with unmodified interferon/ribavirin regimens has not been definitively determined. It is known, however, that the risk of a nonresponse or relapse is increased in subjects who require dosage reduction of PEG-IFN and RBV during therapy (Ferenci et al., 2005; Shiffman et al., 2004). Steps to reduce the need for regimen alterations, including supportive medications (i.e., erythropoietin, filgrastim) and psychiatric evaluation, should be considered as part of any retreatment with PEG-IFN/RBV (Collantes & Younossi, 2005). An assessment of adherence is also important, and adherence-building exercises should be considered. Treatment of alcohol and other substance abuse conditions is critical before beginning retreatment.

Relapse or Nonresponse to Peginterferons

Few studies have looked at treatment of patients who have failed therapy with PEG-IFN-alpha/RBV. Berg (2006a) took patients who relapsed after 24 weeks of PEG-IFN-alpha-2a at 180 μg/week with RBV (800 mg or 1–1.2 g daily) and retreated them for an additional 48 weeks. Treatment was initiated at the same dose (PEG-IFN-alpha-2a/ribavirin) they had received before. The overall SVR was 55%. Fifty-one percent and 64% of subjects infected with HCV genotypes 1 and 2 or 3 achieved an SVR, respectively. Kaiser (2006) compared the efficacy of consensus interferon/RBV with PEG-IFN-alpha-2a/RBV in subjects who had relapsed after 48 weeks of PEG-IFN/ribavirin therapy. Group A received daily consensus interferon (9 μg), and Group B received standard PEG-IFN (180 μg/week) dosing. Both groups received weight-based RBV for a total of 72 weeks. Eighty-three percent of patients were infected with HCV genotype 1. The SVR for Groups A and B were 69% and 44%, respectively, $P < 0.05$. A clinical study is under way to determine the safety and efficacy of daily high-dose consensus interferon/RBV therapy in subjects who are nonresponders to combined PEG-IFN-alpha and ribavirin (ClinicalTrials.gov identifier: NCT00266318).

The "REPEAT" study is a multinational trial to test the value of an induction dose of PEG-IFN-alpha-2a plus RBV compared with standard-dose PEG-IFN-alpha-2a/RBV in patients who are nonresponders to PEG-IFN-alpha-2b/RBV therapy. Patients were randomized to four treatment groups. Groups A and B received PEG-IFN-alpha-2a at 360 μg/week for the first 12 weeks followed by 180 μg/week for a total of either 72 or 48 weeks, respectively. Groups C and D received PEG-IFN-alpha-2a at 180 μg/week for a total of either 72 or 48 weeks, respectively (Marcellin & Jensen, 2005). All groups received RBV (1–1.2 g/day). The results of the first 12 weeks of the trial have been presented. The high-dose induction groups of the study appear to be more effective in lowering HCV RNA concentrations than the standard PEG-IFN dosing arms. Forty-three percent of subjects in Groups A and B had undetectable HCV RNA concentrations at 12 weeks of therapy versus 26% in Groups C and D. Whether these results will translate into an improved SVR awaits completion of the study.

Role of Maintenance Therapy in Patients Who Fail to Respond to Interferon-Based Therapy

The HALT-C, EPIC, and COPILOT trials are three randomized controlled studies whose goal is to determine whether maintenance therapy with PEG-IFN-alpha will decrease the progression of disease in subjects with chronic HCV infection and advance disease compared with no therapy or placebo (Shiffman et al., 2004; Poynard et al., 2005; Curry et al., 2005). Patients had failed prior unmodified interferon/RBV therapy. Whether maintenance therapy will be of value in preventing disease progression awaits the final results of these studies.

Experimental Therapy

It is clear that we have gone about as far as we can go with PEG-IFN/RBV therapy for patients with chronic HCV infection. New modalities that attack the hepatitis C virus at different sites are under development. The serine protease inhibitors are being investigated for their ability to inhibit hepatitis C viral replication. Agents under development include VX-950 and SCH 503034 (Reesink et al., 2005; Zeuzem et al., 2006a). Reesink (2005) performed a phase 1B trial with VX-950 in healthy adults and patients infected with HCV genotype 1. Patients assigned to the 750-mg q8h group experienced the greatest decrease in HCV RNA concentrations after 14 days of treatment (median drop of 4.4 $\log_{10}$). The NS3 protease inhibitor (SCH 503034) was administered to HCV genotype-infected patients who had failed therapy with PEG-IFN-alpha-2b with or without RBV. Patients received varying doses of SCH 503034 alone or with PEG-IFN-alpha-2b for a period of 14 days using a three-way crossover design. A 2- to 3-week washout period occurred between each dosing regimen. Interestingly, undetectable HCV RNA concentrations were found in 4 of 10 patients after 14 days of PEG-IFN plus 400 mg of SCH 503034. These inhibitors will need to be used in combination with other HCV-active antiviral agents due to the emergence of resistant serine protease if used alone (Lin et al., 2006, Zeuzem et al., 2006). Combination phase II studies with PEG-IFN-alpha-2a and 2b are under way (ClinicalTrials.gov Identifier NCT00336479; ClinicalTrials.gov Identifier NCT00160251).

Valpoitabine (NM283) is a viral RNA polymerase inhibitor that is under study in HCV treatment-naïve and nonresponders to prior interferon-based therapies (Pockros et al., 2006; Dieterich et al., 2006). Pockros (2006) outlined the 24-week effects on HCV RNA baseline concentrations of valopicitabine alone, in combination with PEG-IFN-alpha-2a compared with PEG-IFN-alpha-2a/RBV. Patients were infected with HCV genotype 1 and were nonresponders to PEG-IFN/RBV therapy. Subjects were randomly assigned to five treatment groups, including PEG-IFN-alpha-2a at

180 µg/week plus RBV (1–1.2 g/day) (A), PEG-IFN-alpha-2a at 180 µg/week and various doses of valopicitabine (B–D), or valopicitabine monotherapy (E). Subjects in the 800-mg/day valopicitabine/PEG-IFN-alpha-2a group had a greater mean decline of HCV RNA (3.32 $\log_{10}$ IU/mL) versus PEG-IFN-alpha-2a/RBV or valopicitabine monotherapy, 2.31 $\log_{10}$ IU/mL and 0.54 $\log_{10}$ IU/mL, respectively. Dieterich (2006) administered valopicitabine/PEG-IFN-alpha-2a to treatment-naïve HCV genotype 1 infected patients and reported a mean decrease of 4.5 $\log_{10}$ IU/mL in subjects after 8 weeks of combination therapy.

A Toll-like receptor 9 agonist (CPG 10101-CPG) is being investigated in phase II studies. The drug acts by stimulating the immune system. McHutchison (2006) has reported on the first 12 weeks of CPG monotherapy, CPG in combination therapy with PEG-IFN-alpha-2b with or without RBV or with RBV alone, compared with PEG-IFN-alpha-2b plus RBV. Subjects were infected with HCV genotype 1 and had relapsed after PEG-IFN /RBV therapy. At 12 weeks, 57% and 86% of patients receiving PEG-IFN/RBV or PEG-IFN/RBV plus CPG had achieved an EVR, respectively ($P = 0.21$).

Interest in improving the tolerability of PEG-IFN/RBV therapy has led to the development of the drug viramidine. Viramidine is a pro-drug of RBV that is being investigated for use with interferon due to its lower incidence of anemia. Benhamou (2006) reported on the results of a phase III study comparing PEG-IFN-alpha-2b combined with weight-based RBV (1–1.2 g/day) or viramidine (V) (600 mg twice daily). An SVR was achieved in 52% (RBV) and 38% (V) of patients in each group (intent-to-treat) and did not meet the non-inferiority efficacy endpoint. Anemia was found in 24% and 5% of subjects in the RBV and V groups, respectively. The authors noted an increase in SVR with increasing mg/kg dose of V. Anemia was not increased proportionally. Revision of the standard dose of viramidine is likely to result from this study.

A variety of other agents are under investigation for the management of treatment-naïve or nonresponders/relapsers to interferon-based therapies. Most of the agents discussed here are in phase II or early phase III studies, and whether they will have a role in the management of HCV-infected individuals will be determined by results from controlled studies. Due to concerns about viral resistance when these agents are used alone, combination regimens including PEG-IFN will likely be necessary.

Children

The management of chronic HCV-infected children and adolescents with PEG-IFN-based therapy has not been extensively studied. A trial by González-Peralta (2005) recorded an overall SVR rate of 46% (54/118) in children treated for 48 weeks with unmodified interferon/RBV. An SVR was realized in 36% and 84% of children infected with HCV genotypes 1/2 or 3, respectively. Interestingly, Schwarz (2003) reported an SVR of 38% (5/13) children receiving PEG-IFN-alpha-2a (180 µg/1.73 $m^2 \times$ patient's body surface area) monotherapy for 48 weeks. Ninety-two percent (12/13) of patients were infected by HCV genotype 1. Wirth (2005) treated

62 children and adolescents with chronic HCV infection with PEG-IFN-alpha-2b (1.5 µg/kg/week)/RBV (15 mg/kg daily). Subjects infected with HCV genotype 1 were treated for 48 weeks, while patients with HCV genotype 2 or 3 were given the option of receiving 24 weeks of therapy. The overall SVR was 56%. The SVR rate in patients with HCV genotypes 1 and 2 or 3 were 48% and 100%, respectively. Utilizing the same dosing regimen as Wirth (2005), Hasan (2006) found an SVR rate of 75% (9/12) in a small group of adolescents infected with HCV genotype 4. Although these studies do not show the superiority of PEG-IFN/RBV regimens over unmodified interferon/ribavirin treatments, the once-weekly dosing of the peginterferon may appeal to busy parents/children.

Liver Transplantation

Use of PEG-IFN therapy in the post-liver transplant population is an evolving practice with new clinical information being published at a rapid pace. As more liver transplantation programs emerge and grow, and more high-risk transplants are performed, recurrent hepatitis C in the recipients will likely rise, with the potential for graft loss (Lauer & Walker, 2001; National Institutes of Health Consensus Development Conference Statement: Management of Hepatitis C, 2002). High doses of immunosuppressive agents immediately post-transplantation present a particular problem in terms of tolerability of treatment with PEG-IFN and RBV. For this reason, treatment is recommended to be held until months after transplantation, with two scenarios being explored most often. The first is the use of PEG-IFN as preemptive or "prophylactic" treatment before clinical onset of recurrent hepatitis C, relatively early (e.g., one month) post-transplant in high-risk patients with high viral loads. The second is the use of PEG-IFN for treatment of recurrent hepatitis C, later post-transplant (e.g., six months). Much of the literature describing this use has been published for unmodified interferon therapies (Singh et al., 1998; Sheiner et al., 1998; Firpi et al., 2002; Berenguer Prieto et al., 2004; Giostra et al., 2004; Bizollon et al., 2003). Although interferon-based therapies carry theoretical risks of inducing graft rejection, recent trials do not support the association (Kuo & Terrault, 2006; Chalasani et al., 2005).

In a study of the two treatment scenarios mentioned above, preemptive treatment and active therapy of recurrent hepatitis C in patients following liver transplantation, Chalasani et al. (2005) enrolled 54 patients within 3 weeks of an orthotropic liver transplantation (OLT) for "prophylaxis" or preemptive therapy and enrolled 67 patients 6 to 60 months after transplantation for active treatment of recurrent HCV disease. In either treatment group, patients were randomized to treatment with 180 µg/week of PEG-IFN-alpha-2a or no antiviral treatment for 48 weeks with a 24-week followup post-therapy. In the preemptive arm, patients who received PEG-IFN had a significantly greater drop in HCV RNA concentrations at weeks 4 and 24 than untreated patients ($P < 0.003$ and $P < 0.02$, respectively). Likewise, patients in the treatment arm on PEG-IFN-alpha-2a had significantly lower viral loads than untreated patients at each scheduled post-baseline assessment ($P < 0.001$). Unfortunately, only 2 treated patients in the prophylaxis trial (8%) and 3 in the treatment trial (12%) achieved an SVR. Acute rejection rates were similar in the treated and untreated groups in both the prophylaxis (12% vs. 21%; $P < 0.5$) and treatment (12% vs. 0%; $P < 0.1$) trials.

A preemptive-only, randomized comparative study of single therapy or combination therapy with RBV of unmodified interferon-alpha-2b 3 MU three times a week versus PEG-IFN-alpha-2b 1.5 µg/kg once weekly was conducted by Shergill (2005). Therapy was initiated two to six weeks' post-transplantation and continued for 48 weeks. Fifty-one patients were treated, but with a high rate of dose reductions (85%) and drug discontinuations (37%) due to adverse events and intolerability. Rates of viral suppression at 48 weeks (ETR) were low for both groups, 4.5% for single therapy with either interferon product versus 22.7% for combination therapy with either product ($P = 0.093$). The difference was not statistically significant due to the low total number of patients. Even lower comparative rates were recorded for SVR at 4.5% for monotherapy and 18.2% for combination therapy. Comparisons could not be performed regarding unmodified versus PEG-IFN products due to the small number of patients. The authors concluded that while combination therapy was better than monotherapy, SVR rates are far less than in non-transplantation HCV-infected patients, and a majority of patients did not tolerate the drug therapy well. In a post hoc analysis, the authors concluded that the best candidates for treatment are patients with better liver function (MELD score) prior to treatment and those who undergo living donor transplantation.

A French study investigated the benefit of active treatment of recurrent HCV disease post-liver transplantation. Dumortier (2004) treated 20 patients with 12 months of combination therapy with PEG-IFN-alpha-2b and RBV. They evaluated virological and biochemical responses to treatment. Patients were started on low-dose treatment at 0.5 µg/kg/week of PEG-IFN plus 400 mg of RBV daily. Therapy was initiated at least 28 months' post-transplantation, and dosing was escalated as tolerated to a maximum dose of PEG-IFN of 1 µg/kg/week and 1,200 mg of RBV daily. A high rate of adverse events, at 20%, led to drug discontinuation. Dose reductions of PEG-IFN to 0.5 µ/kg/week were required in 37.5% of the remaining patients, and dose reductions of RBV were required in 87% of the remaining patients. Using intent to treat, 55% of the patients had a virological response at 12 months, with an SVR of 45%. Like most studies, virological outcomes were less for genotype 1 than other genotypes (64% vs. 100%, respectively, $P < 0.05$). Five of the patients in this study had a mild acute rejection episode.

Neff (2004) examined the benefits of treatment of recurrent HCV in liver transplant recipients using combination therapy with PEG-IFN-alpha-2b (1.5 µg/kg/wk) and RBV (400–600 mg/day) therapy for at least 48 weeks. The retrospective review identified 57 patients, who were divided into patients who were treatment-naïve versus those who had received interferon-based therapy pre-transplant and were nonresponders to at least six months of combination therapy. Undetectable HCV RNA concentrations were attained in eight (27.6%) treatment-naïve patients and six (21%) treatment-experienced patients at the end of 48 weeks of therapy. An SVR was achieved in 75% (6/8) and 33% (2/6) of treatment-naïve and treatment-experienced patients, respectively. Ribavirin or PEG-IFN dose reductions were required in both groups. Up to 69% of treatment-naïve patients had dose reductions.

A Spanish study conducted by Planas (2005) treated patients with evidence of recurrent HCV infection after liver transplantation with PEG-IFN-alpha-2b at 1.5 µg/kg/week plus weight-based ribavirin (10.6 mg/kg/day). The 30 patients were treated at a median of 43 months' post-transplantation. Treatment duration was dependent upon HCV genotype, 48 weeks for genotypes 1 and 4 or 24 weeks for genotypes 2 and 3. An end-of-treatment response was measured at 63%, with an SVR measured at 47%. Dose reductions were required in 40% of patients.

Other small case series include open-label reviews of patients with recurrent HCV after liver transplantation by Biselli (2005), Beckebaum (2004), Mukerjee (2003), and Oton (2005). Patients were treated with combination therapy including PEG-IFN plus RBV. All four are observational studies designed to examine the benefits of combination therapy with PEG-IFN-alpha-2b plus RBV. In the study by Biselli (2005), the dose chosen for PEG-IFN-alpha-2b was 1.0 µg/kg/week and RBV at 600 mg/day. Therapy was continued for at least six months. Nine patients (45%) had an end-of-treatment response, and an SVR was attained in 60% of treatment-naïve patients versus 30% of previously treated nonresponders. A drop-out rate of 45% was observed at six months. In the study by Beckebaum (2004), 12 patients were treated with 3 months of unmodified interferon, followed by 9 months of PEG-IFN–alpha-2b at 1.5 µg/kg/week plus RBV (10–12 mg/kg). An ETR was observed in 33% of patients. A disappointing 42% had no response after six months of treatment. The study by Mukherjee (2003) followed 39 patients receiving PEG-IFN-alpha-2b at 1.5 µg/kg/week and RBV at 800 mg/day. In this more standard does study, 17 patients withdrew within the first three months due to drug intolerance. Four had not yet completed 3 months of treatment at the time of publication. Of the 18 patients who completed treatment, 17 (94.4%) or 43.6% of the total enrolled had an early virological response (HCV RNA undetectable at 3 months). An end-of treatment response (ETR) was achieved in 15 (83.3%) or 38.5% of the total enrolled patients at six months. An SVR occurred in 12 patients, with results pending for three other subjects. Oton (2005) reviewed outcomes in 21 treatment-naïve patients who received PEG-IFN-alpha-2b at 1.5 µg/kg/week plus RBV at a weight-based dose of at least 10.6 mg/kg/day. The time from liver transplantation was at least 1.7 years. All patients were infected with HCV genotype 1. Treatment was intended for at least 48 weeks. Two patients dropped out, one due to intolerance, the other due to non-drug-related cholangitis. Necessary dose adjustments were high, at 14% for PEG-IFN and 32% for RBV. Fourteen patients (66.7%) had an ETR, with an SVR reported in 42.8% of patients, demonstrating one of the highest response rates among these observational studies.

Initial results of a randomized, multicenter study investigating the efficacy of PEG-IFN-alpha-2a in patients with recurrent HCV infection following liver transplantation were presented in abstract form (Vogel et al., 2002). The time of therapy initiation varied widely, within 6 to 60 months of liver transplantation. All patients were treatment-naive. Patients were assigned to receive either PEG-IFN at 180 µg/week or placebo. The majority of patients in both groups had a viral load >1 million IU/mL at baseline and HCV genotype 1. Treatment was continued for 48 weeks with a 24-week followup that was still pending. At 48 weeks of therapy, 35% of patients receiving active treatment with PEG-IFN and none of the patients receiving placebo had an undetectable HCV RNA level.

The published experience for preemptive therapy or treatment of recurrent HCV infection in liver transplant patients demonstrates fairly low success rates with a reasonably high amount of drug intolerance. Larger prospective dose-ranging studies would help define the ideal approach. In the meantime, initiation of therapy with PEG-IFN plus RBV at lower doses with titration upward as tolerated, and careful monitoring and support with growth factors, appear to be a reasonable approach. The pharmaco-economic benefit of PEG-IFN-based therapies in this patient population has yet to be defined.

Human Immunodeficiency Virus (HIV) Co-infection

In the United States and Europe, between 15% and 30% of all people infected with HIV are also infected with HCV (Sulkowski et al., 2000; Sherman et al., 2002; Greub et al., 2000; Sulkowski et al., 2002). Rates are higher in people who contract the disease through intravenous drug abuse (Garfein et al., 2000) than among men who contract the disease through sexual contact with other men. However, there have been spikes in the incidence in the population of men who have sex with men (MSM), likely linked to the resurgence of unprotected sex (Sulkowski & Thomas, 2003; Rauch et al., 2005; Ghosn et al., 2004). Co-infected patients often have lower CD4 counts, a more rapid progression of liver fibrosis, and increased mortality over people who are singly infected (Nunez et al., 2003; Brau, 2003; Kramer et al., 2005; Sulkowski et al., 2002; Tedaldi et al., 2003). Although there may be some disagreement among researchers, co-infection with HCV does not appear to affect progression or response to therapy of HIV disease (Sulkowski et al., 2002; Sullivan et al., 2006; Hershow et al., 2005).

Data are growing exponentially on the treatment and management of patients with HCV/HIV co-infection. The publication of two well–designed, large, randomized studies has expanded the level of evidence for treating co-infected patients (Table 4). The AIDS Pegasys RBV International Co-infection Trial (APRICOT) evaluated the efficacy and safety of unmodified interferon-alpha-2a (3 MU 3 times/week) plus RBV (800 mg/day) versus PEG-IFN-alpha-2a at 180 ○g/week plus placebo versus PEG-IFN plus RBV (800 mg/day) (Torriani et al., 2004). Patients were young, with a mean age of 40 years, mostly male, white and primarily infected with HCV genotype 1 (60%). Inclusion criteria were compensated liver disease, a $CD4^+$ cell count ≥ 100 cells/mL, and stable HIV disease (with or without HIV therapy). Sustained virological response rates were lower than for mono-infected patients, and the highest SVR response occurred in the group treated with PEG-IFN plus RBV (40%). However, the SVR rate for unmodified interferon plus RBV was only 12%. The SVR rate for patients receiving PEG-IFN plus placebo (20%) was also lower than the combination with RBV. Like mono-infection trials, patients infected with HCV genotype 1 were more treatment refractory. For these patients, the best chance for achieving an SVR was observed when subjects received the combination of PEG-IFN plus RBV (29% vs. 14% and 7%) versus the PEG-IFN plus placebo or unmodified interferon plus RBV groups, respectively.

Table 4 An Overview of Trial Design/Results for Treatment of HIV and HCV Co-infected Patients with Pegylated Interferon

Study Design	APRICOT (Torriani et.al. 2004)	PRESCO (Ramos 2006)	ACTG 5071 (Chung 2004)	RIBAVIC (Carrat 2004)
Sites	International (19 countries)	Spain (1 site)	US (21 sites)	France (71 centers)
Number of patients	868	98	133	412
Outcome measures	Clinical/Safety	Clinical/Safety	Clinical/Safety	Clinical/Safety
Arm 1	PEG-IFN alpha-2a 180 mcg weekly + RBV 800/d	PEG-IFN alpha-2a 180 mcg weekly + RBV 800–1200 mg/d	PEG-IFN alpha-2a 180 mcg weekly + RBV 600 mg × 4wks then 800 mg × 4 wks then 1000 mg/d	PEG IFN alpha-2b 1.5 mcg/kg weekly + ribavirin 400 mg twice daily
Arm 2	PEG-IFN alpha-2a 180 mcg weekly + placebo	None	PEG-IFN alpha-2a 180 mcg weekly + placebo	IFN alpha-2b 3MU three times a week +ribavirin 400 mg twice daily
Arm 3	IFN alpha-2a 3MU three times weekly + RBV 800/d	None	IFN alpha-2a 6MU three times weekly × 12 weeks, then 3mu three times weekly + RBV 600 mg × 4wks then 800 mg × 4 wks then 1000 mg/d	None
Treatment duration (weeks)	48	12 (interim results)	48	48
		RESULTS		
SVR genotype 1	PEG-IFN + RBV 29% PEG-IFN + placebo 14% IFN + RBV 7%	NA	PEG IFN + RBV 14% IFN + RBV 6%	PEG IFN + RBV 17% IFN + RBV 6%
Overall SVR (week 72)	PEG-IFN + RBV 40% PEG-IFN + placebo 20% IFN + RBV 12%	EVR: 83%	PEG IFN + RBV 27% IFN + RBV 12%	PEG IFN + RBV 27% IFN + RBV 20%

The RIBAVIC trial (Carrat et al., 2004), a multicenter, randomized, parallel group, open-label trial, examined the safety and efficacy of PEG-IFN-alpha-2b plus RBV versus unmodified interferon-alpha-2b plus RBV in HIV-HCV co-infected patients (Table 4). The study was conducted in 71 French centers. Dosing for PEG-IFN-alpha-2b was 1.5 µg/kg/week versus interferon-alpha-2b 3 MU three times a week, both arms with RBV at 800 mg/day. An SVR was attained at a higher rate for the PEG-IFN regimen than unmodified interferon (27% vs. 20%, = 0.047), respectively. Different rates of SVR were even more pronounced when comparing patients infected with genotype 1 who were assigned to PEG-IFN-alpha-2b versus unmodified interferon (17% vs. 6%, respectively, $P = 0.006$). When comparing SVR rates for genotypes 2 and 3, there was not a statistically significant difference in SVR rates between those treated with PEG-IFN product (44%) versus unmodified interferon (43%).

Another multicenter randomized trial by Chung (2004) evaluated the efficacy of unmodified interferon-alpha-2a or PEG-IFN-alpha-2a, both with escalating doses of RBV (600–1,000 mg/day) (Table 4). Patients treated with the PEG-IFN-alpha-2a/RBV regimen experienced a higher SVR rate versus unmodified interferon/RBV (27% and 12%). An SVR was achieved in only 6% (unmodified interferon group) and 14% (PEG-IFN group) of patients infected with HCV genotype 1, respectively.

The PRESCO trial (Ramos et al., 2006), in progress, examines the benefit of weight-based doses of ribavirin (1,000–1,200 mg/day) compared to the standard RBV dose (800 mg/day) utilized in the APRICOT trial (Table 4). Data up to week 12 have been presented in abstract form. For HCV genotype 1, an EVR (2 $\log_{10}$ decrease in HCV RNA concentrations at 12 weeks of therapy) was observed in 78% of the study subjects. This compares to 63% in APRICOT and 80% of subjects in the HCV mono-infection trial (Fried et al., 2002). Differences were less pronounced in patients with genotypes 2 and 3. Whether these results will translate into a better SVR rate awaits the conclusion of the study.

Voight (2006) performed a prospective, uncontrolled, multicenter trial in Germany, exploring the efficacy and safety of combination PEG-IFN-alpha-2b (1.5 µg/kg/week) plus RBV (800 mg daily) for 48 weeks for HCV genotypes 1 and 4 and 24 weeks for genotypes 2 and 3 in HIV/HCV co-infected patients. An end-of-treatment response was found in 52% of patients overall, but an SVR response occurred in only 25% of subjects. This rate was lowest for those with genotype 1 or 4 (18%) and highest for genotype 2 or 3 (44%). Discontinuation rates were high (30%).

Soriano (2004) investigated whether an early virological response at 12 weeks (EVR- undetectable HCV RNA or ≥ 2 $\log_{10}$ decrease of HCV RNA concentrations) could be used to make treatment decisions in HCV patients co-infected with HIV. Of the 89 co-infected patients, an EVR occurred in 58% of patients, and 32.6% attained an SVR. The negative and positive predictive value for achieving an SVR was 100% and 56%, respectively. From these results, consideration of treatment discontinuation can be entertained after 12 weeks of therapy if there is no EVR regardless of infecting HCV genotype in HCV/HIV co-infected patients receiving PEG-IFN-based therapies.

Table 5 Common and Clinically Significant Side Effects of Pegylated Interferon Therapies [(pegylated IFN alpha-2a- Pegasys^{TM}) (pegylated IFN alpha-2b-Peg Intron^{TM})]

Appearance-Related/Skin	Hematologic Effects
Alopecia	Neutropenia (both lymphocytic and neutrophilic)
Eczema	Anemia (more pronounced with co-treatment with ribavirin)
Pruritis	Thrombocytopenia
Rash	
Dry skin	
Cardiovascular	**Neurologic/Behavioral**
Angina	Aggression
Cardiomyopathy	Anxiety
Hypertension	Bipolar symptoms
Myocardial Infarction	Depression
	Dizziness
Endocrine	Headache
Diabetes	Hearing loss
Hypo/hyperthyroidism	Impaired concentration
	Mania
Gastrointestinal	Neurocognitive impairments
Colitis	Suicidal ideation
Decreased appetite	Vision changes
Diarrhea	
Hepatic decompensation	
Asterixis	
Elevated liver	
transaminases (ALT/AST)	
Elevated bilirubin	
Decreased albumin	
Decreased total protein	
Encephalopathy	
Increased prothrombin time	
Increased INR	
Nausea	
Upper abdominal pain	
Vomiting	
General	
Influenza-like symptoms	
Cough	
Fatigue	
Fever	
Myalgia	
Rigors	
Pneumonitis	

up to 27% of those treated with unmodified interferon. The primary concern about dose reduction is that while it may allay adverse events, it can also allow for incomplete suppression or eradication of the virus, as described by Fried (2000). The length of therapy also plays a role in the frequency and severity of adverse events. Drug discontinuation secondary to adverse drug events was recorded at higher rates in patients treated for 48 weeks (15%) versus those treated for 24 weeks

Table 6 Treatment Discontinuation of Standard IFN Versus PEG-IFN Due to Clinical Adverse Events or Lab Abnormalities[a]

Trials and Comparative Arms	Standard IFN Group	PEG-IFN Group
Mono-Infection Trials	**Discontinuation Rates (%)**	
Zeuzem et al. (2000)(PEG-IFN alpha-2a vs. unmodified IFN alpha-2a)	7	10
Heathcote et al. (2000)(PEG-IFN alpha-2a 90 mcg/wk vs. 180 mcg/wk vs. unmodified IFN alpha-2a)	10	90 mcg wk 11180 mcg wk 14
Fried et al. (2002)(PEG-IFN alpha-2a vs. unmodified IFN alpha-2a)	10	10
Pockros et al. (2004)(PEG-IFN alpha-2a 135 mcg/wk vs. 180 mcg/wk vs. unmodified IFN alpha-2a)	11	135 mcg/wk 9180 mcg/wk 10
Hadziyannis et al. (2004)(PEG-IFN alpha-2a + ribavirin 800 mg/day vs. PEG-IFN alpha-2a + ribavirin 1000 mg/day)	NA	Ribavirin 800 mg/d16 Ribavirin 1000 mg/d15
Manns et al. (2001)(PEG-IFN alpha-2b vs. unmodified IFN alpha-2b)	13	1.5 mcg/kg wk 140.5 mcg/kg wk 13
HIV/HCV Co-infection Trials		
ACTG 5071, Chung et al. (2004)(PEG-IFN alpha-2a vs. unmodified IFN alpha-2a)	12	12
RIBAVIC, Carrat et al. (2004)(PEG-IFN alpha-2b vs. unmodified IFN alpha-2b)	17	16
APRICOT, Torriani et al. (2004)(PEG-IFN alpha-2a vs. unmodified IFN alpha-2a)	14	12

[a] Discontinuation rates were similar between PEG-IFN and standard interferon. Higher doses of ribavirin were associated with slightly higher rates of discontinuation. Additionally, the discontinuation rate was slightly higher among patients co-infected with HIV.

(5%). Drug discontinuation related to adverse drug events overall are compared in Table 6 for unmodified interferon versus PEG-IFN-based therapies from the larger randomized trials. The primary reasons for drug discontinuation were hematologic abnormalities and neurocognitive disorders. A higher rate of discontinuation is noted in the patients with HIV co-infection. Dose reductions are only slightly higher with longer treatment courses, 30% versus 33% as noted in the Food and Drug Administration briefing document (2002).

Constitutional symptoms resembling influenza-like illness have been noted in up to 51% of patients treated with PEG-IFNs (Zeuzem et al., 2000; Heathcote et al., 2000). Symptoms are often transient, and patients tend to develop tolerance over time. In the larger comparative trials, influenza-like symptoms occurred more frequently in patients treated with unmodified interferon plus RBV than with PEG-IFN plus RBV; 50% compared with 42% ($P = 0.02$) (Fried et al., 2002; Zeuzem et al., 2000).

Neuropsychiatric disorders, such as anxiety and depression, affect up to 26% of patients during treatment with PEG-IFN products. A warning exists for all interferon products regarding the potential new onset or exacerbation of neuropsychiatric disorders, including depression, suicidal ideation, increased addictive behaviors, bipolar symptoms, mania, and aggressive behaviors. This warning is especially emphasized in patients with underlying neuropsychiatric disorders, particularly if they are not being treated or monitored for these conditions. A hypothesis for this adverse event was studied by Schwaiger (2003), wherein serotonergic activity in patients treated with PEG-IFN was measured and greatly reduced in those treated with the drug. Byrnes (2005) reported on a pre and post-evaluation of 10 patients treated with PEG-IFN to evaluate the potential confounding effect of HCV on neuronal function. They performed baseline brain MRI and 1H–MRS (spectroscopy), measurements of neurotransmitters, in addition to neuropsychological evaluation (neuropsychometric tests, Beck's depression inventory, and quality-of-life and self-reported cognitive dysfunction questionnaires) prior to, and 12 weeks following, initiation of PEG-IFN. The study demonstrated that as HCV viral load was reduced, there was a reduction in inflammatory markers, but depression scores increased and cognitive function declined. A history of depression is directly correlated with a higher incidence of drug-related psychiatric events during PEG-IFN therapy (Castera et al.,2004). However, there are some differences relative to unmodified interferon products. In the trial by Fried et al. (2002), depression was observed less frequently in the PEG-IFN-treated group compared with those treated with unmodified IFN-alpha-2b (22% vs. 30%, $P = 0.01$). Other trials confirmed this pattern (Zeuzem et al., 2000; Heathcote et al., 2000), although significantly morbid depression has been described in patients receiving either product. In parallel, the dose finding trial by Pockros et al. (2004) demonstrated a similar rate of depression for patients treated with PEG-IFN product versus unmodified interferon. Neurocognitive decline, such as decreased alertness, attention deficits, decreased vigilance, and impaired short-term memory, was demonstrated in a cohort of 70 patients treated with PEG-IFNs by Krauss (2005). Overall, neuropsychiatric disorders, including depression, mark the highest-ranking reason for drug withdrawal, particularly in those patients treated for 48 weeks versus 24 weeks (FDA brief, 2002).

Bone marrow suppression, resulting in low cell counts, such as anemia and neutropenia, is another common treatment-limiting adverse event of PEG-IFN therapies. This results from direct inhibition of progenitor cell proliferation in the bone marrow by interferon (Ganser et al., 1987). Initial laboratory data should be recorded as a baseline for the absolute neutrophil count, hemoglobin, and platelets. Patients with low counts at baseline (a neutrophil count <1,500 cells/mm^3 or a baseline hemoglobin <10g/dL or a baseline platelet count <90,000 cells/mm^3) should be treated cautiously with lower starting doses of PEG-IFN and careful monitoring of cell lines during therapy. Severe, persistently low counts may result in infectious complications including opportunistic infections. Therapy with PEG-IFN should be interrupted or discontinued if counts do not respond to dose modifications.

Hematologic abnormalities are the most common reason for dose modifications in many of the trials (Fried et al., 2002; Chung et al., 2004; Torriani et al., 2004; Pockros et al., 2004; Carrat et al., 2004). Drug-related neutropenia can begin early, with a nadir in the first four weeks of therapy. Rates of neutropenia have been reported to be higher in patients treated with PEG-IFN products than with unmodified interferon in randomized trials (Fried et al., 2002; Chung et al., 2004; Torriani et al., 2004; Carrat et al., 2004; Heathcote et al., 2000). Neutropenia defined as grade 4 (ANC < 500/mL) has been documented in multiple trials. Dose reductions are employed in up to 11% of the patients treated with the PEG-IFN versus 7% of patients receiving unmodified interferon. In an analysis of the effects on hematologic cell lines, Schmid (2005) analyzed 133 patients undergoing treatment with unmodified interferon, unmodified interferon plus RBV, PEG-IFN-alpha-2a plus RBV, or PEG-IFN-alpha-2b plus RBV. Leukopenia was common to all arms. The maximum decrease in neutrophils and lymphocytes occurred in patients treated with PEG-IFN products, 55% for PEG-IFN-alpha-2a and 52% for PEG-IFN-alpha-2b versus 37% for unmodified IFN ($P < 0.05$). As an exception, the trial by Heathcote (2000) demonstrated similar rates of grade 4 neutropenia among groups (3% with unmodified interferon, 3% and 1% with the PEG-IFN-alpha-2a at 90 and 180 µg/wk). Dose reductions for PEG-IFN-alpha-2a to 135 µg/week from 180 µg/week and to 0.75 µg/kg/week from 1.5 µg/kg/week for PEG-IFN-alpha-2b are recommended when the ANC falls below 750. Treatment discontinuation is recommended if the ANC falls below 500.

Thrombocytopenia has been observed, particularly early in treatment (within 2 weeks). Similar comparative rates of thrombocytopenia in the trial by Fried (2002) were observed between patients receiving PEG-IFN-alpha-2a plus RBV, or PEG-IFN-alpha-2a plus placebo as compared to unmodified interferon plus RBV at 4% versus 6% versus <1%, respectively. Severe thrombocytopenia, classified as grade 3, or a platelet count between 20,000 and 50,000/mL, occurred equally between unmodified and PEG-IFN products in the study by Zeuzem (2000). One HIV/HCV co-infection study (Chung et al., 2004) described two patients with grade 4 thrombocytopenia (<20,000/mL), one in the unmodified and one in the PEG-IFN-treated groups. However, the study by Heathcote et al. (2000) demonstrated a higher incidence of thrombocytopenia in patients receiving PEG-IFN-alpha-2a, 26% versus 7% of all patients treated with unmodified interferon ($P < 0.001$, $P = 0.04$, respectively). In an observational crossover study, Homincik (2003) examined the effects of one dose of unmodified interferon-alpha-2a followed by weekly PEG-IFN-alpha-2a for 36 weeks on both platelet count as well as platelet activity. Decreases in platelet count were noted uniformly, but platelet activity was relatively unchanged. In the analysis of drug-related hematologic effects of six prospective trials by Schmid (2005), median decreases in platelet counts were similar in the patients treated with unmodified interferon alone (25%) versus those treated with PEG-IFN products plus RBV, 21% for PEG-IFN-alpha-2a and 23% for PEG-IFN-alpha-2b. An unusual case of late-onset (six months after drug discontinuation) drug-related autoimmune thrombocytopenia (PEG-IFN-alpha-2b plus RBV) was reported by Elefsiniotis (2006). Drug-related thrombocytopenia was not associated with increased bleeding episodes in any of the major trials, but when thrombocytopenia is severe, the risk of bleeding rises proportionally. The presence of cirrhosis of the liver adds to this risk. Dose reductions for thrombocytopenia have been required in up to 21% of patients with cirrhosis versus 6% for the overall population in a monotherapy study group and 16% for cirrhotics versus 4% for the overall population in the combination study group (FDA Brief, 2002). Recovery of platelet counts occurs by four weeks after treatment is completed or discontinued. Dose reductions to 90 µg/week from 180 µg/week for PEG-IFN-alpha-2a are recommended if the platelet count falls below 50,000. Therapy should be discontinued if the platelet count falls below 25,000.

Drug-related anemia has also been well documented, particularly in patients co-treated with RBV. Data relative to erythropoietin levels in treated patients indicate that patients cannot maintain erythropoietin levels adequate to offset treatment-related anemia (Balan et al., 2003). Hemoglobin values appear to decrease similarly when comparing PEG-IFN to unmodified interferon; in one study, they fell by a maximum of 3.7 g/dL for PEG-IFN-alpha-2a plus RBV, compared to 3.6 g/dL for unmodified interferon-alpha-2b plus RBV. However, when examining rates relative to the percent of patients whose hemoglobin falls below 10 g/dL, Schmid (2005) demonstrated higher rates among patients treated with PEG-IFN-alpha-2a with RBV (29%) versus 6% of patients treated with unmodified interferon-alpha-2b with RBV ($P < 0.05$). The common denominator remains the additive anemic effects mediated by ribavirin. The proposed mechanism is direct hemolysis by RBV as demonstrated by De Franceschi (2000). A handful of patients with HIV co-infection have had observed grade 4 anemia, Hgb less than 6.5 g/dL (Chung et al., 2004). All three hematologic abnormalities have been documented at higher rates in patients with a lower body mass index (BMI), perhaps reflecting the use of fixed doses of PEG–IFN-alpha-2a. High RBV blood levels and concomitant treatment with zidovudine have also been identified as predictors of treatment-related anemia (Rendon et al., 2005). Treatment should be discontinued if the Hgb falls below 8.5 g/dL.

New onset cardiovascular dysfunction has been documented rarely (21 total reported cases) in patients treated with unmodified interferon (Kuwata et al., 2002; Cohen et al., 1990; Sonnenblick et al., 1990). These conditions include cardiomyopathy, myocardial infarction, angina, and hypertension. A baseline EKG and serial monitoring in patients with pre-existing cardiovascular disease are recommended. Risk reduction measures should also be considered in patients at a high risk for cardiovascular complications. A recent case report by Condat (2006) details a case of fatal cardiomyopathy in a patient receiving PEG-IFN/RBV therapy.

Ocular changes have been described with unmodified interferon treatment of hepatitis C and newly with PEG-IFN-alpha-2b (Farel et al., 2004; Ahmed et al., 2003; Schulman et al., 2003). In the case series review by Farel (2004), patients co-infected with HIV and HCV who were treated with PEG-IFN-alpha-2b plus RBV underwent serial ophthalmologic exams every three months, revealing pathology in 35% of the patients. These included cotton wool spots, cataracts, and decreased red-green color vision. The color vision changes were reversed in one patient 10 weeks after discontinuation of combination therapy. The other patient had sustained visual changes from weeks 10 through 23 of therapy. Visual changes spontaneously resolved. None of the patients had diabetes or hypertension. Speculation of the underlying mechanism for this side effect lies in an observational study by Sugano (1998) wherein the authors measured plasma levels of complement 5a that revealed a drug-induced increase in plasma levels of complement 5a. Complement 5a is purported to deposit in the retinal vasculature, leading to local capillary rupture. Baseline and a followup ophthalmologic exam are recommended based on these data.

Interferon treatment for chronic HCV infection may exacerbate or result in hepatic decompensation, particularly in patients co-infected with HIV and receiving highly active antiretroviral therapy (HAART). A baseline liver biopsy and serial laboratory panels for liver injury (serum AST/ALT, total bilirubin) and synthetic function (albumin, PT, INR, total protein) are recommended. An analysis of risk factors for development of hepatic decompensation was performed for the APRICOT trial by Torriani (2004). All patients had advanced cirrhosis as a predisposing factor. Additional risk factors were co-treatment with didanosine and evidence of cholestasis at baseline. Adverse events include potentially acute liver transaminase flares, elevations in bilirubin, and evidence of decompensation (e.g., encephalopathy, elevated bleeding times, asterixis).

Thyroid disorders have been linked to unmodified interferon and PEG-IFN therapy (Imigawa et al., 1995; Fonseca et al., 1991; Preziati et al., 1995; Lisker-Melman et al., 1992). Induction of anti-thyroid antibodies and autoimmune thyroiditis is the primary mechanism by which hypothyroidism is induced by interferon products. A baseline TSH and a followup TSH are recommended in the Veteran's Health Administration guidelines (2005). Some patients require supplementation with levothyroxine.

Allergy or drug hypersensitivity is a rare event linked to PEG-IFN use. Like other drugs, the reaction is not predictable and may manifest itself as an acute IgE-mediated event, with urticaria, hypotension, and anaphylaxis. Anti-IFN antibodies have been positive in a handful (4.8%) of patients in the larger trials (Fried et al., 2002), but no notable sequelae were observed as a result. Skin rashes have been documented in multiple trials. A severe skin rash was described in a case series by Jessner (2002) wherein three patients using PEG-IFN developed a delayed rash requiring drug discontinuation. Two of the three patients were switched to unmodified interferon therapy with resolution and without recurrence of the rash. The third patient was not rechallenged.

Other notable case reports include a documented interstitial pneumonitis with adult respiratory distress syndrome (Abi-Nassif et al., 2003). The patient described had no previous history of pulmonary disease but, upon presentation, required admission to the intensive care unit. The patient died 26 days after admission, of bacteremia and fungemia. The mechanism of this potentially related AE is unknown but may be secondary to a localized autoimmune response to the medication.

Appearance-related side effects include alopecia, dry skin, and injection site reactions. Decreased appetite leading to weight loss and anorexia are also reported. Drug-induced eczema and other skin disorders secondary to interferon therapies have been well documented (Moore et al., 2004; Shen et al., 2005; Dalekos et al., 1998; Dereure et al., 2002). Hyperpigmentation has been described of the tongue and skin, particularly in dark-skinned non-Caucasian patients (Gurguta et al., 2005). Other notable dermatological diseases noted with interferon therapy include cutaneous sarcoidosis and psoriasis, as described and reviewed recently by Hurst and Mauro (2005) and Ketikoglou (2005), respectively.

Adverse events attributed to PEG-IFNs can be difficult to manage, requiring dose adjustments and potentially drug discontinuation. A summary of discontinuation rates in some of the major trials, including the HIV co-infection trials, appears in Table 6. Discontinuation rates were notably higher in the co-infection trials due to a higher frequency of drug-related adverse events. The addition of preemptive agents or drugs to treat or palliate adverse drug effects while continuing PEG-IFN is an option. These include granulocyte cell-stimulating factor or granulocyte-macrophage colony-stimulating factor for neutropenia (Lebray et al., 2005), oprelvekin (NEUMEGA®) for thrombocytopenia, erythropoeitin for anemia (Afdhal et al., 2004; Sulkowski et al., 2005; Shiffman et al., 2005), and selective serotonin reuptake inhibitors (Krauss et al., 2002) for depression.

Teratogenicity

Peginterferons are classified as pregnancy category C when used without RBV. The addition of RBV increases the severity of this classification to category X (PEGASYS® and PEG-INTRON® package inserts). Little to no collaborative data are available on the safety of PEG-IFNs in humans or animals during pregnancy. In Rhesus monkeys, unmodified interferon administered at 20 to 500 times the human weekly dose was associated with no teratogenic events. No information is currently available on the distribution of PEG-IFNs into breast milk, and as such, there are no published data on the safety of the drug in the breastfed child.

Conclusion

The pegylation of the interferon molecule represents a significant breakthrough in the management of chronic HCV infection when compared with unmodified interferon. The therapeutic effect is especially obvious when the pegylated interferons are combined with RBV. In addition, pegylation of interferon allows for less frequent administration, which may improve adherence to therapy. Despite extensive research, efforts to improve the SVR rate for subjects infected with genotypes 1 and 4 above the current 40–50% and 61–69% levels remains elusive. The likelihood of attaining an SVR in subjects infected with HCV genotype 2 or 3 with combination therapy with PEG-IFN/RBV regimens is excellent. However, new data indicate that response to therapy of genotype 3 may be less than that of genotype 2. More research is under way to address the issue of difficult-to-treat patients including subjects who do not respond to treatment or relapse after an end-of-treatment response. Other difficult-to-treat patients include African Americans as well as subjects undergoing liver transplantation. New guidelines to manage HCV co-infection in patients with HIV disease have been developed. Improved attention to adverse events has allowed patients to continue therapy and complete more of therapy, thus improving the chance of achieving an SVR. The combination of PEG-IFN products with new agents that attack hepatitis C in a variety of different sites appear promising, and we await the completion of phase III studies.

References

Abi-Nassif, S., Mark, E.J., Fogel, R.B., Hallisey, R.K., Jr. (2003). Pegylated interferon and ribavirin-induced interstitial pneumonitis with ARDS. *Chest*, 124: 406–410.

Afdhal, N.H., Dieterich, D.T., Pockros, P.J. (2004). Epoetin alfa maintains ribavirin dose in HCV-infected patients: A prospective, double blind, randomized controlled study. *Gastroenterology*, 126: 1302–1311.

Afdhal, N., Jacobson, I., Brown, R., Freilich, B., Santoro, J., Griffel, L., Bass C. (2006). The effect of liver fibrosis and cirrhosis on SVR in 4913 patients with hepatitis C: Results from the Win-R trial. *Gastroenterology*, 130 (4 Suppl 2): A-771; abstract 655.

Ahmed, F., Jacobson, I.M., Chen, S.T., Cofrancesco, S., Cooley, J., Demicco, M., Freilich, B., Hudes, B., Jensen, J., Levin, A., Lyche, K., McCone, J., Monsour, H.P., Peine, C., Pimstone, N., Rosenfield, T., Strauss, R., Stokes, K., Terrault, N., Tsai, N., Wasserman, R.B., Woolf, G.M., Brown, R.S., Brass, C. (2003). Serious ophthalmologic events during pegylated interferon and ribavirin therapy for chronic hepatitis C: Observations from the WIN-R trial. *Hepatology*, 38 (4 Suppl 1): 734–735A.

Alberti, A., Clumeck, N., Collins, S., Gerlich, W., Lundgren, J.D., Palu, G., Reiss, P., Thiebaut, R., Weiland, O., Yazdanpanah, Y., Zeuzem, S. (The ECC Jury) (2005). Short statement of the first European consensus conference on the treatment of chronic hepatitis B and C in HIV-coinfected patients. *Journal of Hepatology*, 42: 615–624.

Algranati, N.E., Sy, S., Modi, M. (1999). A branched methoxy 40 Kda polyethylene glycol (PEG) moiety optimizes the pharmacokinetics (PK) of peginterferon α-2a (PEG-INF) and may explain its enhanced efficacy in chronic hepatitis C (CHC). *Hepatology*, 30: 190A; abstract 120.

Arora, S., O'Brien, C., Zeuzem, S., Shiffman, M.L., Diago, M., Tran, A., Pockros, P.J., Reindollar, R.W., Gane, E., Patel, K., Wintfeld, N., Green, J. (2006). Treatment of chronic hepatitis C patients with persistently normal alanine aminotransferase levels with the combination of peginterferon α-2a (40 Kda) plus ribavirin: Impact on health-related quality of life. *Journal of Gastroenterology and Hepatology*, 21: 406–412.

Asahina, Y., Izumi, N., Enomoto, N., Uchihara, M., Kurosaki, M., Onuki, Y., Nishimura, Y., Ueda, K., Tsuchiya, K., Nakanishi, H., Kitamura, T., Miyake, S. (2005). Mutagenic effects of ribavirin and response to interferon/ribavirin combination therapy in chronic hepatitis C. *Journal of Hepatology*, 43: 623–629.

Bailon, P., Palleroni, A., Schaffer, C.A., Spence, C.L., Fung, W.J., Porter, J.E., Ehrlich, G.K., Pan, W., Xu, Z.X., Modi, M.W., Farid, A., Berthold, W., Graves, M. (2001). Rational design of a potent, long-lasting from of interferon: A 40 kDa branched polyethylene glycol-conjugated interferon alpha-2a for the treatment of hepatitis C. *Bioconjugates and Chemistry*, 12: 195–202.

Balan, V., Schwartz, D., Wu, G.Y., Muir, A.J., Ghalib, R., Jackson, J., Keeffe, E.B., Rossaro, L., Burnett, A., Goon, B.L., Bowers, P.J., Leitz, G.J. (2005). Erythropoietic response to anemia in chronic hepatitis C patients receiving combination pegylated interferon/ribavirin. *American Journal of Gastroenterology*, 100: 299–307.

Bani-Sadr, F., Carrat, F., Pol, S., Hor, R., Rosenthal, E., Goujard, C., Morand, P., Lunel-Fabiani, F., Salmon-Ceron, D., Piroth, L., Pialoux, G., Bentata, M., Cacoub, P., Perronne, C. for the ANRS Hc02-Ribavic Study Team. (2005). Risk factors for symptomatic mitochondrial toxicity in HIV/hepatitis C virus-coinfected patients during interferon plus ribavirin-based therapy. *Journal of AIDS*, 40(1): 47–52.

Barril, G., Quiroga, J.A., Sanz, P., Rodrìguez-Salvanés, F., Selgas, R., Carreño, V. (2004). Pegylated interferon-α2a kinetics during experimental haemodialysis: Impact of permeability and pore size of dialysers. *Alimentary and Pharmacology Therapy*, 20: 37–44.

Beckebaum, S., Cicinnati, V.R., Zhang, X., Malago, M., Dirsch, O., Erim, Y., Frilling, A., Broelsch, C.E., Gerken, G. (2004). Combination therapy with peginterferon alpha-2B and ribavirin in liver transplant recipients with recurrent HCV infection: Preliminary results of an open prospective study. *Transplant Proceedings*, 36: 1489–1491.

Benhamou, Y., Pockros, P., Rodriguez-Torres, M., Gordon, S., Shiffman, M., Lurie, Y., Afdhal, N., Lamon, K., Kim, Y., Murphy, B. (2006). The safety and efficacy of viramidine® plus pegylated interferon alpha-2b versus ribavirin plus pegylated interferon alpha-2b in therapy-naïve patients infected with HCV: Phase 3 results. *Journal of Hepatology*, 44 (Suppl 2): S273; abstract 751.

Berenguer, M., Prieto, M., Palau, A., Carrasco, D., Rayón, J.M., Calvo, F., Berenguer, J. (2004). Recurrent hepatitis C genotype 1b following liver transplantation: Treatment with combination interferon-ribavirin therapy. *European Journal of Gastroenterology and Hepatology*, 16: 1207–1212.

Berg, C., Goncales, F.L., Bernstein, D.E., Sette, H., Rasenack, J., Diago, M., Jensen, D.M., Graham, P., Cooksley, G. (2006a). Re-treatment of chronic hepatitis C patients after relapse: Efficacy of peginterferon-alpha-2a (40 kDa) and ribavirin. *Journal of Viral Hepatology*, 13: 435–440.

Berg, T., VonWagner, M., Nasser, S., Sarrazin, C., Heintges, T., Gerlach, T., Buggisch, P., Goeser, T., Rasenack, J., Pape, G.R., Schmidt, W.E., Kallinowski, B., Klinker, H., Spengler, U., Martus, P., Alshuth, R., Zeuzem, S. (2006). Extended treatment duration for hepatitis C virus type 1: Comparing 48 versus 72 weeks of peginterferon-alfa-2a plus ribavirin. *Gastroenterology*, 130: 1086–1097.

Bernfort, L., Sennfält, K., Reichard, O. (2006). Cost-effectiveness of peginterferon alfa-2b in combination with ribavirin as initial treatment for chronic hepatitis C in Sweden. *Scandinavian Journal of Infectious Diseases*, 38: 497–505.

Biselli, M., Andreone, P., Gramenzi, A., Lorenzini, S., Loggi, E., Bonvicini, F., Cursaro, C., Bernardi, M. (2006). Pegylated interferon plus ribavirin for recurrent hepatitis C infection after liver transplantation in naive and non-responder patients on a stable immunosuppresive regimen. *Digest of Liver Diseases*, 38: 27–32.

Bizollon, T., Ahmed, S.N.S., Radenne, S., Chevallier, M., Chevallier, P., Parvaz, P., Guichard, S., Ducerf, C., Baulieux, J., Zoulim, F., Trepo, C. (2003). Long term histological improvement and clearance of intrahepatic hepatitis C virus RNA following sustained response to interferon-ribavirin combination therapy in liver transplanted patients with hepatitis C virus recurrence. *Gut*, 52: 283–287.

Bodenheimer, H.C., Lindsay, K.L., Davis, G.L., Lewis, J.H., Thung, S.N., Seeff, L.B. (1997). Tolerance and efficacy of oral ribavirin treatment of chronic hepatitis C: A multicenter trial. *Hepatology*, 26 : 473–477.

Bonny, C., Roche, C., Fontaine, H., Poynard, T., Héxode, C., Larrey, D., Marcellin, P., Bourlière, M., Bronowicki, J.P., Merle, P., Zarski, J.P., Nicolas, C., Randl, K., Bommelaer, G., Abergel, A. (2005). Eficacy of interferon (standard or pegylated) plus ribavirin in naive patients with hepatitis C virus genotype 5. A French national study. *Journal of Hepatology*, 42 (Suppl 2): 200; abstract 549.

Brau, N. (2003). Update on chronic hepatitis C in HIV/HCV-coinfected patients: Viral interactions and therapy. *AIDS*, 27: 2279–2290.

Brennan, B., Morrison, R., Hagedorn, C., Marbury, T.C., Sulkowski, M., Grippo, J., Gries, J-M. (2005). Effect of ethnicity on the pharmacokinetics of ribavirin (COPEGUS®) and peginterferon alfa-2a (40KD) (PEGASYS®) in patients with chronic hepatitis C. *Hepatology*, 42 (Suppl 1): 653A; abstract 1161.

Bressler, B., Wang, K., Gries, J.-M., Heathcote, J. (2005). Pharmacokinetics and response of obese patients with chronic hepatitis C treated with different doses of PEG-IFN alpha 2A (40KD) (PEGASYS®). *Hepatology*, 42 (Suppl 1): 661A; abstract 1183.

Brown, R.S., Jacobson, I.M., Afdhal, N., Freilich, B., Regenstein, F., Flamm, S., Kwo, P., Pauly, M.P., Griffel, L.H., Brass, C.A. (2006). Differences in treatment outcome to antiviral therapy based on genotype and viral load in hepatitis C genotypes 2 and 3 in the Win-R trial. *Gastroenterology*, 130 (4 Suppl 2): A-767; abstract 523.

Bruno, R., Sacchi, P., Maiocchi, L., Zocchetti, C., Ciappina, V., Patruno, S., Filice, G., (2005). Area-under-the-curve for peginterferon alpha-2a and peginterferon alpha-2b is not related to body weight in treatment-naive patients with chronic hepatitis C. *Antiviral Therapy*, 10: 201–205.

Burton, J.R., Klarquist, J., Im, K.A., Belle, S.H., Rosen, H.R. (2006). Stronger baseline HCV-specific immunity is associated with a higher likelihood of a sustained virological response to combination antiviral therapy of chronic hepatitis C. *Gastroenterology*, 130 (4 Suppl 2): abstract S1057.

Buti, M., Casado, M.A., Fosbrook, L., Esteban R. (2005). Financial impact of two different ways of evaluating early virological response to peginterferon α-2b plus ribavirin therapy in treatment-naïve patients with chronic hepatitis C virus genotype 1. *Pharmacoeconomics* 23: 1043–1055.

Buti, M., Sanchez-Avila, F., Lurie, Y., Stalgis, C., Valdéz, A., Martell, M., Esteban, R. (2002). Viral kinetics in genotype 1 chronic hepatitis C patients during therapy with 2 different doses of peginterferon alfa-2b plus ribavirin. *Hepatology*, 35: 930–936.

Byrnes, V., Miller, A., Weinstein, C., Hill, E., Lenkinski, R., Alsop, D., Afdhal, N.H. (2005). Cerebral metabolic and neuropsychiatric effects of pegylated interferon (PIFN) therapy in hepatitis C. *Gastroenterology*, 128 (Suppl 2): A684 [abstract].

Cammà, C., Di Bona, D., Schepis, F., Heathcote, E.J., Zeuzem, S., Pockros, P.J., Marcellin, P., Balart, L., Alberti, A., Craxì, A. (2004). Effect of peginterferon alfa-2a on liver histology in chronic hepatitis C: A meta-analysis of individual patients' data. *Hepatology*, 39: 333–342.

Carrat, F., Bani Sadr, F., Pol, S., Rosenthal, E., Lunel-Fabiani, F., Benzekri, A., Morand, P., Goujard, C., Pialoux, G., Piroth, L., Salmon-Ceron, D., Degott, C., Cacoub, P., Perronne, C. for the ANRS HCO2 RIBAVIC Study Team. (2004). Pegylated interferon alfa-2b vs standard interferon alfa-2b, plus ribavirin, for chronic hepatitis C in HIV-infected patients: A randomized controlled trial. *JAMA*, 292: 2839–2848.

Castera, L., Constant, A, Henry, C., Champenoit, P., Sauve, G., de Ledinghen V., Bernard P.H., Foucher, J., Demotes-Mainard, J., Couzigou, P. (2004). Psychiatric events during peginterferon and ribavirin therapy in chronic hepatitis C (CHC): Results of a prospective study in 98 patients. *Journal of Hepatology*, 40 (Suppl 1): 139 [abstract].

Chalasani, N., Manzarbeitia, C., Ferenci, P., Vogel, W., Fontana, R.J., Voigt, M., Riely, C., Martin, P., Teperman, L., Jiao, J., Lopez-Talavera, J.C. for the Pegasys Transplant Study Group. (2005). Peginterferon alfa-2a for hepatitis C after liver transplantation: Two randomized, controlled trials. *Hepatology*, 41: 289–298.

Chatelut, E., Rostaing, L., Grégoire, N., Payen, J.L., Pujol, A., Izopet, J., Houin, G., Canal, P. (1999). A pharmacokinetic model for alpha interferon administered subcutaneously. *British Journal of Clinical Pharmacology*, 47: 365–371.

Chung, R.T., Andersen, J., Volberding, P., Robbins, G.K., Liu, T., Sherman, K.E., Peters, M.G., Koziel, M.J., Bhan, A.K., Alston, B., Colquhoun, D., Nevin, T., Harb, G., van der Horst, C. for the AIDS Clinical Trials Group A5071 Study Team (2004). Peginterferon alfa-2a plus ribavirin versus interferon alfa-2a plus ribavirin for chronic hepatitis C in HIV-coinfected persons. *New England Journal of Medicine*, 351: 451–459.

ClinicalTrials.gov Identifier NCT00077649. A randomized, double-blind study of the effect of PEGASYS and ribavirin combination therapy on viral kinetics and virologic response in interferon-naïve patients with chronic hepatitis C genotype 1 infection. (Study ID number: NV17318). (http://www.clinicaltrials.gov/ct/show/NCT00077649?order=1). Accessed 9/4/06.

ClinicalTrials.gov Identifier NCT 00255008. SEASON South East Asian study of novel genotypes in hepatitis C infection: Pegylated-interferon and ribavirin therapy (PEGATRON REDIPEN combination therapy (PEG-Intron®) REDIPEN plus REBETOL®) in treatment naïve patients with genotypes 1, 6, 7, 8, 9: A comparison of race and genotype on treatment outcome (http://www.clinicaltrials.gov/ct/show/NCT00255008?order=3). Accessed 9/4/06.

ClinicalTrials.gov Identifier NCT00255034. Effects of 48 weeks versus 24 weeks of therapy with peg-Intron/ribavirin in patients with chronic hepatitis C, genotype 3 (study P04143) (http://www.clinicaltrials.gov/ct/show/NCT00255034). Accessed 7/29/06.

ClinicalTrials.gov Identifier NCT00266318. Study of high dosage CIFN plus RBV for HCV genotype 1 infected patients who are nonresponders to prior therapy (http://www.clinicaltrials.gov/ct/show/NCT00266318?order=2). Accessed 8/13/06.

ClinicalTrials.gov Identifier NCT00336479. A phase 2 study of VX-950 in combination with peginterferon-2a (Pegasys®), with ribavirin (Copegus®) in subjects with genotype 1 hepatitis C who have not received prior treatment. (Study VX05-950-104) (http://www.clinicaltrials.gov/ct/show/NCT00336479?order=1). Accessed 8/24/06.

ClinicalTrials.gov NCT00160251. PEG-Intron/REBETOL vs PEG-Intron/SCH503034 with and without ribavirin in chronic hepatitis C HCV-1 peginterferon alfa/ribavirin nonresponders: A SCH 503034 dose-finding phase 2 study. (Study P03659AMS1) (http://www.clinicaltrials.gov/ct/show/NCT00160251?order=1). Accessed 8/30/06.

Cohen, M.C., Huberman, M.S., Nesto, R.W. (1990). Interferon alfa-induced cardiac dysfunction. *New England Journal of Medicine,* 322: 1469.

Collantes, R.S., Younossi, Z.M. (2005). The use of growth factors to manage the hematologic side effects of PEG-interferon alfa and ribavirin. *Journal of Clinical Gastroenterology,* 39 (Suppl 1): S9–S13.

Condat, B., Asselah, T., Zanditenas, D., Estampes, B., Cohen, A., O'Toole, D., Bonnet, J., Ngo, Y., Marcellin, P., Belazquez, M. (2006). Fatal cardiomyopathy associated with pegylated interferon/ribavirin in a patient with chronic hepatitis C. *European Journal of Gastroenterology and Hepatology,* 18: 287–289.

Conjeevaram, H.S., Fried, M.W., Jeffers, L.J., Terrault, N.A., Wiley-Lucas, T.E., Afdhal, N., Brown, R.S., Belle, S.H., Hoofnagle, J.H., Kleiner, D.E., Howell, C.D. for the Virahep-C Study Group (2006). Peginterferon and ribavirin treatment in African Americans and Caucasian American patients with hepatitis C genotype 1. *Gastroenterology,* 131: 470–477.

Curry, M., Cardenas, A., Afdhal, N.H. (2005). Effect of maintenance PEG-INTRON therapy on portal hypertension and its complications: Results from the COPILOT study. *Journal of Hepatology,* 42 (Suppl 2): 40; abstract 95.

D'Heygere, F., George, C., Nevens, R., Van Vlierberghe, H., Van Der Meeren, O. (2005). Patients infected with HCV-5 present the same response rate as patients infected with HCV-1: Results from the Belgian randomised trial for naive and relapsers (BERNAR-1). *Journal of Hepatology,* 42 (Suppl 2): 203; abstract 558.

Dalekos, G.N., Hatzis, J., Tsianos, E.V. (1998). Dermatologic disease during interferon alpha therapy for chronic viral hepatitis. *Annals of Internal Medicine,* 128: 409–410.

Davis, G.L. (2002). Monitoring of viral levels during therapy of hepatitis C. *Hepatology,* 36 (Suppl 1): S145–S151.

De Franceschi, L., Fattovich, G., Turrini, F., Ayi, K., Brugnara, C, Manzato, F., Noventa, F., Stanzial, A., Solero, P., Corrocher, R. (2000). Hemolytic anemia induced by ribavirin therapy in patients with chronic hepatitis C virus infection: Role of membrane oxidative damage. *Hepatology,* 31: 997–1004.

Dereure, O., Raison-Pevron, N., Larrey, D., Blanc, F., Guilhou, J.J. (2002). Diffuse inflammatory lesions in patients treated with interferon alfa and ribavirin for hepatitis C: A series of 20 patients. *British Journal of Dermatology,* 147: 1142–1146.

Dev, A.T., McCaw, R., Sundararajan, V., Bowden, S., Sievert, W. (2002). Southeast Asian patients with chronic hepatitis C: The impact of novel genotypes and race on treatment outcome. *Hepatology,* 36: 1259–1265.

Diago, M.., Romero-Gomez, M.., Crespo, J., Olveira, A., Perez, R., Barcena, R., Sanchez-Tapias, J.M., Munoz-Sanchez, M. (2006). Pharmacokinetics and pharmacodynamics of induction doses of peginterferon alpha-2a (40 KD) (PEGASYS®) and ribavirin (COPEGUS®) in HCV genotype 1 patients who failed to respond to interferon/ribavirin. *Journal of Hepatology,* 44 (Suppl 2): S210; abstract 566.

DiBisceglie, A.M., Fan, X., Chambers, T., Strinko, J. (2006). Pharmacokinetics, pharmacodynamics and hepatitis C viral kinetics during antiviral therapy: The null responder. *Journal of Medical Virology,* 78: 446–451.

Dieterich, D., Lawitz, E., Nguyen, T., Younes, Z., Santoro, J., Gitlin, N., McEniry, D., Chasen, R., Goff, J., Knox, S., Kleber, K., Belanger, B., Brown, N. (2006). Early clearance of HCV RNA with valopicitabine (NM283) plus peg-interferon in treatment-naive patients with HCV-1 infection: First results form a phase IIb trial. *Journal of Hepatology,* 44 (Suppl 2): S271; abstract 736.

Dominguez, S., Ghosn, J., Valantin, M.A., Schruniger, A., Simon, A., Bonnard, P., Caumes, E., Pialoux, G., Benhamou, Y., Thibault, V., Katlama, C. (2006). Efficacy of early treatment of acute hepatitis C infection with pegylated interferon and ribavirin in HIV-infected patients. *AIDS,* 20: 1157–1161.

Dumortier, J., Scoazec, J., Chevallier, P., Boillot, O. (2004). Treatment of recurrent hepatitis C after liver transplantation: A pilot study of peginterferon alfa-2b and ribavirin combination. *Journal of Hepatology*, 40: 669–674.

Dusheiko, G., Main, J., Thomas, H., Reichard, O., Lee, C., Dhillon, A., Rassam, S., Fryden, A., Reesink, H., Bassendine, M., Norkrans, G., Cuypers, T., Lelie, N., Tefler, P., Watson, J., Weegink, C., Sillikens, P., Weiland, O. (1996). Ribavirin treatment for patients with chronic hepatitis C: Results of a placebo-controlled study. *Journal of Hepatology*, 25: 591–598.

Elefsiniotis, I.S., Pantazis, K.D., Fotos, N.V., Moulakakis, A., Mavrogiannis, C. (2006). Late onset autoimmune thrombocytopenia associated with pegylated interferon alfa-2b plus ribavirin treatment for chronic hepatitis C. *Journal of Gastroenterology and Hepatology*, 21: 622–623.

Elmakhzangy, H., Rekacewicz, C., Shouman, S.I., Mohamed, H.N., Esmat, G., Ismail, A., Rafaat, R., El Hosseiny, M., El Daly, M., El Kafrawy, S., Abdel Hamid, M., Fontanet, A., Pol, S., Mohamed, M.K. (2005). Combined pegylated interferon alfa-2a and ribavirin treatment of chronic hepatitis C in Egypt (ANRS 1211 Trial). *Journal of Hepatology*, 42 (Suppl 2): 203; abstract 559.

Farel, C., Suzman, D.L., McLaughlin, M. (2004). Serious ophthalmic pathology compromising vision in HCV/HIV co-infected patients treated with peg interferon alpha-2b and ribavirin. *AIDS*, 18: 1805–1809.

Ferenci, P., Bergholz, U., Laferl, H., Scherzer, T.M., Maieron, A., Gschwantler, M., Brunner, H., Hubmann, R.R., Datz, C., Bischof, M., Stauber, R., Steindl-Munda, P. (2006). 24 week treatment regimen with peginterrferon alpha-2a (40 KD) (PEGASYS®) plus ribavirin (COPEGUS®) in HCV genotype 1 or 4 "super-responders." *Journal of Hepatology*, 44 (Suppl 2): S6; abstract 8.

Ferenci, P., Fried, M.W., Shiffman, M.L., Smith, C.I., Marinos, G., Gonçales, F.L., Häussinger, D., Diago, M., Carosi, G., Dhumeaux, D., Craxì, A., Chaneac, M., Reddy, K.R. (2005). Predicting sustained virological responses in chronic hepatitis C patients treated with peginterferon alfa-2a (40KD)/ribavirin. *Journal of Hepatology*, 43: 425–433.

Firpi, R., Abdelmalek, M., Soldevila-Pico, C. (2002). Combination of interferon alfa-2b and ribavirin in liver transplant recipients with histological recurrent hepatitis C. *Liver Transplantation*, 8: 1000–1006.

Fonseca, V., Thomas, M., Dusheiko, G. (1991). Thyrotropin receptor antibodies following treatment with recombinant alpha-interferon in patients with hepatitis. *Acta Endocrinology*, 125: 491–493.

Food and Drug Administration Antiviral Drugs Committee Briefing document. Pegylated interferon alfa 2a (Pegasys®). Bethesda, MD (May 16, 2006). http://www.fda.gov/ohrms/dockets/ac/02/briefing/3909b1.htm.

Fried, M.W., Shiffman, M.L., Reddy, K.R., Smith, C., Marinos, G., Gonçales, F.L., Häussinger, D., Diago, M., Carosi, G., Dhumeaux, D., Craxi, A., Lin, A., Hoffman, J., Yu, J. (2002). Peginterferon alfa-2a plus ribavirin for chronic hepatitis C virus infection. *New England Journal of Medicine*, 347: 975–982.

Fried, M.W. (2002). Side effects of therapy of hepatitis C and their management. *Hepatology*, 36 (5 Suppl 1): S237–S244.

Gaglio, P., Choi, J., Zimmerman, D., Heller, L., Brown R.S. (2005). Weight based ribavirin in combination with pegylated interferon alpha 2-B does not improve SVR in HCV infected patients who failed prior therapy: Results in 454 patients. *Hepatology*, 42 (4 Suppl 1): 219A; abstract 59.

Gane, E., Diago, M., Mohankumar, A., Wintfeld, N., Patel, K., Zeuzem, S., Shiffman, M., Reindollar, R. (2005). Health-related quality of live (HRQL) in chronic hepatitis C (CHC) patients with elevated vs. persistently "normal" ALT levels (PNALT): Comparison of data from patients treated with peginterferon alfa-2a (40 KD) plus ribavirin in two multinational trials. *Gastroenterology* 128 (4 Suppl 2): A-747; abstract M956.

Ganser, A., Carlo-Stella, C., Greher, J., Volkers, B., Hoelzer, D. (1987). Effect of recombinant interferons alpha and gamma on human bone marrow-derived megakaryocytic progenitor cells. *Blood*, 70: 1173–1179.

Garfein, R.S., Doherty, M.C., Monterroso, E.R., Thomas, D.L., Nelson, K.E., Vlahov, D. (1998). Prevalence and incidence of hepatitis C virus infection among young adult injection drug users. *Journal of AIDS Human Retrovirology,* 18 (Suppl 1): S11–S19.

Ghosn, J., Pierre-François, S., Thibault, V., Duvivier, C., Tubiana, R., Simon, A., Valantin, M.A., Dominguez, S., Caumes, E., Katlama, C. (2004). Acute hepatitis C in HIV-infected men who have sex with men. *HIV Medicine,* 5: 303–306.

Giostra, E., Kullak-Ublick, G., Keller, W. (2004). Ribavirin/interferon-alpha sequential treatment of recurrent hepatitis C after liver transplantation. *Transplantation International,* 17: 169–176.

Glue, P., Rouzier-Panis, R., Raffanel, C., Sabo, R., Gupta, S.K., Salfi, M., Jacobs, S., Clement, R.P., the Hepatitis C Intervention Therapy Group. (2000a). A dose-ranging study of pegylated interferon alfa-2b and ribavirin in chronic hepatitis C. *Hepatology,* 32: 647–653.

Glue, P., Fang, J.W.S., Rouzier-Panis, R., Raffanel, C., Sabo, R., Gupta, S.K., Salfi, M., Jacobs, S., the Hepatitis C Intervention Therapy Group. (2000b). Pegylated interferon α-2b: Pharmacokinetics, pharmacodynamkics, safety, and preliminary efficacy data. *Clinical Pharmacology Therapy,* 68: 556–567.

González-Peralta, R.P., Kelly, D.A., Harber, B., Molleston, J., Murray, K.F., Jonas, M.M., Shelton, M., Mieli-Verganti, G., Lurie, Y., Martin, S., Lang, T., Baczkowski, A., Geffner, M., Gupta, S., Laughlin, M., for the International Pediatric Hepatitis C Therapy Group. (2005). Interferon alfa-2b in combination with ribavirin for the treatment of chronic hepatitis C in children: Efficacy, safety, and pharmacokinetics. *Hepatology,* 42: 1010–1018.

Greub, G., Ledergerber, B., Battegay, M., Grob, P., Perrin, L., Furrer, H., Burgisser, P., Erb, P., Boggian, K., Piffaretti, J.-C., Hirschel, B., Janin, P., Francioli, P., Flepp, M., Telenti, A. for the Swiss HIV Cohort Study (2000). Clinical progression, survival, and immune recovery during antiretroviral therapy in patients with HIV-1 and hepatitis C virus co infection: The Swiss HIV Cohort Study. *Lancet,* 356: 1800–1805.

Gross, J., Johnson, S., Kwo, P., Afdhal, N., Flamm, S., Therneau, T.M. (2005). Double-dose peginterferon alfa-2b with weight based ribavirin improves response for interferon/ribavirin non-responders with hepatitis C: Final results of "RENEW." *Hepatology,* 42 (4 Suppl 1): 219A–220A; abstract 60.

Gross, J., Johnson, S.M., Therneau, T.M., Kwo, P.Y. (2005a). Double-dose peginterferon alfa-2b plus weight-based ribavirin for re-treatment of African-American non-responders with hepatitis C. *Gastroenterology,* 128 (4 Suppl 2): A-684; abstract 87.

Gupta, S.K., Glue, P., Jacobs, S., Belle, D., Affrime, M. (2003). Single-dose pharmacokinetics and tolerability of pegylated interferon-alpha 2b in young and elderly healthy subjects. *British Journal of Clinical Pharmacology,* 56: 131–134.

Gupta, S.K., Pittenger, A.L., Swan, S.K., Marbury, T.C., Tobillo, E., Batra, V., Sack, M., Glue, P., Jacobs, S., Affrime, M. (2002). Single-dose pharmacokinetics and safety of pegylated interferon-α2b in patients with chronic renal dysfunction. *Journal of Clinical Pharmacology,* 42: 1109–1115.

Gurguta, C., Kauer, C., Bergholz, U., Formann, E., Steindl-Munda, P., Ferenci, P. (2006). Tongue and skin hyperpigmentation during PEG-interferon-alpha/ribavirin therapy in dark skinned non-Caucasian patients with chronic hepatitis C. *American Journal of Gastroenterology,* 101: 197–198.

Hadziyannis, S.J., Sette, H., Morgan, T.R., Balan, V., Diago, M., Marcellin, P., Ramadori, G., Bodenheimer, H., Bernstein, D., Rizzetto, M., Zeuzem, S., Pockros, P.J., Lin, A., Ackrill, A.M. PEGASYS International Study Group. (2004). Peginterferon alpha 2a and ribavirin combination therapy in chronic hepatitis C: A randomized study of treatment duration and ribavirin dose. *Annals of Internal Medicine,* 140: 346–355.

Harris, J.M., Martin, N.E., Modi, M. (2001). Pegylation: A novel process for modifying pharmacokinetics. *Clinical Pharmacokinetics,* 40: 539–551.

Hasan, F., Alsarraf, K., Qabandi, W. (2006). Pegylated interferon alfa-2b plus ribavirin for the treatment of chronic hepatitis C genotype 4 in adolescents: A pilot study. *Gastroenterology,* 130 (Suppl 2): A-840; abstract T1819.

Hasan, F., Asker, H., Al-Khalid, J., Siddique, I., Al-Ajmi, M., Owaid, S., Varghese, R., Al-Nakib, B. (2004). Peginterferon alfa-2b plus ribavirin for the treatment of chronic hepatitis C genotype 4. *American Journal of Gastroenterology*, 99: 1733–1737.

Hassanein, T., Cooksley, G., Sulkowski, M., Smith, C., Marinos, G., Lai, M-Y., Pastore, G., Trejo-Estrada, R., Vale, A.H.E., Wintfeld, N., Green, J. (2004). The impact of peginterferon alfa-2a plus ribavirin combination therapy on health-related quality of life in chronic hepatitis C. *Journal of Hepatology*, 40: 675–681.

Heathcote, E.J., Pockros, P.J., Fried, M.E., Bain, M.A., DePamphilis, J., Modi, M. (1999). The pharmacokinetics of pegylated-40K interferon alfa-2a (PEG-IFN) in chronic hepatitis C (CHC) patients with cirrhosis. *Gastroenterology*, 116: A735; abstract G3190.

Heathcote, E.J., Shiffman, M.L., Cooksley, W.G.E., Dusheiko, G.M., Lee, S.S., Balart, L., Reindollar, R., Reddy, R.K., Wright, T.L., Lin, A., Hoffman, J., DePamphilis, J. (2000). Peginterferon alfa-2a in patients with chronic hepatitis C and cirrhosis. *New England Journal of Medicine*, 343: 1673–1680.

Herrine, S.K., Brown, R.S., Bernstein, D.E., Ondovik, M.S., Lentz, E., Te, H. (2005). Peginterferon α-2a combination therapies in chronic hepatitis C patients who relapsed after or had a viral breakthrough on therapy with standard interferon α-2b plus ribavirin: A pilot study of efficacy and safety. *Digestive Disease Sciences*, 50: 719–726.

Herrmann, E., Lee, J.-H., Marinos, G., Modi, M., Zeuzem, S. (2003). Effect of ribavirin on hepatitis C viral kinetics in patients treated with pegylated interferon. *Hepatology*, 37: 1351–1358.

Hershow, R.C., O'Driscoll, P.T., Handelsman, E., Pitt, J., Hillyer, G., Serchuck, L., Lu, M., Chen, K.T., Yawetz, S., Pacheco, S., Davenny, K., Adeniyi-Jones, S., Thomas, D.L. (2005). Hepatitis C virus coinfection and HIV load, CD_4^+ cell percentage, and clinical progression to AIDS or death among HIV-infected women: Women and infants transmission study. *Clinical Infectious Diseases*, 40: 859–867.

Homoncik, M., Jessner, W., Sieghardt, W. (2003). Effect of therapy with interferon-alpha and pegylated interferon-alpha on platelet plug formation in patients with chronic hepatitis C. *Journal of Hepatology*, 38 (Suppl 2): 145; abstract 497.

Hornberger, J., Dusheiko, G., Lewis, G., Patel, K. (2005). An evaluation of the cost-effectiveness of peginterferon alfa-2a (40KD) (PEGASYS®) plus ribavirin (COPEGUS®) for the first treatment of mild chronic hepatitis C (CHC). *Hepatology*, 42 (4 Suppl 1): 653A; abstract 1162.

Howell, C., Dowling, T., Haritos, M., Terrault, N., Thelma, W.-L., Taylor, M. (2006). Relationship between serum peginterferon and 2,5-OAS kinetics and virologic responses in African Americans and Caucasians patients with chronic hepatitis C, genotype 1, during peginterferon combination therapy. *Gastroenterology*, 130 (4 Suppl 2): abstract 199.

Hui, C.K., Yuen, M.F., Sablon, E., Chan, A.O., Wong, B.C., Lai, C.L. (2003). Interferon and ribavirin therapy for chronic hepatitis C virus genotype 6: A comparison with genotype 1. *Journal of Infectious Diseases*, 187: 1071–1074.

Hultgren, C., Milich, D.R., Weiland, O., Sällberg, M. (1998). The antiviral compound ribavirin modulates the T helper (Th) 1/Th 2 subset balance in hepatitis B and C virus-specific immune responses. *Journal of General Virology*, 79: 2381–2391.

Hurst, E.A., Mauro, T. (2005). Sarcoidosis associated with pegylated interferon alfa and ribavirin treatment for chronic hepatitis C. *Archives of Dermatology*, 141: 865–868.

Imagawa, A., Itoh, N., Hanafusa, T., Oda, Y., Waguri, M., Miyagawa, J. (1995). Autoimmune endocrine disease induced by recombinant interferon-α therapy for chronic active type C hepatitis. *Endocrinology and Metabolism*, 80: 922–926.

Izumi, N., Iino, S., Okuno, T., Omata, M., Kiyosawa, K., Kumada, H., Hayashi, N., Yamada, G., Sakai, T. (2006). High response rates with peginterferon α-2a (40KD) (PEGASYS®) PLUS ribavirin (Copegus®) in Japanese non-responders or relapsers to conventional interferon. *Journal of Hepatology*, 44 (Suppl 2): S216; abstract 582.

Jacobson, I., Brown, R., McCone, J., Black, M., Albert, C., Dragutsky, M., Siddiqui, F., Hargrave, T., Kwo, P., Gitlin, N.l, Lambiasi, L., Galler, G., Araya, V., Freilich, B., Harvey, J., Bass, C., and the WIN-R Study Group. (2004). Weight-based ribavirin dosing improves virologic response in HCV-infected genotype 1 African-American (AA) compared to flat dose ribavirin with peginterferon alfa-2b combination therapy. *Hepatology* 40 (4 Suppl 1): 217A; abstract 125.

Jacobson, I.M., Brown, R.S., Freilich, B., Afdhal, N., Kwo, P., Santoro, J., Becker, S., Wakil, A., Pound, D., Godofsky, E., Straus, R., Bernstein, D., Flamm, S., Bala, N., Araya, V., Davis, M., Monsour, H., Vierling, J., Regenstein, F., Balan, V., Dragutsky, M., Epstein, M., Herring, R.W., Rubin, R., Galler, G., Pauly, M.P., Griffel, L.H., Brass, C.A., The WIN-R Study Group. (2005). Weight-based ribavirin dosing (WBD) increases sustained viral response (SVR) in patients with chronic hepatitis C (CHC): Final results of the WIN-R study, a US community based trial. *Hepatology*, 42 (Suppl 1): 749A; abstract LB03.

Jacobson, I.M., Brown, R.S., Freilich, B., Afdhal, N., Kwo, P., Santoro, J., Becker, S., Wakil, A., Pound, D., Ahmed, F., Griffel, L.H., Brass, C.A. (2006). Stratification of high viral load: Impact on sustained virologic response in the Win-R trial. *Gastroenterology*, 130 (4 Suppl 2): A837; abstract T1806.

Jacobson, I.M., Gonzalez, S.A., Ahmed, F., Lebovics, E., Min, A.D., Bodenheimer, H.C., Esposito, S.P., Brown, R.S., Bräu, N., Klion, F.M., Tobias, H., Bini, E.J., Brodsky, N., Cerulli, M.A., Aytaman, A., Gardner, P.W., Beders, J.M., Spivack, J.E., Rahmin, M.G., Berman, D.H., Ehrlich, J., Russo, M.W., Chait, M., Rovner, D., Edlin, B.R. (2005). A randomized trial of pegylated interferon α-2b plus ribavirin in the retreatment of chronic hepatitis C. *American Journal of Gastroenterology*, 100: 2453–2462.

Jeffers, L.J., Cassidy, W., Howell, C.D., Hu, S., Reddy, K.R. (2004). Peginterferon alfa-2a (40kd) and ribavirin for Black American patients with chronic HCV genotype 1. *Hepatology*, 39: 1702–1708.

Jensen, D.M., Morgan, T.R., Marcellin, P., Pockros, P.J., Reddy, K.R., Hadziyannis, S.J., Ferenci, P., Ackrill, A.M., Willems, B. (2006). Early identification of HCV genotype 1 patients responding to 24 weeks peginterferon α-2a (40 kd)/ribavirin therapy. *Hepatology*, 43: 954–960.

Jessner, W., Kinaciyan, T., Formann, E. (2002). Severe skin reactions during therapy for chronic hepatitis C associated with delayed hypersensitivity to pegylated interferons. *Hepatology*, 36: 361A; abstract 793.

Kaiser, S., Hass, H., Lutze, B., Gregor, M. (2006). Comparison of daily Consensus interferon versus peginterferon alfa-2a extended therapy of 72 weeks for peginterferon/ribavirin relapse patients. *Gastroenterology*, 130: A-784; abstract S1060.

Kamal, S.M., El Tawil, A.A., Nakano, T., He, Q., Rasenack, J., Hakam, S.A., Saleh, W.A., Ismail, A., Aziz, A.A., Ali Madwar, M. (2005). Peginterferon α-2b and ribavirin therapy in chronic hepatitis C genotype 4: Impact of treatment duration and viral kinetics on sustained virological response. *Gut*, 54: 858–866.

Kamal, S.M., Fehr, J., Roesler, B., Peters, T., Rasenack, J.W. (2002). Peginterferon alone or with ribavirin enhances HCV-specific CD4+ T-Helper 1 responses in patients with chronic hepatitis C. *Gastroenterology*, 123: 1070–1083.

Kamal, S.M., Fouly, A.E., Kamel, R.R., Hockenjos, B., Al Tawil, A., Khalifa, K.E., He, Q., Koziel, M.J., El Naggar, K.M., Rasenack, J., Afdhal, N.H. (2006a). Peginterferon alfa-2b therapy in acute hepatitis C: Impact of onset of therapy on sustained virologic response. *Gastroenterology*, 130: 632–638.

Kamal, S.M., Moustafa, K.N., Chen, J., Fehr, J., Moneim, A.A., Khalifa, K.E., El Gohary, L.A., Ramy, A.H., Madwar, M.A., Rasenack, J., Afdhal, N.H. (2006b). Duration of peginterferon therapy in acute hepatitis C: A randomized trial. *Hepatology*, 43: 923–931.

Ketikoglou, I., Karatapanis, S., Elefsiniotis, I., Kafiri, G., Moulakakis, A. (2005), Extensive psoriasis induced by pegylated interferon alpha-2b treatment for chronic hepatitis B. *European Journal of Dermatology*, 15: 107–109.

Khuroo, M.S., Khuroo, M.S., Dahab, S.T. (2004). Meta-analysis: A randomized trial of peginterferon plus ribavirin for the initial treatment of chronic hepatitis C genotype 4. *Alimentary Pharmacology and Therapy*, 20: 931–938.

Kozlowski, A., Harris J.M. (2001). Improvements in protein PEGylation: Pegylated interferons for treatment of hepatitis C. *Journal of Controlled Release*, 72: 217–224.

Kramer, J.R., Giordano, T.P., Souchek, J., El-Serag, H.B. (2005). Hepatitis C co infection increases the risk of fulminant hepatic failure in patients with HIV in the HAART era. *Journal of Hepatology*, 42: 309–314.

Kraus, M.R., Schäfer, A., Faller, H., Csef, H., Scheurlen, M. (2002). Paroxetine for the treatment of interferon-alpha-induced depression in chronic hepatitis C. *Alimentary Pharmacology and Therapy*, 16: 1091–1099.

Kraus, M.R., Schafer, A., Wibmann, S., Reimer, P., Scheurlen, M. (2005). Neurocognitive changes in patients with hepatitis C receiving interferon alfa-2b and ribavirin. *Clinical Pharmacology and Therapy*, 77: 90–100.

Kuehne, F., Bethe, U., Freedberg, K., Goldie, S. (2002). Treatment for hepatitis C virus in HIV-infected patients: Clinical benefits and cost-effectiveness. *Archives of Internal Medicine*, 162: 2545–2556.

Kuo, A., Terrault, N.A. (2006). Management of hepatitis C in liver transplant recipients. *American Journal of Transplantation*, 6: 449–458.

Kuwata, A., Ohashi, M., Sugiyama, M., Ueda, R., Dohi, Y. (2002). A case of reversible dilated cardiomyopathy after alpha-interferon therapy in a patient with renal cell carcinoma. *American Journal of Medicine*, 324: 331–334.

Lam, N.P., Pitrak, D., Speralakis, R., Lau, A.H., Wiley, T.E., Layde, T.J. (1997). Effect of obesity on pharmacokinetics and biologic effect of interferon-alpha in hepatitis C. *Digestive Disease Sciences*, 42: 178–185.

Lamb, M.W., Martin, N.E. (2002). Weight-based versus fixed dosing of peginterferon (40 kDa) alfa-2a. *Annals of Pharmacotherapy*, 36: 933–935.

Lamb, M.W., Marks, I.M., Wynohradnyk, L., Modi, M.W., Preston, R.A., Pappas, C. (2001). 40 KDA peginterferon alfa-2a (PEGASYS®) can be administered safely in patients with end-stage renal disease. *Hepatology*, 34: 326A; abstract 618.

Lauer, G.M., Walker, B.D. (2001). Hepatitis C virus infection. *New England Journal of Medicine*, 345: 41–52.

Lebray, P., Nalpas, B., Vallet-Picard, A., Broissand, C., Sobesky, R., Serpaggi, J., Fontaine, H., Pol, S. (2005). The impact of haematopoietic growth factors on the management and efficacy of antiviral treatment in patients with hepatitis C virus. *Antiviral Therapy*, 10: 769–776.

Lee, S.S., Heathcote, E.J., Reddy, K.R., Zeuzem, S., Fried, M.W., Wright, T.L., Pockros, P.J., Häussinger, D., Smith, C.I., Lin, A., Pappas, S.C. (2002). Prognostic factors and early predictability of sustained viral response with peginterferon alfa-2a (40KD). *Journal of Hepatology*, 37: 500–506.

Lee, W., Dieterich, D. (2004). Challenges in the management of HIV and hepatitis C virus co-infection. *Drugs*, 64: 693–700.

Lin, K., Perni, R.B., Kwong, A.D., Lin, C. (2006). VX-950, a novel hepatitis C (HCV) NS3-4a protease inhibitor, exhibits potent antiviral activities in HCV replicon cells. *Antimicrobial Agents and Chemotherapy*, 50: 1813–1822.

Lindsay, K.L., Trepo, C., Heintges, T., Schiffman, M.L., Gordon, S.C., Hoefs, J.C., Schiff, E.R., Goodman, Z.D., Laughlin, M., Yao, R., Albrecht, J.K., for the Hepatitis Interventional Therapy Group. (2001). A randomized, double blind trial comparing pegylated interferon alfa-2b to interferon alfa-2b as initial treatment for chronic hepatitis C. *Hepatology*, 34: 395–403.

Lisker-Melman, M., Di Bisceglie, A.M., Usala, S.J., Weintraub, B., Murray, L.M., Hoofnagle, J.H. (1992). Development of thyroid disease during therapy of viral hepatitis with interferon alfa. *Gastroenterology*, 102: 2155–2160.

Lutchman, G., Hoofnagle, J.H. (2003). Viral kinetics in hepatitis C. *Hepatology*, 37: 1257–1259.

Luxon, B.A., Grace, M., Brassard, D., Bordens, R. (2002). Pegylated interferons for the treatment of chronic hepatitis C infection. *Clinical Therapy*, 24: 1363–1383.

Macias, J., Castellano, V., Merchante, N., Palacios, R.B., Mira, J.A., Saez, C., Garcia-Garcia, J.A., Lozano, F., Gomez-Mateos, J.M., Pineda, J.A. (2004). Effect of antiretroviral drugs on liver fibrosis in HIV-infected patients with chronic hepatitis C: Harmful impact of nevirapine. *AIDS*, 18: 767–774..

Mangia, A., Santoro, R., Minerva, N., Ricci, G.L., Carretta, V., Persico, M., Vinelli, F., Scotto, G., Bacca, D., Annese, M., Romano, M., Zechini, F., Sogari, F., Spirito, F., Andriulli, A. (2005).

Peginterferon alfa-2b and ribavirin for 12 vs. 24 weeks in HCV genotype 2 or 3. *New England Journal of Medicine*, 353: 2609–2617.

Manns, M.P., McHutchison, J.G., Gordon, S.C., Rustgi, V.K., Shiffman, M., Reindollar, R., Goodman, Z.D., Koury, K., Ling, M., Albrecht, J.K. (2001). Peginterferon alfa-2b plus ribavirin compared with interferon alfa-2b plus ribavirin for initial treatment of chronic hepatitis C: A randomized trial. *Lancet*, 358: 958–965.

Marcellin, P., Jensen, D. (2005). Retreatment with PEGASYS® in patients not responding to prior peginterferon alfa-2b/ribavirin (RBV) combination therapy-efficacy analysis of the 12-week induction period of the REPEAT study. *Hepatology*, 42 (4 Suppl 1): 749A; abstract LB04.

Martin, N.E., Modi, M.W., Reddy, R. (2000a). Characterization of pegylated (40KDA) interferon alfa-2a (PEGASYS) in the elderly. *Hepatology*, 32: 348A; abstract 755.

Martin, P., Mitra, S., Farrington, K., Martin, N.E., Modi, M.W. (2000b). Pegylated (40 KDA) alfa-2a (PEGASYS) is unaffected by renal impairment. *Hepatology*, 32: 370A; abstract 842.

Mauss, S., Berger, F., Goelz, J., Jacob, B., Schmutz, G. (2004a). A prospective controlled study of interferon-based therapy of chronic hepatitis C in patients on methadone maintenance. *Hepatology*, 40: 120–124.

Mauss, S., Valenti, W., DePamphilis, J. (2004b). Risk factors for hepatic decompensation in patients with HIV/HCV co infection and liver cirrhosis during interferon-based therapy. *AIDS*, 18: 21–25.

McHutchison, J.G., Ghalib, R., Lawitz, E., Kwo, P., Freilich, B., Muir, A., Masciari, F., Morris, M.L., Himes, J.L., Al-Adhami, M., Bacon, B.R. (2006). Early viral response to CPG 10101, in combination with pegylated interferon and/or ribavirin, in chronic HCV genotype 1 infected patients with prior relapse response. *Journal of Hepatology*, 44 (Suppl 2): S269; abstract 730.

Meier, V., Bürger, E., Mihm, S., Saile, B., Ramadori, G. (2003). Ribavirin inhibits DNA, RNA, and protein synthesis in PHA-stimulated human peripheral blood mononuclear cells: Possible explanation for therapeutic efficacy in patients with chronic HCV infection. *Journal of Medical Virology*, 69: 50–58.

Modi, M.W., Fried, M., Reindollar, R.W., Rustgi, V.R., Kenny, R., Wright, T.L., Gibas, A., Martin, N.E. (2000a). The pharmacokinetic behavior of pegylated (40KDA) interferon alfa-2a (PEGASYS) in chronic hepatitis C patients after multiple dosing. *Hepatology*, 32: 394A; abstract 939.

Modi, M.W., Fulton, J.S., Buckmann, D.K., Wright, T.L., Moore, D.J. (2000b). Clearance of pegylated (40 KDA) interferon alfa-2a (PEGASYS^{TM}) is primarily hepatic. *Hepatology*, 32: 371A; abstract 848.

Moore, M.M., Elpern, D.J., Carter, D.J. (2004). Severe, generalized nummular eczema secondary to interferon alfa-2b plus ribavirin combination therapy in a patient with chronic hepatitis C virus infection. *Archives in Dermatology*, 140: 215–217.

Mukherjee, S., Lyden, E., McCashland, T.M., Schafer, D.F. (2005). Interferon alpha 2b and ribavirin for the treatment of recurrent hepatitis C after liver transplantation: Cohort study of 38 patients. *Journal of Gastroenterology and Hepatology*, 20: 198–203.

Mukherjee S., Rogge J., Weaver L., and Schafer D.F. (2003). Pilot study of pegylated interferon Alfa-2b and ribavirin for recurrent hepatitis C after liver transplantation. *Transplant Proceedings*, 35: 3042–3044.

National Institutes of Health Consensus Development Conference Statement: Management of hepatitis C. (2002). *Hepatology*, 36 (Suppl 1): S3–S20.

Neff, G.W., Montalbano, M., O'Brien, C.B., Nishida, S., Safdar, K., Bejarano, P.A., Khaled, A.S., Ruiz, P., Slapak-Green, G., Lee, M., Nery, J., De Medina, M., Tzakis, A., Schiff, E.R. (2004). Treatment of established recurrent hepatitis C in liver-transplant recipients with pegylated interferon-alfa-2b and ribavirin therapy. *Transplantation*, 78: 1303–1307.

Neumann, A.U., Lam, N.P., Dahari, H., Gretch, D.R., Wiley, T. E., Layden, T.J., Perelson, A.S. (1998). Hepatitis C viral dynamics in vivo and the antiviral efficacy of interferon-alpha therapy. *Science*, 282: 103–107.

Neumann, A.U., Zeuzem, S., Ferrari, C., Lurie, Y., Negro, F., Germanidis, G., Esteban, J., Hellstrand, K., Pellegrin, B., Soulier, A., Haagmans, B., Abrignani, S., Colucci, G.,

Schalm, S.W., Pawlotsky, J.-M.. (2002). DITTO-HCV early viral kinetics report novel decline patterns in Gen 1 but not Gen 2-3 patients treated with PEG-interferon-alfa-2a and ribavirin. *Journal of Hepatology*, 36 (Suppl 1): 121; abstract 430.

Nevens, F., Van Vlierberghe, H., D'Heygere, F., Delwaide, J., Adler, M., Henrion, J., Henry, J.P., Hendlisz, A., Michielsen, P., Bastens, B., Brenard, R., Van Der Meeren, O. (2005). Peginterferon α-2a (40 kDa) plus ribavirin is as effective in patients relapsing after conventional interferon based therapy as in naïve patients: Results from the Bernar-1 trial. *Journal of Hepatology*, 42 (Suppl 2): 214; abstract 588.

Nguyen, M.D., Trinh, H.N., Garcia, R.T., Ning, J., Keeffe, B. (2003). Evaluation and outcomes of combination therapy with interferon or peginterferon plus ribavirin in 67 Southeast Asian patients with hepatitis C genotype 6, 7, 8, and 9. *Hepatology*, 38 (4 Suppl 1): 645A; abstract 1014.

Nunez, M., Benito, J.M., Lopez, M. (2003). Hepatitis C virus increases lymphocyte apoptosis in HIV-infected patients. In 43rd Interscience Conference on Antimicrobial Agents and Chemotherapy, Chicago, IL, September [abstract H-1717].

Oton, E., Barcena, R., Garcia-Garzon, S., Moreno-Zamora, A., Moreno, A., Garcia-Gonzalez, M., Blesa, C., Foruny, J.R., Ruiz, P. (2005). Pegylated interferon and ribavirin for the recurrence of chronic hepatitis C genotype 1 in transplant patients. *Transplant Proceedings*, 37: 3963–3964.

Pawlotsky, J.-M., Dahari, H., Neumann, A.U., Hezode, C., Germanidis, G., Lonjon, I., Castera, L., Dhumeaux, D. (2004). Antiviral action of ribavirin in chronic hepatitis C. *Gastroenterology*, 126: 703–714.

Pawlotsky, J.-M., Hezode, C., Pellegrin, B., Soulier, A., won Wagner, M., Brouwer, J.T., Missale, G., Germanidis, G., Lurie, Y., Negro, F., Esteban, J., Hellstrand, K., Ferrari,C., Zeuzem, S., Schalm, S.W., Neumann, A. (2002). Early HCV genotype 4 replication kinetics during treatment with peginterferon alfa-2a (PEGASYS)-ribavirin combination. A comparison with HCV genotypes 1 and 3 kinetics. *Hepatology*, 36: 291A; abstract 511.

Peginterferon alfa-2a, Pegasys [package insert]. Nutley, NJ: Roche Pharmaceuticals; 2004.

PEG-INTRON(R) Powder for Injection, peginterferon alfa-2b [package insert]. Kenilworth, NJ, Schering Corporation; 2005.

Planas, J.M., Gonzalez, E.R., Grana, E.B., Botella, A.G., Peinado, C.B., Poza, J.L., Garrido, M.J., Turrion, V.S., Martinez, V.C. (2005). Peginterferon and ribavirin in patients with HCV cirrhosis after liver transplantation. *Transplant Proceedings*, 37: 2207–2208.

Pockros, P., O'Brien, C., Godofsky, E., Rodriguez-Torres, M., Afdhal, N., Pappas, S.C., Lawitz, E., Bzowej, N., Rustgi, V., Sulkowski, M., Sherman, K., Jacobson, I., Chao, G., Knox, S., Pietropaolo, K., Brown, N. (2006). Valopicitabine (NM283), alone or with PEG-interferon, compared to PEG interferon/ribavirin (pegIFN/RBV) retreatment in hepatitis C patients with prior non-response to pegIFN/RBV: Week 24 results. *Gastroenterology*, 130 (4 Suppl 2): A-748; abstract 4.

Pockros, P.J., Carithers, R., Desmond, P., Dhumeaux, D., Fried, M.W., Marcellin, P., Shiffman, M.L., Minuk, G., Reddy, K.R., Reindollar, R.W., Lin, A., Brunda, M.J., for the PEGASYS® International Study Group. (2004). Efficacy and safety of two-dose regimens of peginterferon alpha-2a compared with interferon alpha-2a in chronic hepatitis C: A multicenter, randomized controlled trial. *American Journal of Gastroenterology*, 99: 1298–1305.

Poynard, T., Schiff, E., Terg, R., Goncales, F., Flamm, S., Diago, M., Reichen, J., Moreno, R., Tanno, H., McHutchison, J., Fainboim, H., Berg, T., Mattos, A., Burak, K., Mukhopadhyay, P., Bedossa, P., Griffel, L., Burroughs, M., Brass, C., Albrecht, J. (2005). Sustained virologic response (SVR) with PEG-Interferon-alfa 2b/ribavirin weight based dosing in previous interferon/ribavirin HCV treatment failures; week 12 virology as a predictor of SVR in the EPIC[3] trials. *Gastroenterology*, 128 (Suppl 2): A681; abstract 5.

Preziati, D., La Rosa, L., Covini, G., Marcelli, R., Rescalli, S., Persani, L. (1995). Autoimmunity and thyroid function in patients with chronic active hepatitis treated with recombinant interferon alpha-2a. *European Journal of Endocrinology*, 132: 587–593.

Ramos, B., Nunez, M., Rendon, A. (2006). High ribavirin doses and early virological response in HCV/HIV-coinfected patients. Program and abstracts of the 2nd International Workshop on HIV and Viral Hepatitis Co infection; January 12-14, Amsterdam, The Netherlands. Abstract 36.

Rasenack, J., Zeuzem, S., Feinman, V., Heathcote, E.J., Manns, M., Yoshida, E.M., Swain, M.G., Gane, E., Diago, M., Revicki, D.A., Lin, A., Wintfeld, N., Green, J. (2003). Peginterferon α-2a (40 kD) [PEGASYS®] improves HR-QOL outcomes compared with unmodified interferon α-2a [Roferon®-A] in patients with chronic hepatitis C. *Pharmacoeconomics*, 21: 341–349.

Rauch, A., Rickenbach, M., Weber, R. (2005). Association of unsafe sex and increased incidence of hepatitis C infection in HIV-infected men who have sex with men. Program and abstracts of the 12th Conference on Retroviruses and Opportunistic Infections; February 22-25, Boston, MA. Abstract 943.

Reddy, K.R., Wright, T.L., Pockros, P.J., Shiffman, M., Everson, G., Reindollar, R., Fried, M.W., Purdum, P.P., Jensen, D., Smith, C., Lee, W.M., Boyer, T.D., Lin, A., Pedder, S., DePamphilis, J. (2001). Efficacy and safety of pegylated (40-kd) interferon α-2a compared with interferon α-2a in noncirrhotic patients with chronic hepatitis C. *Hepatology*, 33: 433–438.

Reesink, H.W., Zeuzem, S., Weegink, C.J., Forestier, N., van Vliet, A., van de Wetering de Rooij, J., McNair, L.A., Purdy, S., Chu H-M., Jansen, P.L. (2005). Final results of a Phase 1B, multiple-dose study of VX-950, a hepatitis C virus protease inhibitor. *Hepatology* 42 (4 Suppl 1): 234A; abstract 96.

Rendon, A.L., Nunez, M., Romero, M., Barreiro, P., Martin-Carbonero, L., Garcia-Samaniego, J., Jimenez-Nacher, I., Gonzalez-Lahoz, J., Soriano V. (2005). Early monitoring of ribavirin plasma concentrations may predict anemia and early virological response in HIV/hepatitis C virus-coinfected patients. *Journal of AIDS*, 39: 401–405.

Rhodes, S., Erlich, H., Im, K.A., Wang, J., Li J., Bugawan, T., Jeffers, L., Tong, X., Su, X., Lee, L.J., Liang, T.J., Yang, H. (2006). Association between HLA allele a*02, B*58, Dpb*1701 and sustained virologic response to pegylated interferon and ribavirin therapy in chronic HCV genotype 1-infected, treatment naïve Caucasian and African Americans. *Gastroenterology*, 130 (4 Suppl 2): A-771; abstract 656.

Sakai, T., Iino, S., Okuno, T., Omata, M., Kiyosawa, K., Kumada, H., Yamada, G., Hayashi, N. (2006). High response rates with peginterferon alpha-2a (40KD) (PEGASYS®) plus ribavirin (COPEGUS®) in treatment-naïve Japanese chronic hepatitis C patients: A randomized, double-blind, multicentre, phase III trial. *Journal of Hepatology*, 44 (Suppl 2): S224; abstract 605.

Sánchez-Tapias, J.M., Diago, M., Escartín, P., Enríquez, J., Romero-Gómez, M., Bárcena, R., Crespo, J., Andrade, R., Martínez-Bauer, E., Pérez, R., Testillano, M., Planas, R., Solá, R., García-Bengoechea, M., Garcia-Samaniego, J., Munoz-Sánchez, M., Moreno-Otero, R.; TeraViC-4 Study Group. (2006). Peginterferon-alfa 2a plus ribavirin for 48 versus 72 weeks in patients with detectable hepatitis C virus RNA at week 4 of treatment. *Gastroenterology*, 131: 451–460.

Schmid, M., Kreil, A., Jessner, W., Homoncik, M., Datz, C., Gangl, A., Ferenci, P., Peck-Radosavljevic, M. (2005). Suppression of haematopoiesis during therapy of chronic hepatitis C with different interferon α mono and combination therapy regimens. *Gut*, 54: 1014–1020.

Schulman, J.A., Liang, C., Kooragayala, L.M., King, J. (2003). Posterior segment complications in patients with hepatitis C treated with interferon and ribavirin. *Ophthalmology*, 110: 437–442.

Schwaiger, M., Pich, M., Granke, L. (2003). Chronic hepatitis C infection, interferon alpha treatment and peripheral serotonergic dysfunction. *Hepatology*, 38 (Suppl 1): 300A; abstract 301.

Schwarz, K., Mohan, P., Narkewicz, M., Molleston, J.P., Te, H.S., Hu, S., Sheridan, S., Lamb, M., Pappas, S.C., Harb, G. (2003). The safety, efficacy and pharmacokinetics of peginterferon alfa-2a (40KD) in children with chronic hepatitis C. *Gastroenterology*, 124 (Suppl 1): A-700; abstract 215.

Seeff, L. (2002). Natural history of chronic hepatitis C. *Hepatology*, 36 (Suppl 1): S35–S46.

Sheiner, P.A., Boros, P., Klion, F.M., Thung, S.N., Schluger, L.K., Lau, J.Y., Mor, E., Bodian, C., Guy, S.R., Schwartz, M.E., Emre, S., Bodenheer, H.C., Miller, C.M. (1998). The efficacy of prophylactic interferon alfa-2b in preventing recurrent hepatitis C after liver transplantation. *Hepatology*, 28: 831–838.

Shen, Y., Pielop, J., Hsu, S. (2005). Generalized nummular eczema secondary to peg interferon alfa-2b and ribavirin combination therapy for hepatitis C infection. *Archives of Dermatology*, 141: 102–103.

Shepherd, J., Brodin, H., Cave, C., Waugh, N., Price, A., Gabbay, J. (2004). Pegylated interferon α-2a and -2b in combination with ribavirin in the treatment of chronic hepatitis C: A systematic review and economic evaluation. *Health Technology Assessment*, 8: 1–125.

Shergill, A., Khalili, M., Bollinger, K., Roberts, J., Ascher, N., Terrault, N. (2005). Applicability, tolerability and efficacy of preemptive antiviral therapy in hepatitis C infected patients undergoing liver transplantation. *American Journal of Transplantation*, 5: 118–124.

Sherman, K.E., Rouster, S.D., Chung, R.T., Rajicic, N. (2002). Hepatitis C virus prevalence among patients infected with human immunodeficiency virus: A cross-sectional analysis of the US adult AIDS Clinical Trials Group. *Clinical Infectious Diseases*, 34: 831–837.

Sherman, M., Bain, V., Villeneuve, J-P., Myers R.P., Cooper, C., Martin, S., Lowe, C. (2004). The management of chronic viral hepatitis: A Canadian consensus conference 2004. *Canadian Journal of Gastroenterology*, 18: 715–728.

Shiffman, M.L., Diago, M., Tran, A., Pockros, P., Reindollar, R., Prati, D., Rodriguez-Torres, M., Lardelli, P., Blotner, S., Zeuzem, S. (2006). Chronic hepatitis C in patients with persistently normal alanine transaminase levels. *Clinical Gastroenterology and Hepatology*, 4: 645–652.

Shiffman, M.L., DiBisceglie, A.M., Lindsay, K.L, Morishima, C., Wright, E.C., Everson, G.T., Lok, A.S., Morgan, T.R., Bonkovsky, H.L., Lee, W.M., Dienstag, J.L., Ghany, M.G., Goodman, Z.D., Everhart, J.E., and the HALT-C Trial Group (2004). Peginterferon alfa-2a and ribavirin in patients with chronic hepatitis C who have failed prior treatment. *Gastroenterology*, 126: 1015–1023.

Shiffman, M.L., Price, A., Hubbard, S. (2005). Treatment of chronic hepatitis C virus (HCV) genotype 1 with peg interferon alfa-2b (PEGIFN), high weight based dose ribavirin (RVN) and epoetin alfa (EPO) enhances sustained virologic response (SVR). Program and abstracts of the 56th Annual Meeting of the American Association for the Study of Liver Diseases; November 11-15, San Francisco, CA. Abstract 55.

Siebert, U., Sroczynski, G., Rossol, S., Wasem, J., Ravens-Sieberer, U., Kurth, B.M., Manns, M.P., McHutchison, J.G., Wong, J.B., German Hepatitis C Model (GEHMO) Group, International Hepatitis Interventional Therapy (IHIT) Group. (2003). *Gut*, 52: 425–432.

Singh, N., Gayowski, T., Wannstedt, C.F., Shakil, A.O., Wagener, M.M., Fung, J.J., Marino, I.R. (1998). Interferon-alpha for prophylaxis of recurrent viral hepatitis C in liver transplant recipients: A prospective, randomized, controlled trial. *Transplantation*, 65: 82–86.

Sonnenblick, M., Rosenmann, D., Rosin, A. (1990). Reversible cardiomyopathy induced by interferon. *BMJ*, 300: 1174–1175.

Soriano, V., Puoti, M., Sulkowski, M. (2004b). Care of patients with hepatitis C and HIV co-infection. *AIDS*, 18: 1–12.

Soriano, V., Nunez, M., Camino, N., Maida, I., Barreiro, P., Romero, M., Martin-Carbonero, L., Garcia-Samaniego, J., Gonzalez-Lahoz, J. (2004a). Hepatitis C virus-RNA clearance in HIV-co-infected patients with chronic hepatitis C treated with pegylated interferon plus ribavirin. *Antiviral Therapy*, 9: 505–509.

Soriano, V., Perez-Olmeda, M., Rios, P., Nunez, M., Garcia-Samaniego, J., Gonzalez-Lahoz, J. (2004b). Hepatitis C virus (HCV) relapses after anti-HCV therapy are more frequent in HIV-infected patients. *AIDS Research Human Retroviruses*, 20: 351–354.

Strader, D.B., Wright, T., Thomas, D.L., Seeff, L.B. (2004). Diagnosis, management, and treatment of hepatitis C. *Hepatology*, 39: 1147–1171.

Su, X., Rhodes, S., Yee, L., Im, K., Tong, X., Howell, C., Haritos, M., Liang, T.J., Taylor, M.W., Yang, H. (2006). Association of interferon signaling pathway and interferon stimulated genes with response to pegylated interferon and ribavirin therapy in HCV infected Caucasian and African American patients. *Gastroenterology*, 130 (4 Suppl 2): A-748; abstract 2.

Sugano, S., Suzuki, T., Watanabe, M., Ohe, K., Ishii, K., Okajima, T. (1998). Retinal complications and plasma C5a levels during interferon alpha therapy for chronic hepatitis C. *American Journal of Gastroenterology*, 93: 2441–2444.

Zeuzem, S., Herrmann, E, Lee, J.-H., Fricke, J., Neumann, A.U., Modi, M., Colucci, G., Roth, K. (2001). Viral kinetics in patients with chronic hepatitis C treated with standard or peginterferon α2a. *Gastroenterology,* 120: 1438–1447.

Zeuzem, S., Hultcrantz, R., Bourlier, M., Goeser, T., Marcellin, P., Sanchez-Tapias, J., Sarrazin, C., Harvey J., Brass C., Albrecht, J. (2004b). Peginterferon alfa-2b plus ribavirin for treatment of chronic hepatitis C in previously untreated patients infected with HCV genotypes 2 or 3. *Journal of Hepatology,* 40: 993–999.

Zeuzem, S., Pawlotsky, J.-M., Lukasiewicz, E., von Wagner, M., Goulis, I., Lurie, Y., Gianfranco, E., Vrolijk, J-M., Esteban, J.I., Hexode, C., Latgging, M., Negro, F., Soulier, A., Verheij-Hart, E., Hansen, B., Tal, R., Ferrari, C., Schalm, S.W., Neumann, A.U., for the DITTO-HCV Study Group. (2005). International, multicenter, randomized, controlled study comparing dynamically individualized versus standard treatment in patients with chronic hepatitis C. *Journal of Hepatology,* 43: 250–257.

Zeuzem, S., Sarrazin, C., Wagner, F., Rouzier, R., Forestier, N., Gupta, S., Hussain, M., Shah, A., Cutler, D., Zhang, J. (2006a) The HCV NS3 protease inhibitor SCH 503034 in combination with PEG-IFN α-2b in the treatment of HCV-1 PEG-IFN α-2b non-responders: Antiviral activity and HCV variant analysis. *Journal of Hepatology,* 44 (Suppl 2): S35; abstract 78.

Subject Index

Printed in the United States of America